Origins and Phylogeny of Rices

Origins and Phylogeny of Rices

N. M. NAYAR
Emeritus Scientist, Jawaharlal Nehru Tropical Botanic Garden & Research Institute, Pacha-Palode 695 562, Trivandrum, India

ELSEVIER

Amsterdam • Boston • Heidelberg • London
New York • Oxford • Paris • San Diego
San Francisco • Singapore • Sydney • Tokyo
Academic Press is an imprint of Elsevier

Academic Press is an imprint of Elsevier
The Boulevard, Langford Lane, Kidlington, Oxford, OX5 1GB, UK
225 Wyman Street, Waltham, MA 02451, USA

First published 2014

Notices

Knowledge and best practice in this field are constantly changing. As new research and experience broaden our understanding, changes in research methods, professional practices, or medical treatment may become necessary.

Practitioners and researchers must always rely on their own experience and knowledge in evaluating and using any information, methods, compounds, or experiments described herein. In using such information or methods they should be mindful of their own safety and the safety of others, including parties for whom they have a professional responsibility.

To the fullest extent of the law, neither the Publisher nor the authors, contributors, or editors, assume any liability for any injury and/or damage to persons or property as a matter of products liability, negligence or otherwise, or from any use or operation of any methods, products, instructions, or ideas contained in the material herein.

British Library Cataloguing in Publication Data
A catalogue record for this book is available from the British Library

Library of Congress Cataloguing in Publication Data
A catalogue record for this book is available from the Library of Congress

ISBN: 978-0-12-417177-0

For information on all Academic Press publications visit our website at store.elsevier.com

Printed and bound in the United States

Transferred to Digital Printing, 2014

CONTENTS

FOREWORD

We owe a deep debt of gratitude to Dr N. M. Nayar for taking the trouble to write an authoritative book on the Origins and Phylogeny of Rices. Dr Nayar is one of our most eminent and well informed rice experts. He has also devoted his life to unraveling the phylogeny of crops from a global perspective. His book covers all the important rice growing areas and throws much light on the origins of the Asian and West African rices.

I would like to highlight some aspects of the yield revolution in rice which we have witnessed during the last sixty years, starting with the *indica–japonica* hybridization program started at the Central Rice Research Institute, Cuttack in the early 1950s.

The yield of indica rice remained stagnant for a long time. In contrast, the japonica varieties of rice cultivated in Japan and also in northern latitudes yielded two to three times more than the indica rices even before World War II. The reason for this is the ability of the japonica varieties to utilize more nutrients and convert them into grains. For example, the rice plant requires about 20 kg of nitrogen and appropriate quantities of other nutrients for producing a yield of 1 tonne of rice. The indica varieties which were cultivated before World War II had thin straw and therefore tended to lodge when mineral fertilizer was applied. To overcome this problem of lodging, an indica–japonica hybridization program was initiated in 1952 at the Central Rice Research Institute, Cuttack, at the insistence of the late Dr K Ramaiah. I worked on this project for some time at Cuttack. For a variety of reasons the indica–japonica hybridization program did not yield the anticipated results. However, a few good varieties like ADT-27 in Tamil Nadu and Mashuri in Malaysia came out of this program. In the mid-1960s semi-dwarf rice varieties like Taichung Native 1 from Taiwan, which had the Dee-Gee-Woo Gen gene for dwarfing, became available. This made the indica–japonica hybridization program less important from the point of view of breeding fertilizer responsive varieties of rice. Soon, very high yielding rice varieties like IR8 became available from the International Rice Research Institute (IRRI) in the Philippines. In 1968, the rice revolution began in all the indica rice growing countries. In India for example, rice production rose from 20 million tonnes in 1947 to over 100 million tonnes in 2012.

The importance of rice cultivation will grow with the onset of the era of climate change. Rice is much more resilient to climatic factors as will be evident from the fact that rice is cultivated under a wide range of altitudes and latitudes, beginning with below sea level farming in the Kuttanad region of Kerala to nearly 3,000 meters altitude in the Himalayas and J&K. There are also over 150,000 varieties of rice available globally. The IRRI Gene Bank has over 110,000 accessions. Therefore, rice will prove to be an important climate saviour crop and its importance to human food security will grow. It will also play an important role in nutrition, since already varieties rich in iron, zinc, and vitamin A are available from the biofortification programs in progress in different countries. MSSRF has iron-rich varieties of rice developed by genetic modification. Therefore there are uncommon opportunities for combining yield and quality in rice varieties.

At the moment, the semi-dwarf varieties capable of responding to good soil fertility and water management are mainly grown in irrigated or high rainfall/lowland conditions. Hybrid rice was introduced in China in the 1970s based on a male sterile line identified on Hainan Island. It is generally believed that hybrids yield 15–20 percent more than varieties. A major problem in popularizing hybrid rice is seed production. Unless seed yield is increased to about 3 tonnes per hectare, the net gain from the cultivation of hybrid rice will be poor. China has most of its irrigated area under hybrid rice because they have produced a wide range of high yielding hybrids and also developed efficient hybrid seed production techniques. In India, which has the largest area under rice (nearly 45 million hectares), there are several hybrids available in the market, but they are yet to become popular because of poor cooking quality. Culinary qualities are as important as yield, as will be clear from the high premium paid to Basmati rice. Pusa basmati 1121 fetches a very high price in the national and international markets.

The regions where there are significant gaps between the potential and actual yields of rice are South Asia and West Africa. The NERICA rices developed in West Africa have, to some extent, helped to improve the yield per hectare. In India, the average yield of rice is less than 50 percent of the potential. The monsoon and the market are two major determinants of the economic success of rice farming. Both require greater attention.

The yield gap may be due to technological, ecological, or economic or social reasons. A "bridging the yield gap movement" should be launched with concurrent attention to the following five components of successful rice farming:

- Soil healthcare and enhancement
- Water management including the application of System of Rice Intensification (SRI) techniques

- Technology and inputs
- Credit and insurance
- Assured and remunerative marketing

Genetic engineering is largely being used for breeding rice varieties resistant to biotic and abiotic stresses. Varieties like Golden Rice rich in vitamin A have also been produced by genetic modification. Many of these attributes can also be transferred to rice varieties through molecular markers. There are inherent problems with reference to public acceptance of GM rice. Therefore, it will be better to achieve the same results through molecular marker-based selection.

Irrigation water will be a great constraint in the coming decades and centuries. That is why there is interest in techniques like SRI which help to reduce irrigation water need by about 50 percent. There are also experiments for raising more crop per drop of water. There are great opportunities for improving water use efficiency. For example, when I was at the IRRI, I held a joint discussion with WHO on preventing rice fields becoming breeding grounds for mosquitoes. We concluded that one way of controlling mosquito breeding in rice fields is alternate drying and wetting. This does not affect yield. The rice plant does not need standing water all the time.

There are large numbers of farming systems being developed with rice as the principal crop. The rice–wheat rotation in the Punjab has led to the depletion of the water table. Therefore, it will be appropriate to have a 3 or 5 year rotation in which pulses and millets like iron-rich *bajra* are included. Current research in rice farming systems aims to integrate the principles of ecology, economics, employment, energy requirement, and social and gender equity.

The initial work which led to the rice revolution was done in China and Taiwan (China). China gave to the rice world both semi-dwarf and photo-insensitive varieties as well as hybrids. The Chinese material was taken and developed further at the IRRI. The best work on dwarf basmati rices has been done at the Indian Agricultural Research Institute, New Delhi. The IRRI has played an important role in germplasm collection, conservation and distribution and human resource development.

National and international research systems should develop symbiotic relationships. The stronger the national research system, the greater will be the opportunity for deriving benefits from international research. Therefore, the foundation for sustainable progress is the existence of a strong multi-disciplinary national research system. Also, it is important to ensure that public good research receives priority. This means that good varieties should have precedence over hybrids, since farmers will have to buy the hybrid seeds every year.

The rice revolution in India, as well as the other countries, has its roots in synergy among scientific know-how, political do-how, and farmers' enthusiasm. Public policy should bring about synergy between technology and public policy. For example, in the rice–wheat rotation areas of the Punjab and northwest India, free electricity should not be provided for pumping groundwater. Subsidies of this kind can be called "subsidies for ecocides", i.e., ecological suicide. Since rice is going to be a climate resilient and climate smart crop, it is essential that sustainable rice farming using the ever-green revolution pathway (i.e., increase in productivity in perpetuity without associated ecological harm) is promoted. More than subsidies, services are urgently needed, as for example, the provision of appropriate farm machinery on the basis of a custom-hire approach. Public policies in the flood-prone plains of Assam should aim to make the non-flood season (i.e., from Nov–May), the main rice farming season; all that this will require is the provision of funds for shallow tube wells. This will help to lower the water table during the *rabi* and *boro* seasons, thereby allowing more absorption and storage of water during the *kharif* season. Such an approach has been referred to by Roger Revelle as the Ganges Water Machine, which can ensure water for raising good rice crops, and at the same time help to prevent floods.

There is a saying that rice is life in many parts of Asia. In the emerging era of climate change, rice will probably gain greater importance in National Food Security Systems, since this crop is characterized by enormous resilience in terms of its ability to grow under different latitudes and altitudes. For example, the farmers of the Kuttanad area of Kerala started cultivating rice below sea level over 150 years ago. This below-sea-level rice farming system has recently been recognized by FAO as a Globally Important Agricultural Heritage Site.

The lives and livelihoods of over 2 billion women and men depend upon the rice crop. This book brings out clearly the history of this unique plant, its botany and speciation, and genetics and cytogenetics. In Chapters 5 and 6, the origins of the African and Asian rices are dealt with in detail. I find the analysis of the time and place of origin very interesting and original. This book is a must for all interested in the phylogeny of the genus *Oryza* and the interrelationships among different *Oryza* species. Information from both conventional taxonomy and molecular taxonomy is summarized. The available archaeological evidence is also summarized. Hardly any other book brings together, so many aspects of *Oryza* species in an integrated manner. In my view, this book is a must not only for research scholars, teachers, and scientists, but also to all those interested in the conservation of rice genetic diversity and its sustainable utilization in improving the productivity and profitability of rice cultivation.

It is estimated that there may be over 150,000 strains of rice in the world. The gene bank of the International Rice Research Institute has fortunately collected over 110,000 strains. The Global Gene Vault at Svalbard is also maintaining for posterity a wide range of variability occurring in rice and in other crops.

I hope this excellent book by Dr Nayar will stimulate greater interest in the conservation and sustainable and equitable use of rice genetic resources. At the same time, I am confident that the book will generate greater interest in our rice heritage including the cultural and culinary aspects associated with this unique crop.

Professor M.S. Swaminathan
Founder Chairman,
M. S. Swaminathan Research Foundation,
Chennai, India

PREFACE

Wheat and rice are the most important food crops of humankind. Both of them contribute more than one-fifth of a human's calorie requirements. But, rice is the more important food crop because it forms the staple food for more humans than all others.

The last monograph on the origin of rice was published by Oka (1988). The late 1980s was the period when molecular biology studies were beginning to be taken up in rice. The first papers on electrophoretic studies had been published by then.

Since then, there has been literally an explosive interest in studies on rice molecular evolution and phylogeny. The complete sequences of rice published by two groups in 2002 have added to this interest, especially in genomics (Goff et al. 2002, Yu et al. 2002). Archaeological studies were also then picking up in pace, both in west and north Africa, and east and southeast Asia.

After Oka's (1988) monograph, there have only been some short reviews and perspectives on rice origins and diversity, in addition to a couple of multi-authored and edited volumes. Incidentally, Oka's work was primarily a review of only the work done by him and his colleagues. The present work has been prepared with this in the background. This may also be the first attempt to review and synthesize the information on rice phylogeny.

The work on the present publication was begun in 2010. Its preparation was an uphill task mainly because of my working mostly from a location where facilities for intellectual interaction and library were unavailable. Consequently, the help and support of several scientists and institutions working in the US, Europe, and Japan had to be sought to secure copies of some rare publications and suggestions and advice on certain matters. I am grateful to all of them for their support. Their names are given at the end. Part of this work was done while on a two-year Fellowship awarded by the Ministry of Science and Technology, Government of India. Some financial assistance was received from the Kerala State Council of Science, Technology, and Environment also. I thank all the agencies for their assistance.

Rice science is in a state of flux, especially on rice domestication and differentiation. There are sharp differences and contrasting viewpoints on most aspects of rice origin and phylogeny. A sincere attempt has been made here to bring out these diverse viewpoints and opinions objectively.

The volume covers diverse areas such as archaeology, molecular biology, botany, and geology. Hence, it is expected to be consulted by students and specialists from all these disciplines. For their benefit, an appendix and abbreviations used in the volume have been added to provide explanatory information. The definitions of most terms and phenomena covered in the text have been given in the text itself. They have been taken mostly from FAO (2001): *Glossary of Biotechnology for Food and Agriculture,* FAO Rome, and King et al. (2006): *A Dictionary of Genetics,* 7th edition, Oxford University Press.

In the present volume, several early classical references in rice biology (taxonomy and genetics, mainly) have been cited. This has been done mainly to highlight their fundamental and basic contributions and also collate the present data with them to help us arrive at a better perspective and more definitive conclusions. Since students and scholars in other disciplines may be referring to this book, I have attempted to make every chapter as autonomous as possible. Towards this, I have restated also certain fundamental aspects of rice more than once to provide some background information even at the cost of repetition. I expect that this would be appreciated.

I thank profusely Dr M S Swaminathan, the eminent agriculture and food security scientist, for readily writing the Preface to the book. He had earlier served, among several distinguished positions that he held, as Director General, Indian Council of Agricultural Research and International Rice Research Institute. I thank also the following scientists and institutions for providing copies of some publications (other than their own publications) or found time to comment on my queries: C. Connah, N. Ellstrand, P. Heslop-Harrison, S. MacEachern, K. Olsen, R. Singh, National Science Museum Tokyo, University of Kyushu Faculty of Agriculture, and Botanical Museum and Botanical Garden Berlin-Dahlem. Dr Rajendra Singh was particularly helpful in collecting and supplying photocopies of many references from time to time for the period 1972–1990 of those available in the IARI library New Delhi.

It is hoped that this volume will be well received. Comments and suggestions are most welcome. They will be greatly appreciated.

N.M. Nayar
nayarnm1@bsnl.in
nayar.nm@gmail.com

ABBREVIATIONS

AFLP	Amplified fragment length polymorphism
AMOVA	Analysis of molecular variance
ANOVA	Analysis of variance
APG	Angiosperm Phylogeny Group
BCE	Before the Common Era
BEP	Bambusoideae, Ehrhartoideae, and Pooideae
BLAST	Basic local alignment search tool
BP/bp	Before Present
BIL	Backcross inbred line
Cal BCE	Calibrated BCE
Cal BP	Calibrated BP
CE	Common (Christian) Era
CIAT	Centro Internacional de Agricultura Tropical
cpDNA	Chloroplast DNA
FBS	Food Balance Sheet
GM	Genetically modified
IND	Inland Niger Delta
IRRI	International Rice Research Institute
ISSR	Inter-simple sequence repeat
ITS	Internal transcriber spacer
Ka/ka/kya	1,000 years ago
LD	Linkage disequilibrium
MITE-AFLP	Miniature inverted repeat transposable element-AFLP
ML	Maximum likelihood method
MP	Maximum parsimony method
Mtn, Mt	Million tonnes
NJ method	Neighbor joining method
ORF	Open reading frame
PACCMAD	Panicoideae, Arundinoideae, Centrothecoideae, Chloridoideae, Micrairoideae, Aristidoideae, and Danthoninoideae
PCA	Principal component analysis
PCR	Polymerase chain reaction
PCR-AFLP	Polymerase chain reaction – amplified fragment length polymorphism
PIC	Polymorphism information content

QTL	Quantitative trait loci
RAPD	Random amplified polymorphism DNA
rDNA	Recombinant DNA
RFLP	Restriction fragment length polymorphism
RIL	Recombinant inbred line
SINE	Short interspersed nuclear element
SNP	Single nucleotide polymorphism
SSR	Single sequence repeat
STS	Sequence tagged site
UPGMA	Unweighted pair group method with arithmetic averages
WCSP	World Checklist of Selected Plants

Prehistory Timeline

Year	yr
1,000 years	kyr
Uncalibrated radiocarbon years before present	yr bp
Uncalibrated radiocarbon kyr before present	kyr bp
Calibrated radiocarbon years before present	cal BP
Calibrated radiocarbon kyr BC	kyr cal BC
Calibrated radiocarbon kyr AD	kyr cal AD
Historical years BC	BC
Historical years AD	AD

Radiocarbon kyr bp	Radiocarbon kyr cal BP	Calibrated year
0.35	—	AD 1550
0.50	—	AD 1430
1.2–0.7	—	AD 780–1290
1.6–1.7	—	AD 400s
1.7–1.8	—	AD 300s
2.2–2.0	2.3–1.94	360/200 BC – AD 12
2.5–2.2	2.7–2.15	750–360/200 BC
2.5	2.7	775 BC
2.8	2.8	850 BC
3.0–3.1	3.1	1300 (1200)/1150 BC
3.3	2.55–2.45	1600–1500 BC
3.5	3.8	1880–1750 BC
3.7	4.05	2100 BC (2140–1980 BC)
4.0	4.4	2500 BC
4.4	5.0	3100–2920 BC
4.5	5.2	3335–3100 BC

4.9–4.5	5.65	3700–3300 BC
5.2–4.85	5.9–5.6	3950–3650 BC
5.4–4.5	6.25–5.25	4300–3300 BC
5.5–5.4	6.35–6.25	4400–4300 BC
5.72	6.5	4500 BC
5.9–5.5	6.75–6.35	4800–4400 BC
6.0	6.8	4900 BC
6.5	7.35	5400 BC
6.6	7.45	5500–5400 BC
6.9–6.3	7.65–7.15	5700–5200 BC
6.9	7.65	5700 BC
7.0	7.85	5800 BC
7.26	8.0	6100 BC
7.9–7.7	8.75–8.45	6800–6500 BC
8.1	9.0	7000 BC
8.6	9.6	7570 BC
9.4–8.6	10.5–9.6	8420–7570 BC
9.4	10.5	8420 BC
9.8	11.2	9040 BC
10.0	11.5	9250–9150 BC
10.6–10.0	12.55–11.15	10,600–9200 BC
11.7–10.7	13.0–12.65	11,100–10,700 BC
11.7–10.6	13.65–12.55	11,700–10,600 BC
12.0	14.0	12,000 BC
12.7	14.95	13,000 BC

From Hassan (2002) with permission

Geological Time Scale

Era	Period	Epoch	Approximate Time Since Start in Millions Years (Before Present)
Cenozoic	Quarternary	Holocene	0.01
		Pleistocene	1.6
	Tertiary	Pliocene	5
		Miocene	23
		Oligocene	36
		Eocene	57
		Palaeocene	66
Mesozoic	Cretaceous		144
	Jurassic		208
	Triassic		245

Continued

—cont'd

Era	Period	Epoch	Approximate Time Since Start in Millions Years (Before Present)
Palaeozoic	Permean		286
	Carboniferous		320
	Pennsylvanian		320
	Mississipian		360
	Devonian		408
	Silurian		438
	Ordovician		505
	Cambrian		570–600
Precambrian	Proterozoic		2500
	Archaean		4600

CHAPTER 1

Rice in the World

1. IMPORTANCE

Maize, rice, and wheat are the most important food crops of humankind, both in terms of their area under cultivation and their production (FAO 2011) (Table 1.1). In the case of maize, however, only 14% of production is used for food; the rest is used for other purposes (animal feed, industrial purposes, etc.). Its per capita consumption is therefore less than one-third of those of wheat and rice, and in terms of calorie supply for food, only about one-fourth (Table 1.1). At the same time, of the total production, of rice, as much as 80% is used for food, and in wheat, 72% (FAO, Food Balance Sheets (FBS), 2007).

Wheat and rice contribute about the same quantity of energy to humans, although the annual per capita consumption of wheat (65.9 kg) is lower than that of rice paddy (79 kg/cap annually as rice paddy, 53 kg as milled rice). For maize, the corresponding value is only 14.3 kg/cap/year (FAO 2007). Over the decades, the production of wheat has increased by marginally more than that of rice (Table 1.2). Both crops have more than tripled their production in about the last five decades (1961–2009) (Table 1.2), even while the areas under production for the two crops had increased by only about 30% in rice and 10% in wheat. During the same period, the production of maize increased by as much as 400%, while the area under production for the crop increased by about 50%.

The area and production trends of the next three major crops of the world – bananas, cassava, and potato – are also given in Tables 1.1 and 1.2 for comparison. While the per capita food supply of rice increased by about 37% during 1961–2007, that of wheat increased by only about 20%.

2. SPECIAL FEATURES OF RICE PRODUCTION

Certain features in the production and consumption of rice make this cereal crop more important/crucial than wheat in providing food security for the human race. Firstly, rice is grown in as many as 177 countries

Origins and Phylogeny of Rices
http://dx.doi.org/10.1016/B978-0-12-417177-0.00001-2

Table 1.1 Area, Production, and Per Capita Consumption of the Six Major Crops (2007)*.

	Area	Production	Total Per Capita Consumption/Annum	
Crops	**(1000 ha)**	**(1000 t)**	**(in kg)**	**(in kcal)**
Rice, paddy	155060	656254**	52.92	533
Wheat	216705	604465	65.90	530
Maize	158385	787440	16.80	139
Potato	18663	322603	31.70	59
Cassava	18635	224312	16.40	43
Bananas (including plantains)	10237	122034	14.40	27
Total	–	–	–	**2796**

*Source FBS (FAO's Food Balance Sheets) 2007 (the latest year for which FBS data is available)
**Rice as milled equivalent is 4.38 Mt (66.7% conversion factor)
Source: FAOSTAT. Downloaded 30.08.2011

of the world (out of 194 countries). Next, rice possesses the widest adaptability of all the major crop plants. It is grown in latitudes ranging from 53° North to 35° South (Lu and Chang 1980), and in altitudes ranging from 1,500 meters above sea level to more than 4.0 meters below sea level. Thirdly, rice shows the widest variability in terms of a number of characteristics. For instance, its period of maturity varies from 65–70 days to 330–365 days. Some decades ago, the Central Rice Research Institute, Cuttack, India even released an improved cultivar called *Sathari* (meaning 'seventy-er'), indicating its period of maturity as 70 days. More than 90% of the crop is produced and consumed in just one continent, Asia – mostly in the east, southeast, and southern Asia. Only about 6–8% of rice production enters the international trade. In wheat, the equivalent figure is 18–28%.

3. PRODUCTION AND AVAILABILITY OF RICE

Production data for rice across the world and for different continents for the period 1961–2007 are given in Table 1.3. The most recent Food Balance Sheets of the FAO are for 2007 (since updated to 2009). However, country/continent-wise area and production figures are available up to 2011. Here, we use the 2009 data.

Presently (Table 1.3), over 90% of rice production comes from Asia. In broad terms, this position has not changed significantly since 1961. In 1961, the distribution of production in percentage terms across the five continents was: Asia: 92.2%, Africa: 2.0%, Americas: 4.9%, Europe: 0.9%, and Oceania: <0.1%. In 2007, the distribution was Asia: 91.1%, Africa: 3.2%, America:

Table 1.2 Trends in Area and Production of the Six Major Crops (1961–2009)

Crop	1961	1970	1980	1990	2000	2009	CGR* in %
				Area (1000 ha)			
Rice, paddy	115365	132873	144412	146960	154060	158300	5.8**
Wheat	204210	207979	237252	231263	215437	225622	1.7 NS
Maize	105560	113076	125776	131318	137001	158629	7.6**
Potato	22148	20773	18761	17621	20062	18652	−2.9 NS
Cassava	9624	11618	13601	15210	17000	18917	13.2**
Bananas†	4456	5912	6590	7701	9436	10263	16.4**
				Production (1000 t)			
Rice, paddy	215647	316346	396871	518568	599356	685240	22.8**
Wheat	222357	310741	440188	592311	585690	685614	22.4**
Maize	205028	265831	396623	483344	592473	818823	27.2**
Potato	270552	298048	240464	266625	327340	329581	3.9 NS
Cassava	71260	98540	124135	152496	176533	233796	22.6**
Bananas†	21293	31342	36688	46809	65125	97378	28.7**

*CGR = Compound Growth Rate
**Significant at 5%
NS= Not significant
†Includes plantains
Source: FAOSTAT. Downloaded 30.08.2011

Table 1.3 Trends in Production and Consumption of Rice in Various Continents (1961–2007)

	1961	1970	1980	1990	2000	2007	CGR* in %
Rice							
Production (1000 t)							
World	143550	210603	264283	345458	399513	437509	22.2★★
Africa	2875	4860	5741	8469	11656	13940	31.2★★
Americas	7049	10473	15385	15097	21353	22269	22.5★★
Asia	132302	193103	239764	318210	363636	398779	22.0★★
Europe	1231	1986	2962	3048	2122	2399	10.2 NS
Oceania	94	182	430	634	746	122	16.9 NS
Domestic supply (1000 t)							
World	140079	206207	261069	338862	393093	423471	22.1★★
Africa	3283	4865	8104	12006	16374	21616	38.4★★
Americas	6264	8609	12346	16263	20405	21762	26.0★★
Asia	128891	189981	236266	306344	352250	375565	21.2★★
Europe	1571	2663	4167	4011	3638	4102	16.3 NS
Oceania	69	88	185	238	425	426	40.2★★
Rice supply (kg/cap/yr)							
World	38.70	47.70	50.00	54.50	53.80	52.90	5.7★
Africa	9.00	10.60	14.20	16.00	17.50	19.60	15.8★★
Americas	11.80	13.50	16.00	18.30	19.60	19.00	10.4★★
Asia	65.20	78.40	78.70	84.30	79.80	77.90	2.9 NS
Europe	2.10	3.40	5.00	4.40	4.50	4.80	13.8 NS
Oceania	4.60	4.70	8.50	9.80	11.50	12.90	22.8★★
Wheat							
World production (1000 t)	219214	307416	436653	589456	583950	604465	20.8★★
World domestic supply (1000 t)	222911	3304463	439847	544111	576801	612530	19.8★★
World food supply (kg/cap/yr)	54.7	57.7	66.1	67.4	67.6	65.9	4.1★

★CGR = Compound Growth Rate
★★Significant at 5% ★Significant at 10%
Source: FAOSTAT: Food Balance Sheets, 1961–2007. Downloaded 30.08.2011

5.1%, Europe: 0.5%, and Oceania: 0.02%. Production is steadily increasing in Africa. The production and consumption patterns of rice in different continents and in different regions of the continents also show certain interesting trends (see Tables 1.3 and 1.4).

Rice production across the world has increased by about three-fold in 45 years. It has increased by as much as five times in Africa, three times each in the Americas and Asia, two-fold in Europe, and by just one-third in Oceania (Table 1.3). However, during this period, the per capita rice supply within different regions showed a much smaller increase. This is because of the variable or differential rates of population increase in different regions and continents. Domestic supply has increased almost seven-fold in Africa, six-fold in Oceania, 3.5-fold in the Americas, a little less than three-fold in Asia, and in Europe, about two and half times. These figures indicate that Africa continues to remain a major importer of rice, as do Europe and Oceania, though to a lesser extent. However, since the overall consumption and import of rice in these continents remained at relatively low levels, there seems to have been little global impact of this on supply and prices. Only the situation in Europe appears to have shown any change. The influx of Asians and Africans into Europe after the Second World War appears to have contributed to the increased consumption of rice in this continent (Table 1.3).

In the Food Balance Sheets (FBS), the term domestic supply indicates the actual quantity available for direct human consumption. This is calculated by adding imports, if any, to the country's production, and subtracting exports, quantities retained for seed and processing, and losses suffered due to various causes (handling, insect damage, etc.). The data (in Table 1.3) show that Africa, Europe, and Oceania are net importers. While rice consumption (i.e., rice consumed in kg/cap/yr) increased globally by about 1.4 times during 1961–2007 – from 38.7 to 52.9 kg/cap/yr – that of Africa increased by about 2.2 times – from 9.0 to 19.6 kg/cap/yr. The increase in consumption has been nominal in the other continents.

In 2007, about 7.8% of total rice production entered international trade (Table 1.4). In wheat, the equivalent figure was 27.6%. Over the long term, only about 4–8% of annual rice production enters international trade. In wheat, the figure fluctuates between 18 and 28%.

The top rice exporting countries of the world are USA, Thailand, Brazil, Paraguay, and Egypt, and the major importing countries are Bangladesh, Indonesia, Venezuela, and Costa Rica, in addition to West Africa. Both the exporting and importing countries show considerable year to year variation, in both imported and exported quantities. However, USA, Thailand, and Egypt are usually regular exporters.

Table 1.4 Food Balance Sheet of Rice (Milled Equivalent) in Various Regions (2007)

Continent/Region	Production 1000 t	Import 1000 t	Export 1000 t	Domestic Supply 1000 t	Food 1000 t	Rice supply kg/cap/yr
World	437509	29151	34368	423471	348732	52.9
Africa	13940	8127	1372	21616	18908	19.6
Eastern Africa	3314	1397	65	4755	4141	13.7
Middle Africa	363	663	0	1251	1197	10.1
Northern Africa	4624	445	1192	3935	3232	16.0
Southern Africa	2	1042	19	1035	1033	18.3
Western Africa	5637	4579	95	10639	4305	32.8
Americas	22269	4543	5497	21762	17211	19.0
Northern America	6002	1363	3579	3977	2854	8.4
Central America	726	1100	92	1705	1580	10.7
Caribbean	819	1059	9	1903	1612	44.6
South America	14721	1021	1817	14177	11165	29.4
Asia	398779	12516	25290	375565	308741	77.9
Central Asia	448	55	44	473	395	6.7
Eastern Asia	137534	2698	1333	138908	115363	75.0
Southern Asia	137124	2021	9501	128177	116552	72.1
Southeastern Asia	123237	4633	14317	105113	73863	131.0
Western Asia	435	3109	94	2893	2567	13.8
Europe	2399	3687	2007	4102	3519	4.8
Northern Europe	0	767	221	546	501	5.1
Eastern Europe	587	729	113	1260	1162	3.9
North & South Europe	1753	606	1154	1169	986	6.5
Western Europe	58	1585	519	1126	870	4.6

Oceania	122	279	203	426	353	12.9
Australia & New Zealand	109	177	203	310	249	9.9
Melanesia	14	85	0	97	87	48.2
Micronesia	0	4	0	6	5	48.3
Polynesia	0	13	0	13	12	26.9

Source: FAOSTAT: Food Balance Sheets, 2007. Downloaded 01.09.2011
Domestic supply: Actual quantity available for direct human consumption

4. RICE AS A STAPLE FOOD

There does not appear to be a universally used definition of a staple food. An older one from the FAO (1995) is that it is a food that: "is eaten regularly and in such quantities as to constitute the dominant part of the diet and supply a major portion of energy and nutrient needs".

It does not quantify the term, and neither does the development literature. This is understandable because there are regions in the world, particularly in the tropics, where different commodities constitute the major food of the local population during different seasons of the year. For instance, in some islands of Oceania, breadfruit, aroids, sweet potato, and banana constitute the major food during different periods of the year. However, in recent publications, the FAO has considered food items to be staple foods if they contribute a minimum of 10% of the total energy intake. Thus, with a global 19.0% contribution of energy to humans (533 kcal/cap/day; Table 1.1), rice is the most important staple food of humankind. The only other comparable food by this definition is wheat (Table 1.1, 530 kcal/cap/day; 19.0%). The total per capita consumption of food by humans is 2,798 cal/day.

There is much variation in the consumption patterns of rice among different continents of the world, among different regions within the same continent, and also among countries within the same geographical region of a continent (Tables 1.5 and 1.6). Rice forms a staple food (providing a minimum of 10% of the calories consumed) in 68 out of total 193 countries in the world (Table 1.5). This includes also a few countries where consumption is marginally lower than 10%. Among them, there are 11 countries in the world where the annual per capita rice consumption is more than 100 kg (Tables 1.5 and 1.6). They are: Bangladesh (160 kg/year, 69.8% of total calories), Brunei (245 kg/year; 26.6% of total calories), Cambodia (152 kg/year; 64.1% of total calories), Guinea (128 kg/year; 37.5% of total calories), Laos (163 kg/year, 64.2% of total calories), Madagascar (102 kg/year; 48.1% of total calories), Myanmar (157 kg/year; 54.4% of total calories), Philippines (129 kg/year; 49.6% of total calories), Thailand (103.1 kg/year; 40.3% of total calories), and Vietnam (166 kg/year; 57.8% of total calories). For comparison, the equivalent values for three of the developed countries are also given. They are: Japan (56.6 kg/year; 21.6% of total calories), UK (5.6 kg/year; 2.1% of total calories), and the USA (8.3 kg/year; 2.3% of total calories). The overwhelming dependence of the most populous regions of the world on rice is thus obvious. Rice can thus truly be described as the staff of life for humankind.

Table 1.5 Countries Where Rice is Used as Staple Food (2007)*

Continent/Region/ Country	Production (2009) Rice Paddy 000 t	Food Supply Kcals/cap/day Total	Rice	Rice Kg/cap/year
World	–	2796	532(19.0%)	52.9
Africa	–	2445	167(6.8%)	19.6
Eastern Africa	–	2045	138(6.7%)	13.7
Comoros	19	1884	595(31.6%)	59.5
Djibouti	Not available	2291	387(16.9%)	39.2
Madagascar	4738	2160	1038(48.1%)	102.1
Mozambique	180	2067	198(9.6%)	20.3
Seychelles	Not reported	2463	296(12.0%)	29.8
Tanzania	1105	2032	194(9.5%)	18.9
Cameroon	175	2269	223(9.8%)	22.4
Gabon	1	2755	375(13.6%)	37.5
Middle Africa	–	1860	101(5.4%)	10.0
Northern Africa	–	3016	168(5.6%)	16.0
Egypt	4330	3195	388(12.1%)	37.0
Southern Africa	–	2918	181(6.2%)	18.3
Mauritius	Nil	2965	503(17.0%)	49.9
Swaziland	<1	2292	204(8.9%)	20.7
Western Africa	–	2649	324(12.2%)	32.8
Benin	167	2533	326(12.9%)	32.6
Cape Verde	Not reported	2572	462(18.0%)	46.2
Cote d'Ivorie	723	2528	552(21.8%)	61.0
Gambia	100	2385	580(24.3%)	57.6
Ghana	492	2907	273(9.4%)	28.2
Guinea	1499	2561	961(37.5%)	128.2
Guinea-Bissau	177	2306	848(38.5%)	84.7
Liberia	296	2204	734(33.3%)	73.0
Mali	2308	2614	540(20.7%)	54.4
Mauritania	134	2841	302(10.6%)	30.8
Sierra Leone	1027	2170	898(41.4%)	92.4
Togo	110	2161	198(9.2%)	20.3
Americas	Not reported	3216	193(6.0%)	19.0
Northern America	–	3727	88(2.4%)	8.4
Belize	19	2718	314(11.6%)	32.2
Central America	–	3043	106(3.5%)	10.7
Costa Rica	265	2840	297(10.5%)	51.7
Nicaragua	454	2403	399(16.6%)	42.6
Panama	274	2484	587(23.6%)	60.1
Caribbean	–	2561	444(17.3%)	44.6

Continued

Table 1.5 Countries Where Rice is Used as Staple Food (2007)*—cont'd

Continent/Region/ Country	Production (2009) Rice Paddy 000 t	Food Supply Kcals/cap/day Total	Food Supply Kcals/cap/day Rice	Rice Kg/cap/year
Cuba	454	3274	635(19.4%)	64.1
Dominican Republic	567	2295	342(17.1%)	40.3
Haiti	125	1870	392(21.0%)	39.2
Jamaica	<1	2812	285(10.1%)	26.0
Netherlands Antilles	Not reported	3004	341(11.4%)	35.1
St Vincent and the Grenadines	Not reported	2821	288(10.2%)	28.7
South America	–	2886	298(10.3%)	29.4
Bolivia	449	2064	276(13.4%)	27.0
Brazil	11236	3113	338(10.9%)	33.2
Columbia	2412	2685	308(11.5%)	30.3
Ecuador	1706	2301	441(19.2%)	46.8
Guayana	556	2759	783(28.4%)	80.6
Peru	2831	2457	516(21.0%)	48.0
Suriname	227	2492	616(24.7%)	66.4
Asia		2668	783(29.3%)	77.9
Central Asia	–	2783	65(2.3%)	6.7
Eastern Asia	–	2955	782(26.5%)	75.0
China	197212	2981	799(26.8%)	76.8
Japan	10600	2812	607(21.6%)	56.6
N Korea	242	2087	757(36.3%)	77.5
S Korea	7023	3074	825(26.8%)	75.9
Southern Asia	–	2370	715(30.2%)	72.1
Bangladesh	50061	2281	1591(69.8%)	159.7
India	143963	2352	703(29.9%)	70.9
Maldives	Not reported	2685	423(15.8%)	43.3
Myanmar	32682	2440	1342	156.9
Nepal	4024	2360	744(31.5%)	171.3
Sri Lanka	4301	2361	942(39.9%)	97.3
Southeastern Asia	–	2586	1270(49.1%)	131.0
Brunei	1	2968	789(26.6%)	244.9
Cambodia	8245	2268	1453(64.1%)	152.2
Indonesia	66469	2583	1238(48.8%)	125.3
Laos	3071	2240	1439(64.2%)	162.5
Malaysia	2548	2923	755(25.8%)	76.5
Myanmar	33205	2465	1342(54.4%)	156.9
Philippines	15772	2565	1271(49.6%)	129.3
Thailand	31597	2539	1024(40.3%)	103.1
Timor-Leste	113	2066	673(32.6%)	74.6
Vietnam	39989	2816	1629(57.8%)	165.6

Table 1.5 Countries Where Rice is Used as Staple Food (2007)*—cont'd

Continent/Region/ Country	Production (2009) Rice Paddy 000 t	Food Supply Kcals/cap/day Total	Rice	Rice Kg/cap/year
Western Asia	–	3104	140(4.5%)	13.8
Saudi Arabia	Nil	3144	331(10.5%)	31.0
UAE	Not reported	3171	412(13.0%)	41.7
Europe	–	3406	49(1.4%)	4.8
Eastern Europe	–	3323	38(1.1%)	3.9
Northern Europe	–	3401	54(1.6%)	5.1
Southern Europe	–	3510	65(1.9%)	6.5
Western Europe	–	3536	49(1.4%)	4.6
Oceania	–	3182	123(3.9%)	12.9
Australia & New Zealand	–	3216	92(2.9%)	9.9
Melanesia	–	2800	501(17.9%)	48.2
Micronesia	–	2899	479(16.5%)	48.3
Polynesia	–	2919	270(9.2%)	26.9
Fiji	8	3041	491(16.2%)	44.7
French Polynesia	Not reported	2942	367(12.5%)	36.6
Kiribati	Not reported	2899	479(16.5%)	48.3
Solomon Islanda	4	2422	652(26.9%)	66.7
Vanuatu	Not reported	2740	528(19.3%)	52.6

*Computed from FAOs Food Balance Sheets (2007)

5. RICE AND CULTURE

Rice is one of the most ancient crops of humankind. Archaeological remains of rice that are about 10,000 years BP (Before the Present) have been discovered in eastern China, and the Ganga valley in India, and of slightly more recent periods have been found in several other locations in south, southeastern, and eastern Asia (cf., Chapter 6). Remains of rice that are about 2,500 years old have been discovered in West Africa, but the taxonomic identities of some of them are uncertain (cf., Chapter 5). Thus, the antiquity of rice in southern, southeast, eastern Asia, and tropical West Africa is well established. We shall be discussing this aspect in detail in Chapters 5 and 6. In both Asia and tropical West Africa, most field operations of the rice growing systems have involved collective, cooperative, and synchronized community efforts. They include bunding, dewatering, terracing, water management, etc. in most lowlands of these regions, including those that are situated below sea level,

Table 1.6 Rice Calories Supply as Per Cent of Food Used by Country (2000)

Country/Region	Per cent	Country/Region	Per cent	Country/Region	Per cent
World	21	North & Central America	3	Mauritania	17
				Mauritius	18
Asia	32			Nigeria	9
Bangladesh	72	Costa Rica	18	Senegal	34
Brunei	30	Cuba	25	Sierra Leone	46
Cambodia	71	Dominican Republic	19	Tanzania	7
China	30			Togo	10
India	31	El Salvador	4		
Indonesia	49	Haiti	20	Europe	1
Japan	24	Honduras	5	Belgium	1
Korea DPR	33	Jamaica	11	France	1
Korea Republic	31	Nicaragua	19	Germany	1
Laos	66	Panama	17	Netherlands	1
Malaysia	28	Trinidad	8	UK	1
Myanmar	70	USA	2	Ukraine	1
Nepal	39			Greece	2
Philippines	43	Africa	7	Italy	2
Sri Lanka	38	Burkina Faso	6	Russian Federation	2
Thailand	40	Cape Verde	16		
Timor	35	Comoros	29	Spain	2
Vietnam	68	Congo Republic	8		
		Cote d' Ivorie	24	Oceania	4
South America	11	Egypt	12	Australia	3
Bolivia	9	Gabon	13	Fiji	16
Brazil	13	Gambia	17		
Columbia	11	Ghana	5		
Ecuador	17	Guinea	29		
Guyana	21	Guinea Bissau	31		
Peru	20	Liberia	28		
Suriname	30	Madagascar	46		
		Mali	20		

Source: Calculated from FAOSTAT 2000

and in river valleys, terraced uplands etc. (Carnay 2001, Linares 2002, Fields-Black 2008, author's observations, etc.). This has been a unique feature of rice cultivation, which contrasts with those of other major ancient crop plants.

Rice has been and continues to be an integral part of the social, religious, and cultural traditions and customs of most people of southern, southeastern, eastern Asia, and tropical West Africa (Watt 1892, Kumar 1988, Carney 2001, Fields-Black 2008, Prance and Nesbitt 2005, Carney and Rosomoff 2009, Sharma 2010). Readers are invited to go through the above references for further details.

6. RICE AND NUTRITION

There are two cultivated rice species in the world, the universally cultivated Asian rice, *Oryza sativa* L. and the African rice, *O. glaberrima* Steud. The latter species belongs to tropical West Africa. Its commercial cultivation has now all but disappeared and it has been replaced by the higher yielding, more adaptable Asian rice. African rice is now only grown by some isolated farmers mainly for ceremonial purposes (Nayar 2010). Rice is the staple food for almost two-thirds of the human population (FAO 2004). It provides more than half of the food intake for the majority of people in Southeast Asia.

Rice is mostly consumed in its raw form after milling, but in certain regions of the world, after parboiling paddy rice. Hand milling and pounding of rice paddy to remove the husk, and also polishing the grains used to be normal practice until the 1940s in most parts of the Old World, including tropical West Africa. Also, rice, which has a red pericarp, continues to be preferred in a few regions of the world, such as some in the Kerala-Konkan coast and in the southeastern regions of India, southern Bangladesh, and tropical West Africa. African rice also has a red pericarp. Hand-pounded rice, red pericarped rice, and parboiled rice are more nutritious than the predominantly consumed white-pericarped raw, milled, and polished rice (Table 1.7). Efforts are underway to develop rice cultivars with higher nutritional value. There are also some rices that are famously used for medicinal and therapeutic purposes, such as the *njavara* rices of the Malabar Coast in southwest India.

Milling and polishing rice removes considerable quantities of nutrients, including proteins, fats, fiber, iron, and B-vitamins. Rice is consumed in a great variety of forms – as milled rice after boiling, and decanting the water or evaporating the water, as gruel, or as unmilled rice after pounding (beaten rice, like oats) or puffing (like popcorn), and so on. Rice is also consumed after wet and/or dry grinding, alone or in combination with a pulse. It is used for making innumerable snack and breakfast preparations, and is eaten after steaming and frying with various additives, such as grated coconut, sugar or jaggery, as pasta and noodles, and even after deep frying in oil.

Rice is a rich source of dietary energy and a good source of thionin, riboflavin, and niacin (FAO, 1993, 2004). Of the amino acids, it is high in glutamic and aspartic acids, but low in lysine (Juliano 1993). This may explain why the traditional rice diets in Asia and tropical West Africa came to be complemented with fish and/or pulses (Table 1.7). Different cultures have varied preferences for the taste, texture, color, and stickiness of cooked

Table 1.7 Nutritional Content of Rice

Type of Rice	Protein (g/100 g)	Iron (mg/100 g)	Zinc (mg/100 g)	Fiber (g/100 g)
White, polished	6.8	1.2	0.5	0.6
Brown	7.9	2.2	0.5	2.8
Red	7.0	5.5	3.3	2.0
Purple	8.3	3.9	2.2	1.4
Black	8.5	3.5	–	4.9

Source: FAO 1993

rice. For instance, dry flaky rice is used in south Asia, tropical West Africa, and the Middle East, parboiled and red pericarped rice is preferred in south-western and southeastern India and adjoining Bangladesh, and wet sticky (glutinous) rice is preferred in East Asia (FAO 2004).

Rice possesses tremendous variability. This may be primarily because the putative ancestral species of cultivated rice occurs extensively throughout most of Monsoon Asia and tropical West Africa, where most of the rice is grown and both the species share several morphological and genetic attributes.

CHAPTER 2

History and Early Spread of Rice

There are two cultivated rice species, the universally cultivated Asian rice, *Oryza sativa* L. and the African rice, *O. glaberrima* Steud., that is cultivated only in west Africa. Asian rice originated in Asia about 10,000 years ago and African rice in tropical west Africa about 3,000–5,000 years ago. The origins of the two rices will be discussed in Chapters 5 and 6. Rice is one of the oldest food plants of humankind.

1. RICE IN EAST ASIA

i. In Korea and Japan

Rice is assumed to have been taken to Korea and Japan from China (Heu and Moon 2010). The earliest archaeological evidence of rice culture in Korea appears around 4,300 BP. The authors have given details of rice samples obtained from 10 archaeological sites. They range in age from 4,330 ± 80 years BP to 2,000 years BP. The number of rice grains obtained from these sites ranged from 7 to 400, but only 22 and 7 grains were retrieved from the oldest two – both of which are the only ones dated before 4,000 BP. The small number of rice grains obtained from the archaeological sites may suggest that rice was only a minor food item at that period of time, but comparative information *vis-à-vis* other foods is not currently available. Most of the rice grains were shorter and thicker, and looked like japonicas. Some of the sites contained also grains from other crops, including sorghum (*Sorghum bicolor*), millet (*Panicum miliaceum, P. glaucum*), and soybean (*Glycine max*) (Heu and Moon 2010).

In Korea, evidence for the use of agricultural implements began to appear in the Bronze Age (3,000–2,300 BP). Rice cultivation was widely diffused throughout the peninsula and became a common crop by the early Iron Age (2,300–2,100 BP) (Heu and Moon 2010). Another significant finding is the discovery of rice fields in several locations in the Korean Peninsula dating from the 6th century CE onwards.

Rice is believed to have been introduced into the north or northwest of Korea from China. Here, the earliest archaeological evidences for rice cultivation have been obtained from Shandong Banda (4,440 ± 130 BP) and Liaodong (2,945 ± 130 BP), both in northeast China. There are also

Origins and Phylogeny of Rices
http://dx.doi.org/10.1016/B978-0-12-417177-0.00002-4

historical records of the conquest of the Gochosun state of Korea in 2,059 BP by the Han state of China (Tang and Xuan 2010). This could have contributed to the spread of rice cultivation into the Korean Peninsula. Incidentally, the introduction of Chinese civilization into the Korean Peninsula dramatically changed its culture and social structure. It provided also a system of writing (Tan 1995) at a relatively early period in its history, about 2,000 years ago.

Rice was introduced into Japan about 3,000 years BCE, again possibly from China. The archaeological evidence of rice in Japan is scarcer than that reported in Korea. The art of writing also came into vogue somewhat later in Japan, in the early 8th century CE (Kaneda 2010).

Currently, the earliest archaeological evidence of rice in Japan has been obtained from Hikosaki Kaisuka, Okayama City, southeast Honshu Island (Kaneda 2010), which has been dated to about 6,000 years BP. Here, rice opals were found in the soil. Opals are microscopic silica bodies found in motor cells. When plant tissues decay, the silica bodies are released into the soil, where they remain indefinitely. The morphological features of the silica bodies found appeared to be like those of japonica rice. However, the earliest evidence of rice cultivation in Japan relates to a much later date. Rice is generally grown in Japan under irrigated lowland conditions, and the varieties used are temperate japonicas. Rice was possibly first introduced into Japan in Kyushu, its southernmost island. Kyushu faces both Korea and northeastern China. While evidences of farming are indicated in the form of farm implements, fields in various stages of preparation and also artifacts, there was little evidence of actual rice cultivation then (Kaneda 2010). The author has recently reviewed the history of rice in Japan.

The evidences for early rice cultivation found so far in Japan consists of lowland rice fields, several kinds of farm tools, and carbonized rice grains. These have come from the excavations of Itazuke (Fukuoka prefecture) and Nabataka (Saga prefecture), both in northern Kyushu. They have been ascribed to the late Jomon (period 7,500 BCE – 250 BCE, Mesolithic) to Yayoi periods (250 BCE – 250 CE, Neolithic). Excavations conducted in different locations showed signs of progressive improvements having been made in paddy cultivation with the passage of time. These included the use of more specialized and a wider range of farm tools, use of progressively improved systems for land preparation, irrigation, and cultivation, and so on (Kaneda 2010).

Taneka et al. (2010) have indicated that rice cultivation in Japan is only 2,400–3,000 years old. It appears that initially, tropical japonica was introduced more frequently than temperate japonica, and even this was only sporadic. However, rice cultivation had spread up to the northern island of Honshu by 400 BP. In the well-known Jomon site (5,900–3,500 BP) in Sannai Maruyama (northern Honshu), the evidence indicates that the inhabitants subsisted by hunting and gathering. There is also some evidence that the local population did provide some kind of protection from floods to chestnut, wild grapes, and millet, *Echinochloa crusgalli*. No evidence of rice was found here however.

ii. In China

China has been a major rice growing country since very early times. Presently, China ranks first in the world in terms of rice paddy production, at 197.2 Mt annually (2010). The early history of rice cultivation, and its role in the cultural traditions of the country have been outlined by de Condolle (1886) and in several ancient Chinese texts.

Tang and Xuan (2010) have listed 16 archaeological sites, mostly in northeastern China, which are older than 7,000 years, and where rice has been found (Table 6.1). The wild rices, *O. rufipogon*, both annual and perennial, also grow widely in almost all of the regions where archeological rice remains have been obtained. This is reflective of its wide cultivation and use since prehistoric times. Both indica (hsien) and japonica (keng) rices are grown in China. Broadly, the indica rices are grown in the warmer south and southwest regions up to 29° N lat., and japonica rices are grown in the north. Rice and rice cultivation have a long and hoary past in China. Many aspects of rice cultivation, such as terraced farming, irrigation, and transplanting have been practiced in China from very ancient times. Tang and Xuan (2010) have summarized the history and spread of rice cultivation in China.

The first evidence of transplanting rice has come from the Han dynasty (25–200 CE). The technique used was described by the author Jia Sixie in the book, *Quiminyaoshu*, written during the Beiwei dynasty (386–531 CE). Transplanting rice became a common practice in northern China in the sixth century CE (Tang and Vuan 2010).

Rice cultivation is thought to have started in the middle and lower reaches of the Yangtze River (30° N), and then spread gradually both northwards and southwards (Bellwood 2004, Bellwood and Glover 2004, Tang and Xuan 2010). The practice spread to the Huai River (34° N) by 6,000–5,000 years BP, and over the next millennium, its cultivation also spread to both south and west China. However, rice did not appear to have been the dominant food crop of the people in the northern region during the earliest period of rice cultivation. This idea is borne out by the recovery of shellfish, pigs, fruits, legumes, and millets from excavations where the earliest rice remains were obtained (Tang and Xuan 2010, Liu and Chen 2012). Incidentally, millet too has a very ancient origin in China (Zohary et al. 2012).

At present, rice is one of the principal foods of China. The quantity, extent, and manner of consumption vary widely across regions and nationalities within the country. Possibly, rice has the most diversified end uses of all the food plants consumed in China.

2. RICE IN SOUTHEAST ASIA AND OCEANIA

This is a vast region, extending from mainland southeast Asia and the thousands of islands scattered to its east, to the southeastern region of the continent (the Far East) and then beyond to the thousands of islands dotting the Pacific Ocean, and then to Australia and New Zealand.

Southeast Asia is the home of rice even if it was not its homeland. It is in this region that the per capita consumption of rice is the highest in the world (Fig. 2.1, Tables 1.5 and 1.6).

i. In Mainland Southeast Asia

This region extends from Vietnam and Laos in the north, Myanmar in the west, and Indonesia in the south and southeast, up to western New Guinea, and the Philippines in the east.

In this region, the sea level had gone down by more than 100 meters from the present levels during the Last Glacial Maximum (26.5 to 19.2 ka; Clark et al. 2009). This resulted in the formation of the biogeographical region of Sundaland, which included the Malay Peninsula, Sumatra, Java, Borneo, and their surrounding islands. This region was bounded in the east by the Wallace Line. Then deglaciation began to take place from 19–20 ka. Most of this process appears to have been completed by 16 ka (Hanebuth et al. 2000, Lambeck and Chappel 2001). The southeast Asia that we know today may have been created by this time.

After the completion of deglaciation, the sea level in this region is assumed to have risen by 40–50 m in the last 12,000 years (some authors claim up to 90 m), corresponding to the Holocene Period. The earliest humans are now thought to have settled in this region about 50,000 years ago. They were hunter-gatherers and led a semi-nomadic life in the jungles or on the sea shore (Bellwood and Glover 2004).

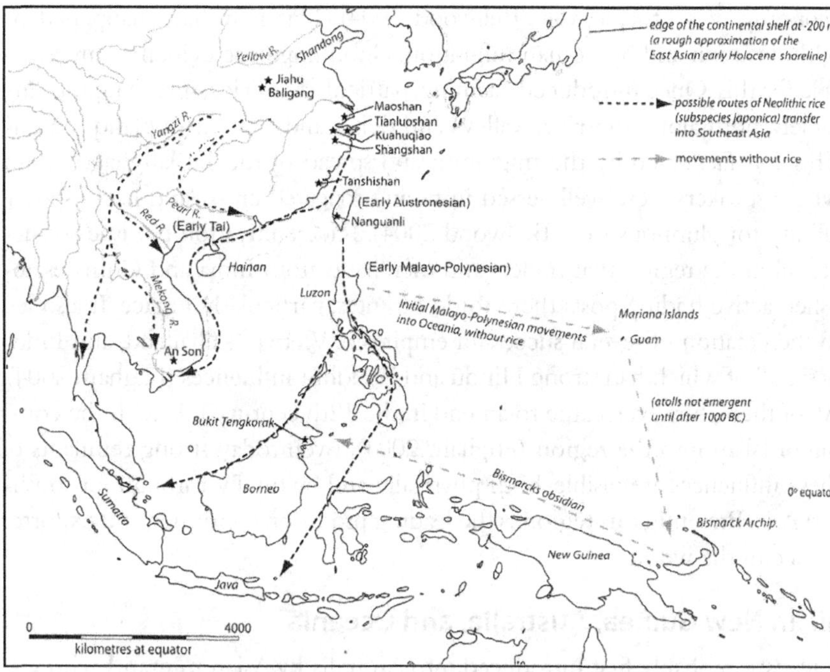

Figure 2.1 ***East Asia and Southeast Asia Showing Possible Spread of Rice from Lower Yangtze River Basin and Some Important Archaeological Sites.***
Adapted from Bellwood (2011) with Permission

From their studies in Thailand, Weber et al. (2010) had concluded that millet and rice were first introduced into the region in only 2,500 BCE. The authors had excavated in the Khao Wong Prachan (KWP) valley in central Thailand, where they had found evidence of human occupation in the region ranging from the second millennium BCE to the later centuries of the first millennium CE. Millets were used more widely than rice during that period. In excavations conducted at three sites in the region, they found carbonized seeds (3,294 in number) of nine millet genera and rice; these were mostly *Setaria italica*, but *Panicum* and *Coix lachryma-jobi* were also discovered. AMS determinations of two rice samples yielded dates of 2,490 ± 40 yrs (conventional)/78–140 cal BCE and 2,550 ± 40 yr (conventional)/690–540 cal BCE. Further details of the rice archaeology of the region are given in Chapter 6.

ii. In Island Southeast Asia

Agriculture involving paddy culture was introduced by people arriving from southern and central China about 5,000 years ago into the hunter-gatherer economy of southeast Asia (Bellwood 2004). This may have happened by both diffusion and by human migration. Good linguistic evidences are available for this. Once introduced, farming, particularly of rice, spread rapidly and widely, both along the river valleys and deltas and also into upland regions. This was facilitated by the migration and spread of the Tai language group, whose speakers were well versed in raising irrigated crops, and had a special affinity for glutinous rice (Bellwood 2004). Rice cultivation became so successful in the region that traders and merchants from India and China established active trading posts there to do business, particularly in rice. It also led to the creation of several successful empires in Vietnam, Thailand, and Indonesia, all of which had strong Hindu and Buddhist influences (Higham 2004). All of them, however, came to an end in the 12th century CE with the coming of Islam into the region (Higham 2004). Even today, strong remnants of these influences are visible, both physically and culturally, throughout southeast Asia. Presently, this region is the leading producer, consumer, and exporter of rice in the world.

iii. In New Guinea, Australia, and Oceania

Rice was probably first introduced into Australia by Asian migrants who visited the northern region in the 15th century to fish and trade with the indigenous people. There is, however, no direct evidence for this (Falvey 2010). The general assumption is that rice was first introduced into southeastern Australia in the 1850s by Chinese gold prospectors. As the gold rush declined, the Chinese became dispersed. Numerous subsequent attempts made by the Europeans to grow rice in southeast Australia failed. Rice cultivation became finally successful in the 1850s in north Queensland, northern Australia, but some time later, rice production gave way to sugarcane in this region. Then in the early 20th century, rice cultivation was again attempted in southeast Australia using irrigated agriculture, and this was a success. It received a fillip after the Second World War. Presently, Australia produces more than 2 Mt rice (japonica), most of which is exported (FAO 2010).

3. RICE IN SOUTH ASIA

i. In Continental South Asia

South Asia includes Afghanistan, Pakistan, Nepal, Sri Lanka, India, Bhutan, Maldives, and Bangladesh. Discussions in this section covering India in the time period up to 1947 will include Pakistan, India, and Bangladesh, unless otherwise specified, since at that time, they were one political entity.

South Asia is a major producer, consumer, and exporter of rice. India and/or China are considered as the centers of origin of Asian rice. Like China, India is also home to one of the ancient civilizations of the world, the Indus Valley or Harappan civilization (Allchin and Allchin 1997). Further, from prehistoric times, travelers and traders from Rome, Mesopotamia, and China used to visit this region by sea and/or land for trade and learning, even though the northern and western borders are naturally protected by the great Himalayan and Hindu Kush mountains. The Hindu Kush Mountains, however, have several mountain passes within them, through which settlers and conquerors have been gaining access to the subcontinent since prehistoric times. The peninsular region lying to the south of the Vindhya Mountains consists of a semiarid plateau, and consequently, the region is basically unsuitable for rice cultivation. However, the Konkan-Malabar coast that lies between the Western Ghats and the Arabian Sea, the Coromandel Coast lying between the Eastern Ghats and the Bay of Bengal, and the valleys and estuaries of numerous rivers that flow through this region provide suitable conditions for raising two, or three crops of rice annually. This is because of the prevailing tropical climate, good rainfall, and availability of irrigation from large numbers of tanks and reservoirs. To the north of the Vindhya mountains is the great Indo-Gangetic plain, which is watered by the three major river systems, the Indus, Ganga-Yamuna, and Brahmaputra. Here, the major cereal crops consist of wheat, rice, sorghum, and millet.

It was widely assumed that indica rice originated in southern Asia, but recently, Fuller and co-workers (summarized in Fuller 2011) proposed a "proto-indica domestication hypothesis" for the origin of indica rices in south Asia, and their subsequent spread. He observed that numerous archaeological findings from the Ganga basin revealed the prevalence of integrated agricultural systems, in existence from 2,200–2,000 BC to the early centuries of the second millennium.

Rice was introduced to the Ganga basin from the Indus region from about 2,400–2,200 BC, and its cultivation was well established there by 2,000 BC. A mixed cultivation system was practiced, using pulses from peninsular India, wheat and barley from west Asia, and rice (Fuller 2011).

Prior to about 2,000 BC, peach, apricot, millet (*Panicum miliaccum*), and cannabis were introduced into southwest Asia from the Yangtze basin of China, via central Asia. According to Fuller, japonica rice might also have been introduced alongside them. It then hybridized with the local wild rice to produce what the author had designated as proto-indica populations. He proposed further that the indica rices developed from these mixed populations. Fuller (2011) has stated that no supporting archaeological evidence for these ideas is available as yet in the form of ceramic types, settlement patterns, or palaeobotany materials..

In contrast to this idea is the "austric farming dispersal hypothesis" put forward by Bellwood (1996) and Higham (2004). According to this, rice was introduced into south Asia in Assam (in northeast India) from the Yunnan province (of southwest China), and from Assam, rice spread westwards to the Ganga basin and beyond.

The middle Ganga plains, lying to the south of present-day Allahabad city, have been considered to be one of the ancient centers of farming (Saraswat 2005). However, some of the earlier series of excavations and consequent claims about the antiquity of rice remains (Sharma and Manda 1980) have been marred by problems in archaeobotanical sampling and analyses, and hence, its chronology has also been questioned. The excavations and analyses carried out more recently in Lahurdewa (Sant Kabir Nagar district, Uttar Pradesh state, India) and elsewhere have attempted to overcome the deficiencies of the earlier excavations (Saraswat and Pokharia 2004, Pokharia 2011).

The lowest layers of excavation in Lahurdewa yielded mixtures of cultivated and wild rice, and also broken pieces of foxtail millet (*Setaria italica*) (Saxena et al. 2006, Tewari et al. 2006). The associated charcoal yielded dates of 6th–5th century BCE, and a husk of cultivated rice yielded an AMS date of 6,409 BCE. Further excavations yielded wild rice phytoliths dated at 10,000 BP, and cultivated rice at 8,300 BP. In another series of excavations conducted at Mahagara, Koldihwa, and Chopani Mande in the same region, Sharma and his co-workers observed the transition from food gathering to farming (Sharma et al. 1980, Sharma 1985). They dated this to 7th–6th century BCE. The authors observed impressions of rice in pottery, and both cultivated and wild rice grains in Mahagara.

The Indus Valley civilization in northwest India (mature period: ca. 2,600 to 1,900 BCE) was a major development in southern Asia, that was urban in its organization (Allchin and Allchin 1997, Thapar 2003). Most of the originally excavated sites are located in present-day Pakistan. After India and Pakistan became independent in 1947 CE, several new sites were excavated in northwest India. The main excavation sites are Harappa, Mohenjo Daro, Rojdi, Rangpura, and Lothal (Allchin and Allchin 1997, Possehl 2002, Singh 2009). While wheat, barley, and sorghum were the staple foods of the people of that period, as many as 11 rice samples, all older than 2,000 years, and consisting mostly of charred grains and husks, and their impressions on pottery have also been discovered in the region (Chowdhary and Ghosh 1953, Ghosh 1961, Chowdhary 1965).

Numerous Neolithic sites have been discovered in Deccan Peninsula. Although millets were the most common food item found in these sites, there were also several examples of rice in the region, dating from the first millennium BCE (Fuller 2002). Then in the Vedic (1,500–500 BCE), protohistoric, and ancient Tamil epochs, there are numerous references to the extensive use of rice in rituals and as food. Presently, south Asia is one of the two major rice producing and consuming regions of the world.

ii. In Sri Lanka

This island nation lies to the southeast of peninsular India. The two countries are separated by the Palk Straits, which are about 50 km wide. Sri Lanka (6.0–9.8°N latitude) is 65,600 km^2 in area with a typical tropical climate. Broadly, the southern part is hilly and the northern part is a vast plain. Rice is the most important crop of the country, occupying about 35% of the net cropped area.

Only a few limited archaeological studies have been conducted in Sri Lanka. Sattar et al. (2010) have given a brief overview of the history of rice in the country. According to the ancient chronicle *Mahavansha*, written in the Pali language during the 5th century CE, Vijaya was the founder of the Sri Lankan community. He came from Kalinga (the present day Orissa State in southeast India) during the 6th century BCE, bringing rice with him, which he distributed among his people. Falvey (2010) has however suggested that "mainland southeast Asia may have been the source of Sri Lankan rice before 543 BC, as the rice here is similar to that of the Malay Archipelago and Indonesia...".

It is possible that at least some rice lots might have come to Sri Lanka from southeast Asia, since we have evidence of people from southeast Asia traveling across the Indian Ocean to Madagascar from more than 2,000 years ago, and that many boats used to stop at Sri Lanka and/or on the Malabar coast to restock their food supplies (Mokhtar 1981, Ray 1993). This is discussed in Chapter 5.

4. RICE IN CENTRAL AND WEST ASIA

This vast region extends all the way from Xinjiang province in west China to Asia Minor in the west, the Russian Federation in the north and the Arabian Sea in the south. The region is marked by very low rainfall and hot and/or extremely cold climates. The staple foods are wheat, barley, and millet. Rice is traditionally used as only a specialty food. Pulao, pudding, and rice gruel are the main forms in which rice is used (Nesbitt et al. 2010). Pulao is a favorite of the region, and its various forms and varieties are widely used here. An exception may be Iran. However, contrary to what has been stated here, the central Asian countries of Azerbaijan, Kazakistan, Kyrgistan, Tajikistan, Turkey, and Turkmenistan grew rice on 330 Mha of land during 2010 (FAOSTAT 2010). In addition, the Russian Federation, which can only grow rice in the central Asian region bordering the Caspian Sea, grew rice on 0.2 Mha and Iran, on 0.6 Mha. The total rice production in the above countries was 91.30 Mt during 2010 (FAOSTAT 2010) (Fig. 2.2).

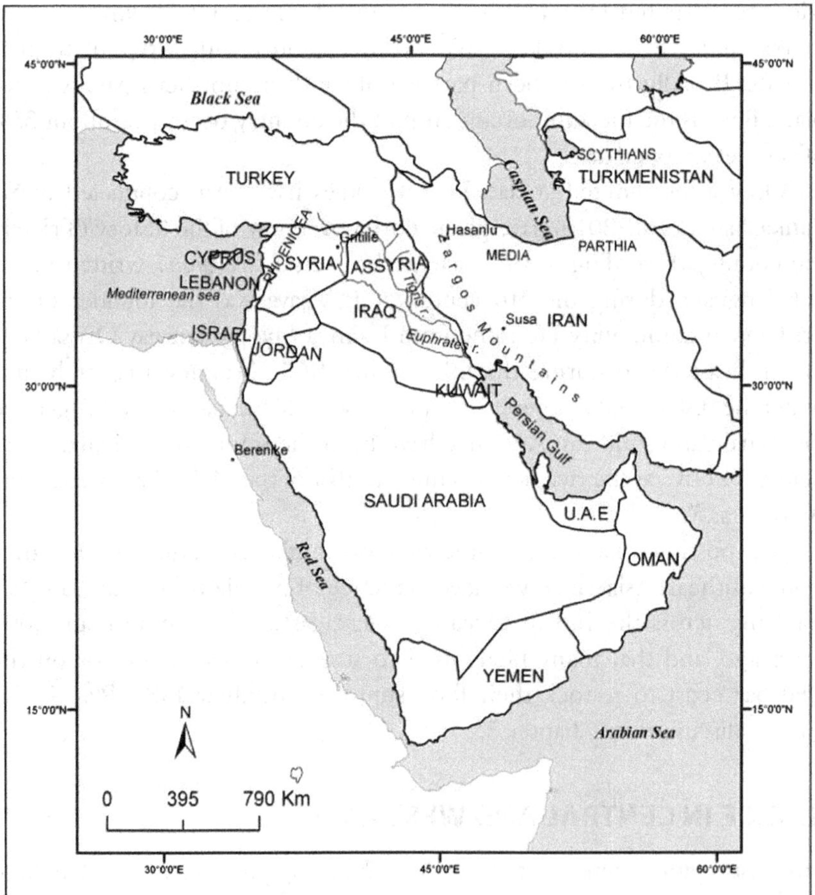

Figure 2.2 ***Southwest and West Asia Showing Places Mentioned in the Text***

In central and western Asia, rice is now grown only in the region adjoining the Caspian Sea, the southern slopes of Zagroes Mountains, and sparingly in the Euphrates-Tigris valley and Turkey. In recent years, the region, especially the Arab countries, has been importing increasing quantities of rice, possibly to meet the needs of the large population of expatriate workers from south Asia and the Philippines.

Very little information is available for the pre-Islamic period about the production and use of rice in this region. It is known that rice used to be imported into the Roman Empire from the Malabar coast of India in prehistoric times (Strabo [64/63 BCE – ca. 24 CE], Periplus [60–90 CE], and so on). Rice dating to 500 BCE has been recovered from the Roman-Egyptian trading ports of Berenike and Myos Hormos on the Red Sea in present-day Egypt (van der Veen 1999, Cappers 2007). Berenike used to act as an entrepot for goods from the west coast of India, the east coast of Africa, and the Roman Empire. From Africa and Asia came rice, spices, timber, ivory, precious and semiprecious stones, and jewelry, and from Rome came gold and silver (Sidebotham 2011) (Fig. 5.5).

The earliest mention of rice cultivation in the region dates to the second century BCE (Nesbitt et al. 2010). It reports the impressions of Zhang Qian, a Chinese envoy of the Han dynasty's Imperial Court, who stated that he saw farmers plowing the fields and growing rice and wheat in the Ferghana valley (now in Uzbekistan). In addition, Potts (1991) reported the recovery of a single charred grain of rice in a pit attributed to Period III (750–590 BCE) at Hasanlu. But it is possible that this came from a commercial sample of rice brought from elsewhere for trade and consumption.

The famous naturalist, Strabo (64/63 BCE – 24 CE) has mentioned about the writings of Aristobolous, one of the Greek chroniclers, in his book, "Geographika". Aristobolous was deputed to write about the expedition of Alexander the Great to India in 326 BCE, and he described in detail the cultivation practices of rice then being followed in the land lying between the Euphrates and Tigris rivers.

The earliest reliable evidence for the use of rice in this region has come from Susa in the South Dez plain in western Iran. This dates to the 1st century CE, and consisted of 373 carbonized grains of a short-grained variety of rice (Wenke, 1975–1976) belonging to the Parthian period (Fig. 2.2). Parthia was an ancient kingdom which lay southeast of the Caspian Sea. Rice might have been cultivated during this period in the South Dez plain, since evidence has been found in the form of rice hull impressions on fired bricks found in several sites dated between ca. 35 BCE and 250 CE

(Nestbitt et al. 2010). Susa was an ancient city of the Elamite, Persian, and Parthian empires of Iran. It was located on Zagros mountain, ca. 250 km east of the Tigris river, between the Kerkhol and Dez rivers (Wikipedia 2012). Potts (1991) has reproduced the account of a chronicler who stated that rice was cultivated in Susiana in the late 4th century BCE. Susiana was an ancient city on the present Jarahi river, east Mesopotamia. It is now known as Ja Neshin (Khuzestan Province, Iran). The Greek historian, Diodorus Siculus (ca. 90–30 BCE), observed that rice might have been introduced into lower Mesopotamia from India during the Achemenid period. In the conflict between Eumenes of Cardia (the head of Alexander's secretariat) and Seleucus in 318/317 BCE, Eumenes had to feed his troops with rice, sesame, and dates, which the region had to produce in abundance due to a shortage of wheat (Nesbitt et al. 2010).

It is apparent from these accounts that, in keeping with the state of crop agriculture of that period (the late prehistoric period), rice was grown in the region, but possibly not as a dominant crop, when considered as a whole, except in central Asia. In prehistoric times, the dominant cereal crops were wheat and barley.

5. RICE IN EUROPE

Europe is currently a very minor producer and consumer of rice, producing only 0.5% and consuming only about 1.0% of the global total (Tables 1.3 and 1.4). The increase in consumption may be attributable to significant immigration of people from Asia and Africa during the 20th century, and also to the changing lifestyles of the native population.

Europe knew of rice long before its cultivation was attempted there. Rice was used in the Roman period as a specialty and health food by the elite population. During the prehistoric and early historic period, and during the Roman Empire, rice was imported into Rome from the Egyptian ports on the Red Sea, which used to act as trans-shipment ports for exchange of goods between Rome, east Africa, the Arab coast, and the west coast of India. Ancient texts such as *Periplus* (50–77 CE), Strabo's *Geography* (ca. 20 CE), and Pliny's *Natural History* (77–79 CE) give accounts of these transactions on the Arabian Sea and the importing of rice from India and elsewhere (Cappers 2007, Sidebotham 2011).

The history of rice in Europe has been summarized by Ferrero and Vidotto (2010). The earliest reference to its use appears to be in the writings of Sophocles (495–406 BCE), a Greek dramatist, who mentioned a cereal *orinda* that was cultivated by people living along the lower Indus river (now Pakistan). The earliest information about the use of rice in Europe appears to have been for medicinal purposes as given in the accounts of the Greek poet and writer Horace (65–8 BCE), who mentioned that rice was then so expensive that only the affluent Romans could afford its use, even for medicinal purposes. In the *De Materia Medica,* the physician, Dioscorides (ca. 40–90 CE) suggested the use of rice infusion against pain and worms in the intestine. There are also some other reports of using rice to treat other ailments (Ferrero and Vidotto 2010).

Flavius Arrianus (95–175 CE) has described in detail the then-ongoing trade between Greece and Egypt. Rice was one of the traded commodities. Cappers (2007) and Sidebotham (2011) have given details of this trade from an archaeological perspective. They have consolidated reports of nine years excavations conducted in the Egyptian Red Sea port of Berenike. Rice was among items that were imported from the Malabar coast of India, along with black pepper, and coconuts (van der Veen 1999) (Fig. 2.3).

Thus, though the use of rice was known in Europe since prehistoric times, especially in the Roman Empire and Greece, its cultivation there only commenced several centuries later, during the historic period. The exact time and place of its introduction to Europe is unknown, but it has been suggested that it was first introduced in the 12th century CE into the Valencia region of Spain (Piqueras and Boira 2005). This was done by the Moors of northwest Africa, who had established a sort of suzerainty over this region during 8th–15th centuries CE. Rice cultivation might have been taken up in Portugal during the reign of D Dinis (1279–1325 CE), possibly in the swampy areas along the Mondego and Tejo rivers (Vasconcellos 1953). The Portuguese had acquired a fair knowledge of rice cultivation from the mid-15th century onwards from the Guinea coast of tropical West Africa, where they were then attempting to establish dominance to obtain people for the slave trade. Rice farming was taken up in France in the delta of the River Rhone sometime later than in the Iberian peninsula (Barbier and Mouret 1993). In Italy, it is unclear when rice farming was first introduced, because it broke up into many regions after the fall of the Roman Empire, and the Arabs and other European powers fought continuously to establish their control and influence in the region. It has been suggested that Charlemagne brought rice from Spain after his battles with the Moors, or that the Venetian merchants introduced it from the East through their extensive trade contacts, and so on.

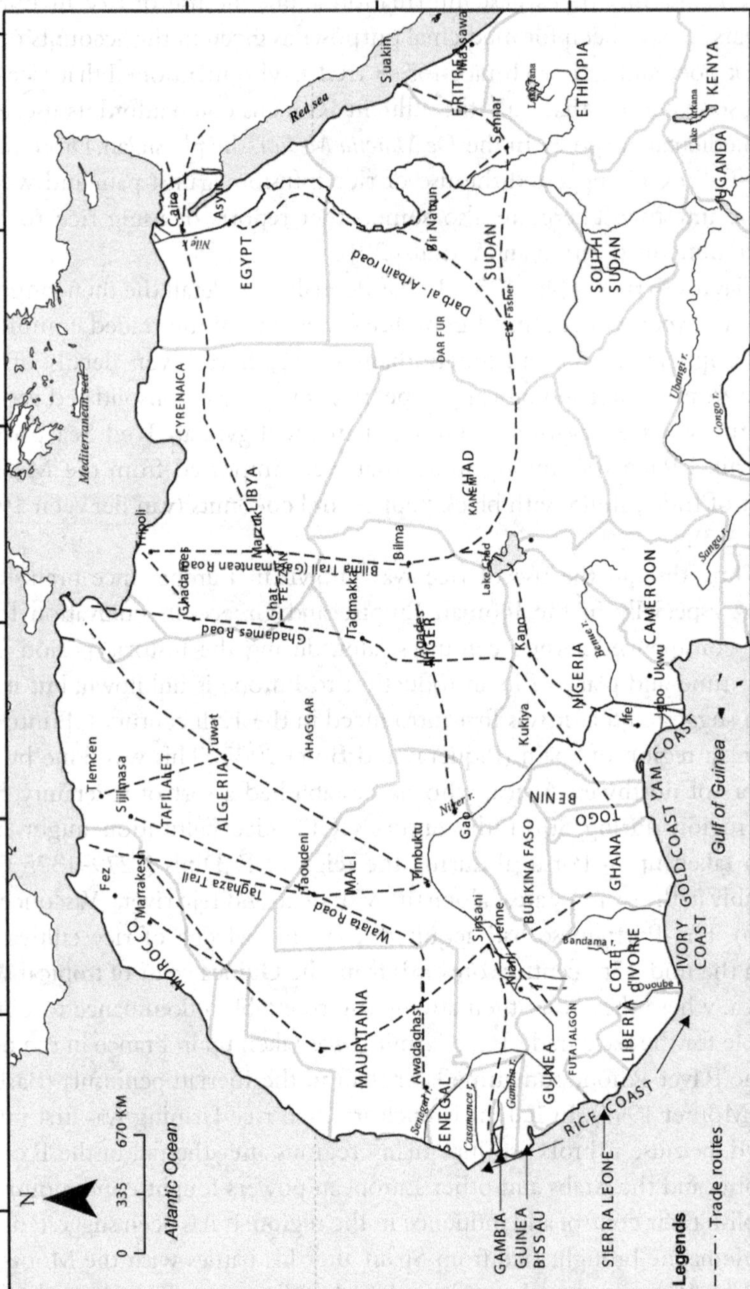

Figure 2.3 *Major Trans-Saharan Trade Routes*

Throughout southern Europe, the cultivation of rice expanded steadily during the 17th to 19th centuries. Opposition to rice cultivation then began to build up alleging that it was responsible for the spread of malaria and other waterborne diseases (Ferroro and Vidotto 2010). As a consequence, several local administrations and governments began to impose restrictions on its cultivation. Later, its cultivation began to expand for a variety of reasons in Europe again after the Second World War. Of late, however, this expansion appears to have slowed down. Presently, Europe imports about half of its requirement for rice (Tables 1.3 and 1.4).

6. RICE IN AFRICA

Africa is the second largest continent on earth after Asia (31.0 Mkm^2 vs. Asia: 44.5 Mkm^2). In population too, it ranks second to Asia (13% global population vs. Asia: 57%). Badawi et al. (2010) have reviewed the recent and current status of the crop in Africa, and Nayar (2010) has reviewed the archaeology and early history of African rice.

In Africa, rice ranks only fifth in area under cultivation after maize, sorghum, millets, and cassava. Rice provides 8.0% of the calories consumed in the continent, and occupies 4.4% of the net cropped area (Nayar 2010). With 13.9 Mt annual production, the continent produces only about half of its annual requirements. The rest is imported (Tables 1.3 and 1.4). Western and eastern Africa and Egypt are the major regions of rice cultivation in Africa (FAOSTAT 2010, Nayar 2010), but in west Africa, rice has been a major component of the culture and customs of the people. They have a rich tradition of rice culture dating from early times.

i. In Northern and Western Africa

The first record of the introduction of Asian rice into Africa appears to be that of Alexander the Great in 332 BCE after his conquest of Egypt, but it did not become established then. Later, the Arabs introduced rice cultivation into Egypt in 639 CE (Erygin and Natalin 1968). Africa began to experience Arab and Muslim influences from 942 CE, beginning initially in Egypt (Badawi et al. 2010). By the time of the reign of Al Mansur, Abbasid Caliph (754–775 CE), rice cultivation had spread to the province of al-Fayyum (Sato 2005). Rice cultivation in this region spread steadily until the 12th century, and after this, it began to spread steadily into lower Egypt. Currently, Egypt is the biggest producer of rice in Africa (Table 1.5). It also exports more than one fourth of its production (4.6 Mt/1.2 Mt) (FBS 2007).

Prehistoric evidence indicates that prior to Alexander's (the Great) introduction of rice in 332 BCE, the Greek geographer and historian Strabo (64/63 BCE – ca. 24 CE) recorded the growing of rice in Cyrenaica in northeast Sudan in ca. 12 CE, but this has not as yet been confirmed (Lewicki 1974). A major archaeological finding has been the recovery of rice (along with broken coconut shells and black pepper) from the Red Sea port of Berenike and Myos Hormos (Egypt), which was a trans-shipment and trading port for goods emanating from eastern Africa, the west coast of India, and Rome, mainly to cater for the demands of the Roman Empire (Fig. 2.3) (Van der Veen 1999, Cappers 2007, Sidebotham 2011). This has been dated to the first century CE. The material originated in the Malabar Coast on the west coast of India. Rice was apparently meant for the use of traders from India, some of whom used to stay in Berenike for extended periods (Fig. 2.4).

In addition to the finds from Berenike, there are three other reports of archaeological findings of rice from Africa, all from west Africa. These are discussed in detail in Chapter 5, so only an outline is presented here. The first of these was the discovery of RJ and SK McIntosh in Jenne Jeno (IND, Inland Niger Delta, Mali Republic) of the chaff of *O. glaberrima* and *O. barthii* (according to the authors). This was dated to 40–50 CE (cf., McIntosh 1995). The second report is from Kursakata and Mege (Nigerian Lake Chad basin), dated to 1,000–800 BCE. These sites were excavated by Neumann and co-workers (cf., Neumann 2003), who identified the remains as wild rice. The third report is from Dia (IND, Mali), and is dated to 799–413 BCE (Bedaux et al. 2001, Murray 2004, 2005). Murray (*loc. cit.*) studied the material in detail and stated that it is the African cultivated rice, *O. glaberrima*, but this conclusion has been questioned (cf. Chapter 5) (Nayar 2012).

The Islamic scholar al-Bakri (1014–1094 CE) provided in 1068 CE the first description of rice cultivation in west Africa along the River Niger (Lewicki 1974). When the famed Moroccan traveler Ibn-Battuta (1304–1368/1369) journeyed along the Niger River during 1341–1343 CE, he too had noticed the abundance of rice in the IND region (Gibbs 2006). Several other Arab travelers and chroniclers of west Africa have also reported seeing rice cultivation in west Africa (IND region) in the 13–14th centuries CE (Lewicki 1974, Gibbs 2006). All of them had traveled to west Africa from north Africa across the Sahara desert.

The Sahara desert is the most defining characteristic of north Africa. It is the biggest hot desert in the world (8.6 Mkm^2) followed by the Arabian desert (2.3 Mkm^2) (Fig. 5.1). It is bounded by the Mediterranean coastal region in the north, the Sahel (a semiarid tropical savanna) in the south, the Atlantic Ocean in the west, and the Red Sea and Ethiopean Highlands in the east. This region of north Africa became a desert only about 5,000 years ago, after the last glacial (8th–6th century BCE), and after the rains came to an abrupt end ca. 4,200 BCE (Phillipson 2005, Collins and Burns, 2007, Wikipedia 2013).

Trans-Saharan trade has been taking place in the region almost from the time of earliest habitation. In the beginning, it used to be sporadic and local or regional in nature. In the Sahara, various evidences of human habitation, such as pottery, cave paintings, etc., including pastoralism have been observed in several scattered locations dating from 7th–6th century BCE. Gradually, the trade began to expand. First, the goods used to be transported as head-load, then by oxen, followed by mules and horses, and finally by camel. Camel was introduced into the region in 3rd–4th century CE (Collins and Burns 2007, Wikipedia 2013).

The trade in the early period used to be mostly carried out by the Berbers, the indigenous people of north Africa, and it used to consist of salt, ivory, gold, animals, meat, and hide. With the introduction of horses and camels, trans-Saharan trade also began to increase. The trade has been taking place in spite of the harsh climate, scarcity of water and food (except in the oases), difficult terrain, and dangers from predatory animals and robbers. It had reached a peak during the 8th–16th centuries CE after the trade was taken over by the Arab Muslims.

These accounts relate to the period before the first European sailors – the Portuguese – set foot on the Guinea coast in west Africa in 1443 (Collins and Burns 2007). Thus, the view widely stated in the literature, that the Europeans introduced rice cultivation into west Africa, needs to be corrected. After the arrival of Europeans, the bulk of the trade shifted to the seagoing routes.

It is possible that when such exchanges of humans and materials were taking place on such a large scale, Asian rice could have constituted one of the items brought to west Africa by these caravan traders. There is also the further possibility that the 'rice' that the early travelers to the IND region had observed was either large stands of *O. barthii* or *O. longistaminata*, the two widely occurring African wild rices, or even the Asian cultivated rice. This should be considered in the context of the writings of the Greek geographer and historian, Strabo (64/63 BCE – ca. 24 CE), who observed rice cultivation in Cyrenaica (northernmost Libya) in ca. 12 CE. Cyrenaica lies to the west of Egypt. Egypt too was growing rice from the mid-7th century CE (Fig. 2.2).

Incidentally, Fezzan, in the southwest of present-day Libya, was the center of a major prehistoric human habitation. This region used to have sources of water, and agricultural activities were therefore possible. Garamantes was a state that existed there in prehistoric times (900 BCE – 500 CE). Its capital was Jarma. Pellig (2007, 2008) conducted a detailed archaeobotanical study of Jarma. She obtained plant remains of over 50 crop plants, covering a broad range of cereals, fruit, and fiber crops from both winter and summer seasons. These did not include any rice material. This region, like Cyrenaica in northeast Libya, was also a center on the trans-Saharan trade routes. It was experiencing severely arid conditions for the last 3,000 years, with an annual precipitation of only 5–12 mm, and temperatures of up to 50°C (Pellig 2008). The rice observed by Strabo is unlikely to have been the African rice, *O. glaberrima*. In her studies, Pellig (2005, 2008) only observed two sub-Saharan plants at Fezzan, pearl millet and sorghum.

It was in this context that Nayar (1973) and Carpenter (1978) proposed that Asian rice and its cultivation technology could have been introduced into west Africa during the 7th or 11th century, respectively, either through one of the several caravan routes across the Sahara from the north or northeast (Nayar 1973), or from the east across the tropical forests of central Africa (Carpenter 1978).

ii. In East Africa and Madagascar

We shall now examine how rice cultivation in east Africa and Madagascar may date back to prehistoric times, or at the latest, early historic periods. The historical aspects of the introduction of rice to the east coast of Africa, including Madagascar, have received only minimal attention from rice scientists. What we know from archaeologists and historians is that the entire eastern coast of Africa from Alexandria (Egypt) in the north to Madagascar and Mozambique in the south have been centers of much human activity since the Roman period (27 BCE – 5th century CE) (Verin 1981, Sherieff 1981, Parker 2008). Our knowledge of the early history of this region has come from several documents, but most importantly from the writings of Strabo (64/63 BCE – ca. 24 CE), Pliny (23–79 CE), the anonymous account, *Periplus Maris Erythracei* (ca. 50–70 CE), and Ptolemy's (90–168 CE) *Geography* (156 CE).

During the Roman period, some ports on the Red Sea in present-day Egypt, notably Berenike and Myos Hormos, used to act as entrepots to receive and transport goods to Rome. The Romans had an insatiable desire for various articles from India, and also elephants and incense (mainly myrrh and frankincense) from east Africa and southern Arabia (Sidebotham 2011). Goods and merchandise used to move in three directions: from the west coast of India, Sri Lanka, and beyond, along the coasts of Iran and Arabia; rice, spices (mostly black pepper), coconut, and textiles; from Africa, timber (mainly teak), ivory, and semiprecious and precious stones (up to Pamba and Zanzibar islands); and gold coins and wine from Rome. Rice has been recovered from the Egyptian ports of Berenike and Myos Hormos that has been dated to the first century CE (Van der Veen 1999, Cappers 2007, Sidebotham 2011). This however appears to have come from the Malabar Coast and was possibly meant for consumption by the Indian traders who stayed in Berenike for extended periods, or visited there for shorter periods of time.

Prehistorically, Madagascar was different from the rest of Africa. The island lies ca. 400 km east of Mozambique in southeastern Africa, in the Indian Ocean (latitude 12–25 °S). With an area of 587,041 km^2, it is the fourth largest island in the world. Over half of the population of about 25 M are of Indo-Malayan descent. The rest of the population consists of Africans and people of mixed descent. The language of Madagascar, Malagasy, is most similar to the Indonesian language (Sheriff 1981, Verin 1981, Blench 2007). Their habits, culture, and agriculture have also been derived for the most part from eastern cultures (e.g., house forms, terraced rice cultivation, ancestral worship, many technological developments [in household articles, musical instruments, farm implements etc.]). Several Asian crop plants are cultivated here, such as Asian rice (*O. sativa*), banana (*Musa paradisiaca*), Asian yam (*Dioscorea alata*), taro (*Colocasia esculenta*), coconut (*Cocos nucifera*), sugarcane (*Saccharum officinarum*), breadfruit (*Artocarpus altilis*), jackfruit (*Artocarpus heterophyllus*) (Murdoch 1959, Verin 1981). There has been considerable interest in the origin and phylogeny of the language and people of Madagascar, Malagasy. It has long been known that there was movement of people from southeast Asia to Madagascar across the Indian Ocean since late prehistoric or early historic times. This was made possible by the trade winds, which allowed a one way voyage to be completed in 40 days (*Periplus*, 50–70 CE). The island of Madagascar lies ca. 400 km from mainland Africa and more than 5,000 km from southeast Asia.

Linguistic studies have shown that more than three-fourths of the basic vocabulary of the present-day Malagasay language has its roots in the language spoken by the ethnic people living south of the Barito River in southern Borneo (Hurles et al. 2005). Hurles and colleagues studied the genetic relationships between four ethnic groups of Madagascar and 10 potential ancestral populations of Island Southeast Asia and the Pacific. They found broadly equal contributions of African and Indonesian people in both the maternal and paternal lineages of Malagasy people, who came most likely from Borneo.

There is some evidence that the ancient voyagers stage-posted in Sri Lanka, and/or peninsular India en route. Migrations across the Arabian Sea might have taken place as early as 2,600 BCE, and they came to an end with the arrival of Islam in present-day Indonesia, which replaced Hindusim in about the 12th century CE (Purseglove 1976, Verin 1981, McPherson 1993, Connah 2001, Ray 2002). Porteres (1955) and Ahmadi et al. (1991) have pointed out the close similarities in the names of some rice landraces of Madagascar and south India.

These findings are indicative of the antiquity of rice in east Africa. In about the last 150 years, there has been substantial emigration of people from south Asia to east Africa. Initially, they were brought in by the British as indentured labor for construction and large-scale farming, and later, for trade and commerce. A consequence of this has been the transfer of various crop plants, including rice, to east Africa by people from southeast and south Asia.

7. RICE IN THE AMERICAS

Rice came to the Americas as part of the Columbian Exchange (Carney 2001, Crosby 2003). There is a widely held belief that the seed and the technology for growing rice were brought to the Americas by the Europeans following the discovery of America by Christopher Columbus in 1492 and the mass migrations of people from Europe that followed it. But of late, the "black rice theory" (cf., Carney 2001) has gained acceptance. This assumes that rice and the technology for growing it were first brought into the Americas by west African slaves, especially those from the Rice Coast of west Africa. The Rice Coast region has a long tradition of rice culture, especially wet rice cultivation (Wood 1974, Littlefield 1981, Carney 2001, Field-Black 2008). The Rice Coast extends from the Gambia river to Cape Palmas (Fig. 5.4).

i. In the USA

Coclanis (2011) has given an account of the history and development of the rice industry in the USA. There were some isolated attempts by the Spaniards to grow rice in Florida in the 16th century and by the British in Virginia in the following century, but they did not succeed. In the 1670s, however, both Europeans and Africans were able to grow rice in South Carolina for their personal use. After this, rice production became a routine and systematic activity by the 1720s, mainly by African slaves for their own consumption. Rice production increased steadily in the tidal backwaters of major rivers throughout the Atlantic coast of North Carolina, Georgia, and Florida.

However, rice cultivation, production, and marketing became established as a commercial venture in the USA only in the early 18th century. By the 1820s, this region had become the center of the rice industry in the USA (Coclanis 2011). About 70,000 ha were used to grow rice by the mid-1850s, producing about 80,000 t. Up to 90% of production was exported to Europe. By the end of the 19th century, rice production had reached almost 500,000 t.

However, great changes began to take place in rice production in the USA in the latter half of the 19th century, after the end of the American Civil War (1861–1865). Rice production got reduced on the Atlantic coast, and rice cultivation was taken up in the coastal region of the Gulf of Mexico, in the southern states of Louisiana and Texas. Soon after that, rice industry began in Arkansas, and later in California in the Sacramento valley (Coclanis 2011). Presently, USA produces more than 6 Mt rice annually. Of this about 60% is exported. USA is presently one of the major exporters of rice in the world (FAOSTAT 2007). This major switchover took place because rice production on the Atlantic coast was a highly labor-intensive operation, with the labor mostly being supplied by African slaves. The American Civil War led to the liberation of the slaves, and consequently, it became uneconomic to produce rice with hired labor.

Rice cultivation in the Deep South started in small enterprises which very soon became organized into large farms of 400 ha or more, and also steadily became increasingly mechanized. It is currently estimated that just two workmen are enough to manage a farm of 400 ha. The rice industry also remains one of the most heavily subsidized operations in the USA (Grisworld 2006).

ii. In South and Central America and the Caribbean

This region produces over 16 Mt rice annually, but consumes ca. 28 Mt (FAOSTAT 2007). Hence, Central and South America is a net importer of rice.

Pereira and Guimaraes (2010) have given a brief history of rice in Latin America. The first rice cultivation in this region occurred in 1580 in the

Magdalena River valley of Columbia. However, commercial rice cultivation only started here in 1908. In Brazil, rice was first introduced in 1761. Precise dates for the introduction into Ecuador, Columbia, and Peru are not available, but, it is assumed to have taken place at almost the same time as in Brazil. Its introduction into Uruguay and Argentina is more recent, taking place only in the late 19th century. Presently, Brazil, Columbia, Peru, Ecuador, and Uruguay are the major producers of rice in South America (Table 1.5).

In South America, most of the early cultivars came from the USA, or from Spain or Italy. After the 1930s, several of these countries set up their own rice R&D programs. Rice research in Central and South America received a major fillip in 1967, with the establishment of the International Center for Tropical Agriculture (CIAT) in Cali, Columbia. In the last three decades or so, all the rice cultivated in the region has been of cultivars developed either locally in their respective countries, or by the CIAT or IRRI.

CHAPTER 3

Phylogeny of the Genus *Oryza* L.

1. INTRODUCTION

Phylogeny is defined as "the relationships of groups of organisms as reflected by their evolutionary history" (King et al. 2006). In this chapter, we shall discuss the evolutionary history of the rice genus *Oryza* and how it relates to higher taxonomic entities and closely related genera.

Cultivated rice consists of two species, African rice, *Oryza glaberrima* Steud. and Asian rice, *O. sativa* L. At present, African rice has been all but replaced in cultivation by Asian rice. The two rices belong to the genus *Oryza* L., tribe Oryzeae Dumort., subfamily Ehrhartoideae Link., family Poaceae Barnhart (synonym: Gramineae L.), order Poales Small, division Monocotyledonae, and class Magnoliopsida (Angiospermae, including Liliopida) (GPWG 2001, APG 2009, Kellogg 2009). In a sense, "angiosperms are a subset of seed plants, those a subset of Euphyllophytes, those in turn a subset of vascular plants, those a subset of land plants, those a subject of Charophyta, in turn a subset of green algae as a whole" (Mabberley 2008).

2. CLASSIFICATION OF ANGIOSPERMAE

The class Magnoliopsida (Angiospermae) includes 413 families referable to 55 orders with 273,000 species in ca. 14,000 genera (Mabberley 2008, the Plant List 2011). There is some disagreement about the actual number of taxa at the lower levels. The Angiosperm Phylogeny Group (APG) (2009) classifies flowering plants into Magnolids (four orders), Monocots (seven orders), Commelinids (five orders), and Eudicots (36 orders), plus a few taxa of uncertain affinity. In addition, four orders (Amborellales, Nymphaeales, Austrobaileyales, and Chloranthales) are sometimes considered as basal Magnolids.

The Monocots include 12 orders and 80 families. The orders are Acorales, Alismatales, Petrosaviales, Dioscoreales, Pandanales, Liliales, Asparagales, Arecales, Commelinales, Poales, and Zingiberales (APG 2009) (according to Mabberley (2008), 11 orders and 86 families). Of these, the last four orders – Arecales, Commelinales, Poales, and Zingiberales, together with the family Dasypogonaceae – are designated as Commelinids. The order Poales includes 16 families including Poaceae (APG 2009).

Origins and Phylogeny of Rices
http://dx.doi.org/10.1016/B978-0-12-417177-0.00003-6

The Angiosperm Phylogeny Group (APG) has been working since 1998. The latest report of the Group was published in 2009 (APG III). A Grass Phylogeny Working Group (GPWG) was established in 1996. It was organized on similar lines to the APG, mainly to combine a series of existing data sets to produce a comprehensive phylogeny for the entire family (GPWG 2001). The family Poaceae Barnhart (Gramineae Juss.) consists of 715 genera and 10,550 species (Mabberley 2008). At the same time, Clayton and Renvoize (1986) have included 651 genera and ca. 10,000 species in the grass family.

3. DIVERSIFICATION OF GRAMINEAE

Stebbins (1981) was possibly the first author to suggest an age for the origin of the grass family. He stated that fossilized silica bodies in the leaf epidermis of grass and fossils of mammals having high-crowned teeth (which is an indication of herbivory) appeared nearly simultaneously in Patagonia (South America) during lower to middle Eocene strata (57–36 Ma).

In North America, the earliest identified grass fossils are the stipoid fruits of the lower phase of the Miocene (23–5 Ma) age, while the oldest mammals with high-crowned teeth are the rhinoceroses of the Miocene age. Stebbins (1981) went on to point out that the record of co-evolution of grasses and herbivores might have spanned most of the Tertiary (66–1.6 Ma) and Quaternary (1.6–0.01 Ma) periods. It began in Patagonia in the middle Eocene epoch (45 Ma) and continued until the retreat of glaciers in the post-Pleistocene (1.64–0.01 Ma) epoch (Fig. 3.1).

Subsequently, fossil records of grass pollen grains were recovered from South America and Africa from the Palaeocene epoch (60–55 Ma) (Jacobs et al. 1999). At present, the earliest known grass fossil is from North America, which dates to the early Eocene (ca. 55 Ma) (Crepet and Feldman (1991). Prior to these two findings, Litke (1968) had discovered bits of fossilized grass leaf epidermis in coal deposits in east Germany, which he had dated to the end of the Eocene epoch (i.e., before 36 Ma). However, Thomasson (1987) referred this as an unidentified grass, and not an Oryzoid (cf., Kellogg 2009). Subsequently, Crepet and Fieldman (1991) obtained unequivocal evidence of grasses in western Tennessee (USA) in the form of spikelets, panicles, and pollen, which have been dated to Palaeocene-Eocene boundary (i.e., just before 57 Ma). They stated that their findings were consistent with an upper Cretaceous (ca. 70 Ma) origin of the grass family. Thus, this appears to be the earliest fossil evidence of grasses obtained so far.

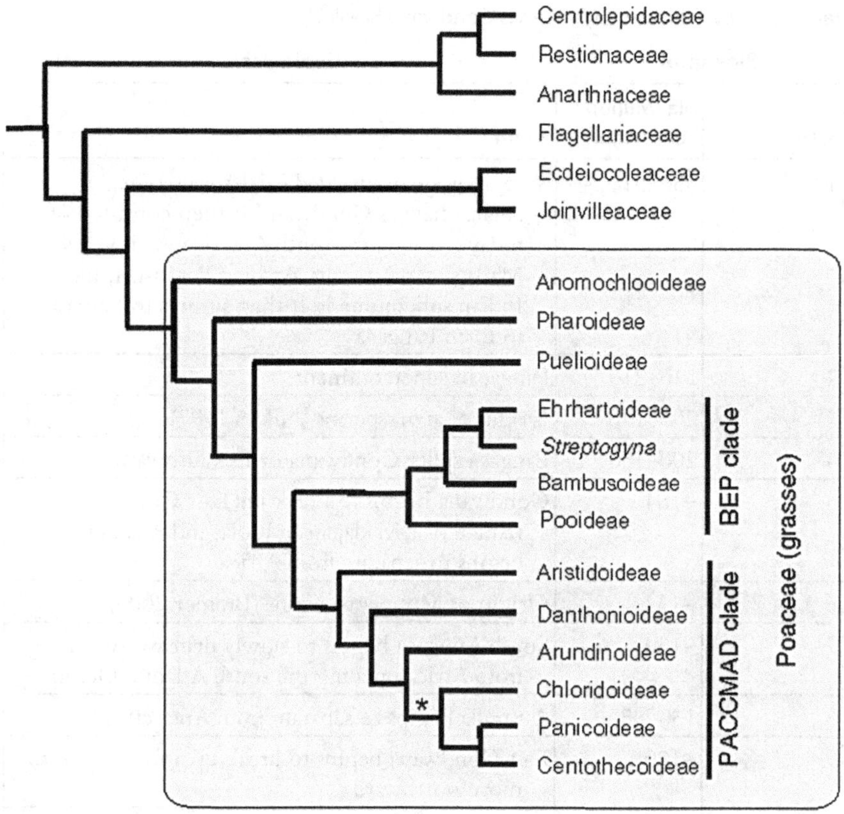

Figure 3.1 ***Phylogeny of the Graminoid Poales.*** Grasses are in the gray box. Relationships among grass subfamilies based on data from Vicentini et al. (2008). The position of subfamily Microairoideae is unresolved within the PACCMAD clade, and is not included. Asterisk indicates a clade that is poorly supported by the available data. Relationships among the grasses and the other poalean families are based on Marchant and Briggs (2007). *Reprinted from Kellogg (2009) with permission.*

The origin and evolution of the grass ecosystems has been described by Jacobs et al. (1999). They ascribed the origin of Poaceae to the late Maastrichian-Palaeocene Epoch (ca. 66–70 Ma) based on fossil evidences, including of pollen from various continents (South America, South Africa, India). Gaut (2002), in his comprehensive review of the evolutionary history of the Gramineae, indicated that the phylogenetic divergence of the grass family took place at ca. 77 Ma (by employing the parametric rate smoothing method of Sanderson [1977]), and that of the subfamilies Ehrhartoideae (to which rice belongs) and Pooideae at 46 Ma. He took into account the fact

Table 3.1 *Oryza* Phylogeny vis-à-vis Gondwana Break Up

Biological **Geological**

S. No.	Ma (Million Years Ago)	Event
1	550–510	Several orogenies (folded earth's crust) finally amalgamate as Gondwana. It then consisted of today's Antarctica, South America, Africa, Madagascar, Australia, Arabian Peninsula, and Indian subcontinent. It then sutures to Laurasia to form Pangaea
2	510–210	Pangaean supercontinent
3	208–146	Origin of angiosperms (Niklas 1997)
4	200–180	Pangaea splits; Gondwana drifts southward
5	~184	Gondwana begins to break up; East Gondwana (Antarctica, Madagascar, India, and Australia) begins to separate from Africa
6	~134	Origin of Monocots begins (Bremer 2000)
7	~130	South America begins to slowly drift westward from Africa opening the south Atlantic Ocean
8	130–85	Australia begins to separate from Antarctica
9	~120	East Gondwana begins to break up. India begins to move northward
10	~110	Open Atlantic Ocean develops
11	91–79	Most Poales diverge (17 families including Poaceae) (Bremer 2000)
12	80	Late Cretaceous: Australia separates from Antarctica
13	~77	Divergence of Poaceae (Gaut 2002)
14	~70	Origin of grass family (Crepet and Fieldman 1991)
15	70–55	Origin of Poaceae (Vicentini et al. 2008)
16	67–55	Grass diversification (Prasad et al. 2005)
17	65–60	Madagascar and present-day Seychelles break off from India: coincides with Cretaceous-Tertiary Extinction event
18	55	Australia–New Guinea begins to separate and move north, and rotate about its axis, but retains some connection with Antarctica for another one million years
19	51.6 ± 7.9	Divergence of BEP and PACCMAD clades (Vicentini et al. 2008)

Table 3.1 *Oryza* Phylogeny vis-à-vis Gondwana Break Up—cont'd

Biological		Geological
S. No.	**Ma (Million Years Ago)**	**Event**
20	45	Indian Plate collides with Asia. Himalayas is formed Tasmania finally separates from Antarctica
21	43.1 ± 7.2	Divergence of Ehrhartoideae (Vicentini et al. 2008)
22	34.5 ± 6.8	Break up of Ehrhartoideae into Ehrharteae and Oryzeae (Vicentini et al. 2008)
23	30	S America separates from west Antarctica and causes major climate change due to opening of the southern sea
24	20.5	Divergence of Oryzeae into *Oryza* + *Leersia* and the rest (*Rhynchoryza* + *Zizaniopsis* + *Luziola* + *Prosphytochloa* + *Polamophila* + *Zizania* + *Hygroyza* + *Chikusichloa*)
25	15	Collision of New Guinea–Australian Plate edge with SW Pacific Plate pushing up New Guinea Highlands
26	14.2	*Leersia* separates from *Oryza*
27	10.2	Differentiation begins within *Oryza*; *O. granulata* separates from other *Oryza* species

that the earliest age of the pollen and vegetative parts of grass fossils has been put at 55–70 Ma. Gaut supported Stebbins' (1981) suggestion that the time of divergence of rice and maize was 50 Ma (cf. Fig. 3.1).

More recently, based on their studies of dinosaur coprolites from central India, Prasad et al. (2005) estimated the time of grass diversification as 65–67 Ma. At that period, India was not connected to any other landmass. After the breakup of East Gondwana, about 120 Ma, the Indian Platelet floated in the Indian Ocean before colliding with Asia about 45 Ma, leading to the formation of the Himalayas. These authors proposed that the radiation would have occurred before the breakup of Gondwana ca. 80 Ma (Table 3.1, Fig. 3.1).

Vicentini et al. (2008) carried out an elaborate study of the grass family to determine the divergence timings and phylogeny of the family. They employed the nuclear gene, *phyB* and chloroplast gene, *ndhF* sequences and used 67 grass species. They took into consideration the earlier published data on fossils (of Linder 1987, Crepet and Fieldman 1991, Jacobs et al. 1991) and coprolites

(Prasad et al. 2005), and employed Kishino's Bayesian relaxed clock method of 2007 for their analysis. Using these, they worked out the time of differentiation of the family, its 13 subfamilies, and orders. The stem node of the family was estimated to have diverged at 55–70 Ma and the two clades, Bambusoideae, Ehrhartoideae, and Pooideaea (BEP), and Panicoideae, Arundinoideae, Centothecoideae, Chloridoideae, Micrairoideae, Aristidoidae, and Danthoninoideae (PACCMAD) separated at 35 Ma. They also calculated the timing of the maize–rice divergence at 57.6 ± 7.9 Ma (which is not wide off from the date of 50 Ma suggested by Stebbins (1981), who had based his inferences on fossil records of grasses and grass-eating ungulates in North and South America, South Africa, and India).

4. CLASSIFICATION OF GRAMINEAE

The grass family is the most cosmopolitan family in the angiosperms. As stated earlier, it includes 715 genera and 10,550 species. It occurs from the equator to the polar regions, and from mountain peaks to the sea. About a fourth of the vegetation cover on the earth consists of grasslands (Clayton 1975). Gramineae also contributes the most to human food security. All cereals, millets, and sugarcane belong to this family. Further, it provides fodder, turf grasses, and bamboos. Bamboos are used in housing and building materials, pulp for paper making, and in making numerous utility articles, especially in tropical Asia (Clayton 1975). Gramineae provides almost two-thirds of the calories consumed by the human race: 21% by rice, 20% by wheat, and the rest by sugarcane, temperate cereals, and millets.

The Grass Phylogeny Working Group (GPWG) (2001) report was prepared by a team of eight scientists. They combined and cladistically analyzed eight data sets (one structural, four plastome, and three nuclear) from 62 grasses and four outgroups. Their study included ca. 0.6% of all species and 8% of genera. The data sets were generated using *ndh*F, *rbc*L, *rpo*C2, *phy*B, ITS2, and GBSSI (or *waxy*), chloroplast restriction site data, and data on 53 morphological characters. Together, these genetic sites gave 7,025 characters. The group members then conducted a parsimony analysis, which produced 2,143 informatic characters. The analysis yielded 12 subfamilies consisting of 11 subfamilies that had been previously described, plus one new subfamily, viz. Danthonioideae. Ehrhartoideae is one of the 11 previously published subfamilies. It includes the rice order, Oryzeae (GPWG 2001). Kellogg (2009) has recently reviewed the evolutionary history of the subfamily Ehrhartoideae.

There have been eight comprehensive classifications of the grass family since 1900 (GPWG 2001) including that by the Grass Phylogeny Working Group. All of them show differences in their subfamilial number, classification and nomenclature, both in subfamilies (7–12 in number) and orders (ca. 45). There are also a few genera that are of uncertain affinity (*Incertae sedis*).

Within Gramineae, most of the grasses fall into the two clades, BEP (Bambusoideae, Ehrhartoideae, and Pooideaea) and PACCMAD (Panicoideae, Arundinoideae, Centothecoideae, Chloridoideae, Aristidoidae, and Danthoninoideae) (Kellogg 2009). The BEP clade, to which rice belongs, is particularly supported by data from *ndh*F, *rpo*C2, and *phy*B. It is characterized by the presence of a single fertile floret per spike (GPWG 2001).

The subfamily Ehrhartoideae, to which *Oryza* belongs, includes three orders, Oryzeae, Ehrharteae, and Phyllorachideae. The last order, Phyllorachideae, is included in the subfamily solely on the basis of morphological characters (GPWG 2001). Ehrhartoideae contains about 120 species (Fig. 3.2).

The Gramineae had been recognized as a distinctive and coherent family long before the term monophyly was applied. They are distinguished by the structure of the embryo, their fruit, and their pollen (Kellogg 2009). Further, the genetics of the family indicates that the whole genome has undergone duplication (Paterson et al. 2004). This might have occurred before the differentiation of the family took place (70.9 ± 9.2 Ma) (Vicentini et al. 2008). Its fruit (caryopsis) is considered to be unique to the family. The embryo is placed laterally, and unlike most monocot embryos, it is highly differentiated, with an obvious root and shoot meristem, leaves, and vascular system. The pollen has only one aperture (GPWG 2001).

The combined analyses carried out by the GPWG (2001) have shown that Poaceae is monophyletic, as are the BEP clade and also the rice tribe, Ehrhartoideae. This conclusion has been based on two characters: the highly differentiated grass embryo and its lateral position, and the *trn*T inversion in the large single copy region of the chloroplast genome. The tribe Ehrhartoideae is strongly supported by molecular data. It is morphologically characterized by the presence of a three-flowered spikelet in which the lower two flowers are sterile.

The present distribution of grasses does not indicate anything about its center of origin (GPWG 2001). However, in a recent study, Bouchenak-Khelladi et al. (2008, 2010) have produced some answers to this question. The authors first developed from a comprehensive sample of Poaceae (298 in number) a phylogenetic tree based on plastid DNA sequences. Then, they created a second tree using a supermatrix of morphological and molecular

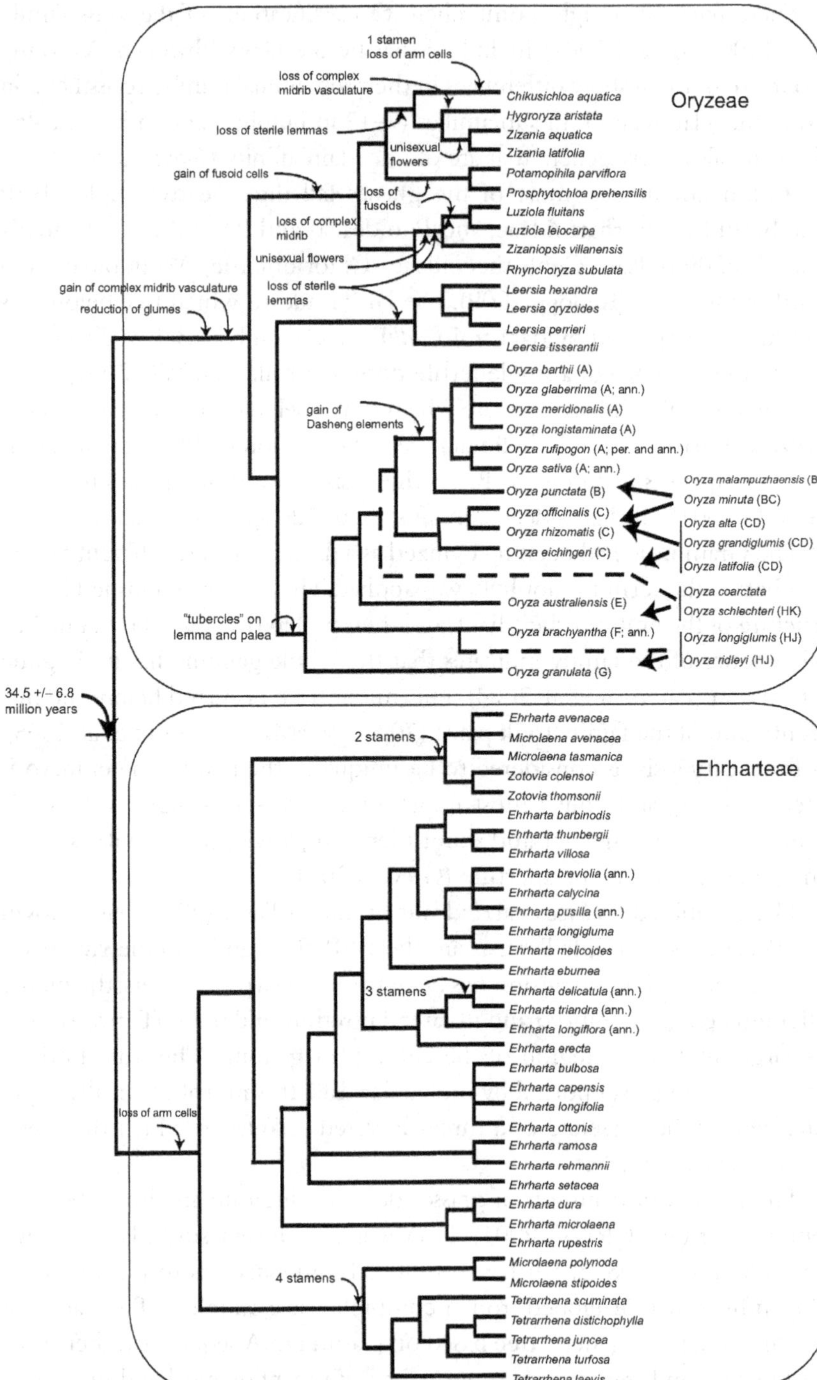
Oryzeae
1 stamen
loss of arm cells
loss of complex
midrib vasculature
loss of sterile lemmas
gain of fusoid cells
unisexual
flowers
loss of
fusoids
loss of complex
midrib
unisexual flowers
gain of complex midrib vasculature
reduction of glumes
loss of sterile
lemmas
Chikusichloa aquatica
Hygroryza aristata
Zizania aquatica
Zizania latifolia
Potamophila parviflora
Prosphytochloa prehensilis
Luziola fluitans
Luziola leiocarpa
Zizaniopsis villanensis
Rhynchoryza subulata
Leersia hexandra
Leersia oryzoides
Leersia perrieri
Leersia tisserantii
Oryza barthii (A)
Oryza glaberrima (A; ann.)
Oryza meridionalis (A)
Oryza longistaminata (A)
Oryza rufipogon (A; per. and ann.)
Oryza sativa (A; ann.)
gain of
Dasheng elements
Oryza punctata (B)
Oryza officinalis (C)
Oryza rhizomatis (C)
Oryza eichingeri (C)
Oryza australiensis (E)
"tubercles" on
lemma and palea
Oryza brachyantha (F; ann.)
Oryza granulata (G)
Oryza malampuzhaensis (B)
Oryza minuta (BC)
Oryza alta (CD)
Oryza grandiglumis (CD)
Oryza latifolia (CD)
Oryza coarctata
Oryza schlechteri (HK)
Oryza longiglumis (HJ)
Oryza ridleyi (HJ)
34.5 +/− 6.8
million years
Ehrharteae
2 stamens
Ehrharta avenacea
Microlaena avenacea
Microlaena tasmanica
Zotovia colensoi
Zotovia thomsonii
Ehrharta barbinodis
Ehrharta thunbergii
Ehrharta villosa
Ehrharta breviolia (ann.)
Ehrharta calycina
Ehrharta pusilla (ann.)
Ehrharta longigluma
Ehrharta melicoides
Ehrharta eburnea
3 stamens
Ehrharta delicatula (ann.)
Ehrharta triandra (ann.)
Ehrharta longiflora (ann.)
Ehrharta erecta
Ehrharta bulbosa
Ehrharta capensis
Ehrharta longifolia
Ehrharta ottonis
Ehrharta ramosa
Ehrharta rehmannii
Ehrharta setacea
Ehrharta dura
Ehrharta microlaena
Ehrharta rupestris
loss of arm cells
4 stamens
Microlaena polynoda
Microlaena stipoides
Tetrarrhena scuminata
Tetrarrhena distichophylla
Tetrarrhena juncea
Tetrarrhena turfosa
Tetrarrhena laevis

data that included all the genera of the grasses (800 entries). Next, they used a likelihood-based method developed by them, which allowed estimation of ancestral polymorphism in both biogeographical and ecological analyses. They determined that the Poaceae originated in Africa as a shade-adapted family. The crown node of the two clades, BEP and PACCMAD, was dated to 57 My in the early Eocene. About 50 My, the grasses had dispersed to all the continents after their Gondwanan origin in the late Cretaceous (144–66 My). The PACCMAD clade adapted to open habitats in the late Eocene (57–36 My), whereas the BEP clade (to which *Oryza* belongs) underwent this adaptation much later, in Eurasia during the Oligocene (36–23 My).

5. SUBFAMILY EHRHARTOIDEAE

As stated before, rice belongs to the subfamily Ehrhartoideae, which in turn belongs to the BEP clade. The other two subfamilies in this clade are Bambusoideae and Poideae. The subfamily Ehrhartoideae includes three tribes, Ehrharteae, Oryzeae, and Phyllorachideae. The last tribe is included solely on the basis of its morphological affinities with the other two tribes (GPWG 2001). The subfamily Ehrhartoideae includes about 13 genera and 120 species (Figs. 3.2 and 3.3).

Vicentini et al. (2008) had found monophyly in the three subfamilies, Ehrhartoideae, Bambusoideae, and Pooideae. In phylogenetic studies, Ehrhartoideae was seen as sister to the genus *Streptogyna,* and both of them were sister to Bambusoideae. They estimated the divergence of Ehrhartoideae from *Streptogyna* at 43.1 $\pm$ 7.2 Ma, and of the two tribes, Oryzeae and Ehrharteae, at 34.5 $\pm$ 6.8 Ma. Kellogg (2009) has created the phylogeny of the subfamily Ehrhartoideae by grafting the phylogeny of tribe Ehrharteae (Verboom 2003) to that of the tribe Oryzeae (from Ge et al. 1999, Guo and Ge 2005) (Figs. 3.2 and 3.3).

There is a difference of opinion about the classification of the tribe Ehrharteae. While Watson and Dallwitz (1992) divided it into four genera

Figure 3.2 ***Phylogeny of Ehrhartoideae.*** Created by Grafting the Phylogeny of Ehrharteae (from Verboom et al. (2003) on to that of Oryzeae (from Ge et al. (1999) and Guo and Ge (2005). Polyploid members of Oryzeae are placed on the right of diploid species according to their genomic constitution. Changes in major morphological characters mapped on the tree using MaClade 4.0, Madison and Madison (2005). *Reproduced from Kellogg (2009) with permission.*

Figure 3.3 Geographic Distribution and Approximate dates of Divergence of Members of Ehrhartoideae. Phylogeny as in Fig. 3.2 except that the Cape species of Ehrharta are represented as a single branch. Luziola peruviana is synonymous with L. leiocarpa following Martinez-Y- Perez et.al. (2008).
Reproduced from Kellogg (2009) with permission.

(*Ehrharta, Tetrarrhena, Microlaena,* and *Zotovia*), Clayton and Renvoize (1986) combined them into a single genus *Ehrharta* containing 35 species, with this being the only genus in the tribe.

Members of the subfamily Ehrhartoideae are widely distributed (cf., Clayton and Renvoize 1986). Most species occur in the Old World, although the common ancestor might have originated in Australia or tropical Africa (Kellogg 2009). The common ancestor of Ehrharteae might also be from Australia and that of Oryzeae from Africa or Australia (Kellogg 2009). In Ehrharteae, the dispersals might have been from Australia to southern Africa, or it might have originated in southern Africa itself, as could be inferred from phylogenetic studies. This is apparent from the fact that more than two-thirds of *Ehrharta* species (ca. 25/30) occur in Africa (Verboom et al. 2003). Fig. 3.3 shows the geographic distribution and approximate dates of divergence of Ehrhartoideae. Its one limitation is that the 20–22 *Ehrharta* species of South Africa are presented as only a single branch. Verboom et al. (2003) have given the phylogeny of the extant *Ehrharta* species.

The ancestral spikelet structure of the subfamily Ehrhartoideae is a three-flowered spikelet in which the lower two flowers are sterile (GPWG 2001). Members of the tribe Ehrharteae exhibit this structure clearly (Kellogg 2009). Many members of the tribe Oryzeae, including *O. sativa,* retain this ancestral structure, but the glumes are minute. In other members of Oryzeae, the glumes are reduced to a cup-like rim. The sterile lemmas may be small or missing. It has been suggested that morphologically the rice spikelet is similar to those of the other members of the subfamily Ehrhartoideae, in having three flowers with the uppermost flower fertile and the other two reduced to scale-like sterile lemmas (Bor 1960). This was suggested first by Stapf in 1917. The genus *Oryza* too has six stamens, like its ancestors.

6. TRIBE ORYZEAE

This tribe consists of 12 genera and ca. 70 species, distributed in the tropical and warm temperate regions of the world (Clayton and Renvoize 1986). More than half the genera and species of the subfamily Ehrhartoideae are included in this tribe (GPWG 2001). It is divided into two groups – those having "strictly one flower", and those with "two sterile lemmas below the fertile floret". The first group contains *Leersia* Swartz (18 species), *Chikusichloa* Koidzumi (three species), *Hygroryza* Nees (one species), *Zizania* L. (three species), *Zizaniopsis* Doell & Ascherson (five species), and *Luziola* de Jussieu (11 species). The genera having two sterile lemmas below the fertile floret are: *Oryza* L. (22–24 species), *Porteresia* Tateoka (1 species), *Rhyncoryza* Baillon (1 species), *Potamophila* Robert Brown (1 species), *Maltebrunia* Kunth (3–5 species), and *Prosphytochloa* Schweickerdt (1 species).

The basic chromosome number of the order is $x = 12$, except in *Zizania,* which has $x = 15$.

The phylogeny of the tribe has been studied by Ge and his coworkers (Ge et al. 1999, Ge et al. 2001, Bao and Ge 2004, Guo and Ge 2005). They used five DNA sequences from chloroplast (*matK, trnL*), mitochondrial (*nad*1), and nuclear (*Adh*2, *Gpa*1) genes, and observed that the tribe Oryzeae is monophyletic, and that *Oryza* and *Leersia* are sister clades. Sister to this clade is another one containing *Prosphytochloa, Potamophila, Chikusichloa, Rhynchoryza, Zizania* and *Zizaniospsis,* and *Luziola.* Their analysis also suggested that *Porteresia coarctata* should be restored to the genus *Oryza* as *O. coarctata.* This aspect will be discussed in Chapter 5.

This sisterly relationship between *Oryza* and *Leersia* may be seen in the context of the taxonomic division of the Oryzeae tribe by Clayton and Renvoize (1986). These authors divided Oryzeae into two groups of six species, each based on the "strict presence" of only one flower and those having one flower subtended by two sterile lemmas. Following this classification, they placed *Oryza* and *Leersia* in two different groups, as *Leersia* has only one floret and *Oryza,* one floret and two sterile lemmas (Figs. 3.4 and 3.5).

The divergence of the subfamily Ehrhartoideae into the tribes Oryzeae and Ehrharteae at 34.5 ± 6.8 Ma, happened in the late Eocene–early Oligocene epoch (Vicentini et al. 2008). The tribe Oryzeae diverged into two clades at 20.5 Ma, with one clade containing the genera *Leersia* and *Oryza,* and the other containing the other six genera (Fig. 3.2). The genus *Leersia* then diverged from *Oryza* at 14.2 Ma. The first diversification from the genus *Oryza* was of *O. granulata* at 10.2 Ma (Guo and Ge 2005). To estimate the divergence points of the tribe Oryzeae and its major lineages, a simple molecular clock assumption was employed, based on the premise (Stebbins 1981, Kellogg 1998, Gaut 2002) that maize and rice diverged ca. 50 Ma. Relative rate tests were implemented using various software packages (Hyphy, RR Tree), and rates of nucleotide substitution in different genes were estimated using MEGA 2.1 (Guo and Ge 2005).

Some other authors have also estimated the divergence points of Oryzeae; for instance, Clayton (1975) made some estimations based on phytogeographic evidence and computer-simulation studies of distribution, Second (1985) used isozyme electrophoresis data, and Chang (1985) based his on Gondwana break up. Since all the above authors have attributed the

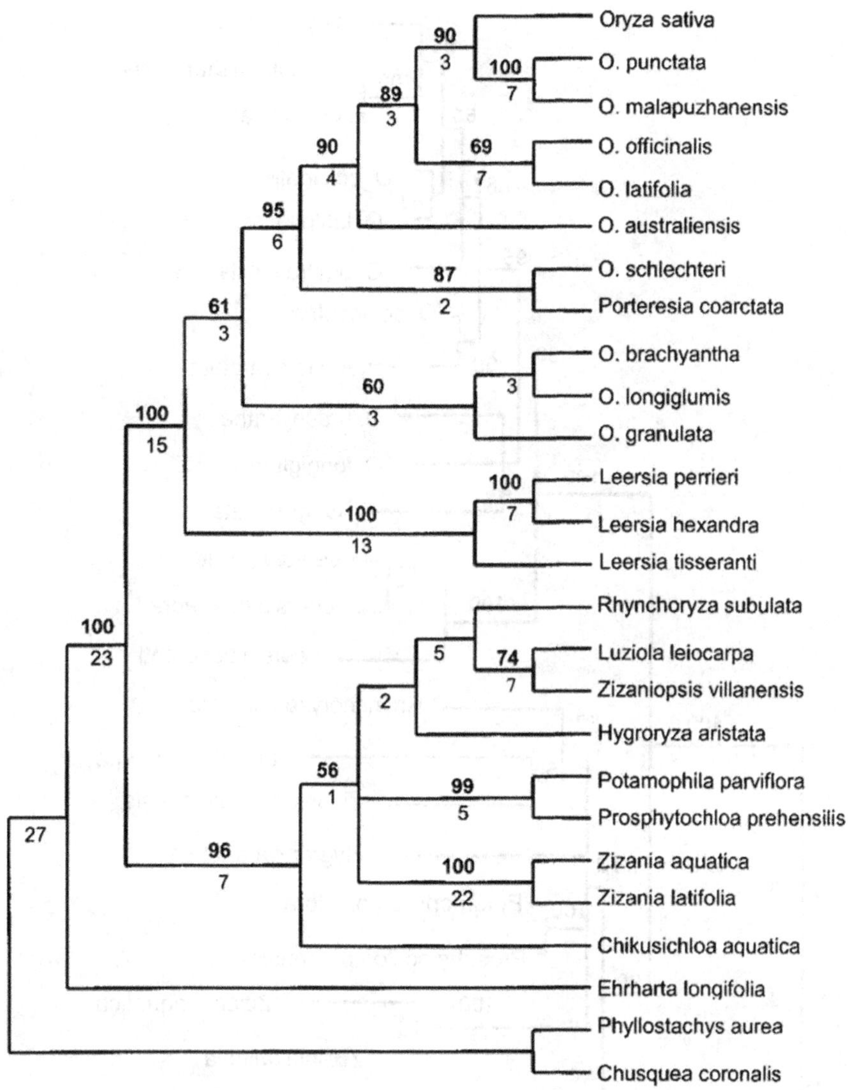

Figure 3.4 ***Phylogeny of the Tribe Oryzeae Based on* matK *Sequence data.*** A. Strict consensus of three equally most parsimonious trees of 23 Oryzeae species and three outcrops (Ehrharta longifolia, Chusquea coronalis , and Phyllostachys aurea) generated from sequences of matK gene of chloroplast DNA. Tree length = 476, consistency index = 0.790, retention index = 0.819. Numbers above branches represent bootstrap values above 50 % and those below branches are nucleotide substitutions.
Reproduced from Ge et al. (2002) with permission.

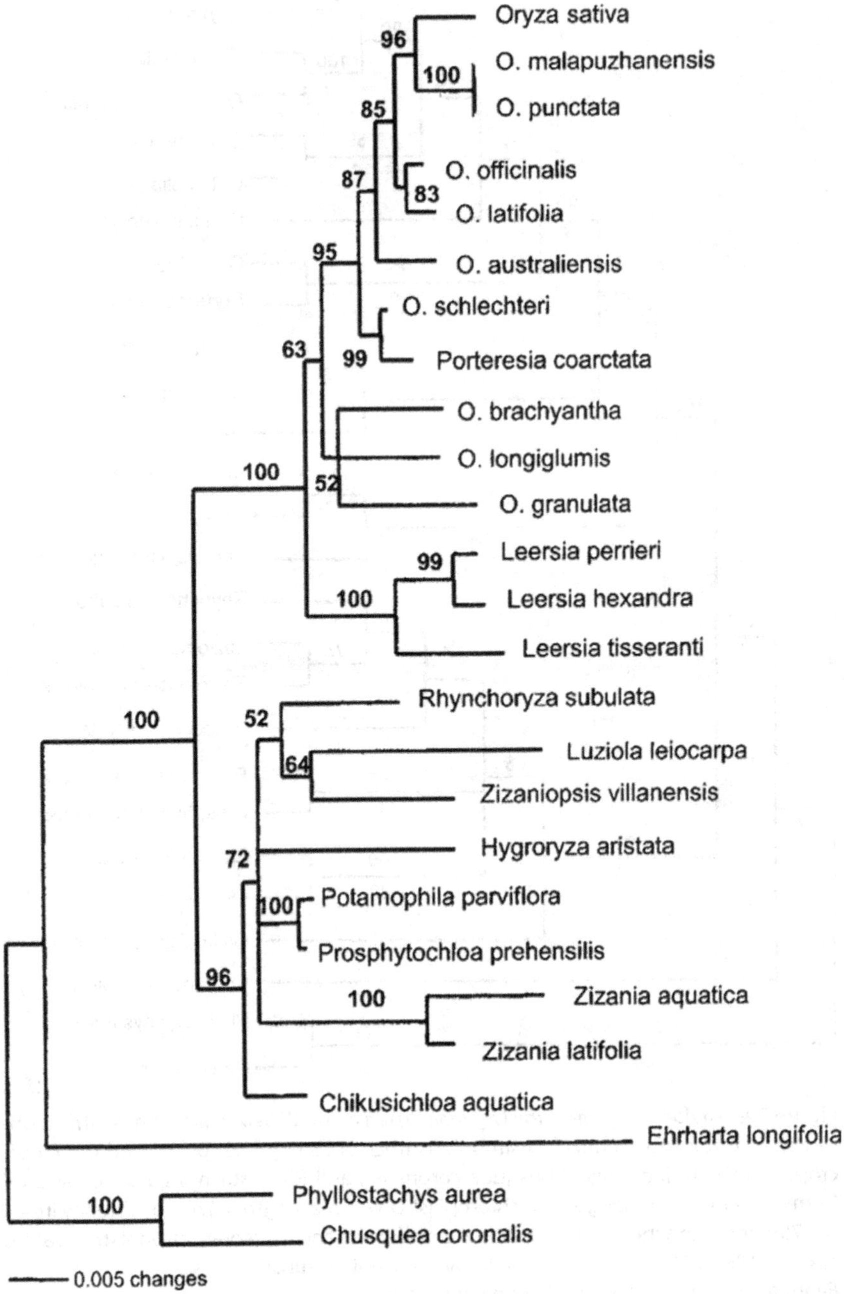

Figure 3.4 Cont'd B. Neighbor joining phylogram based on Kimura two parameter distance model. Numbers of branches are bootstrap values from 1000 replicates. Branch lengths are proportional to distance.

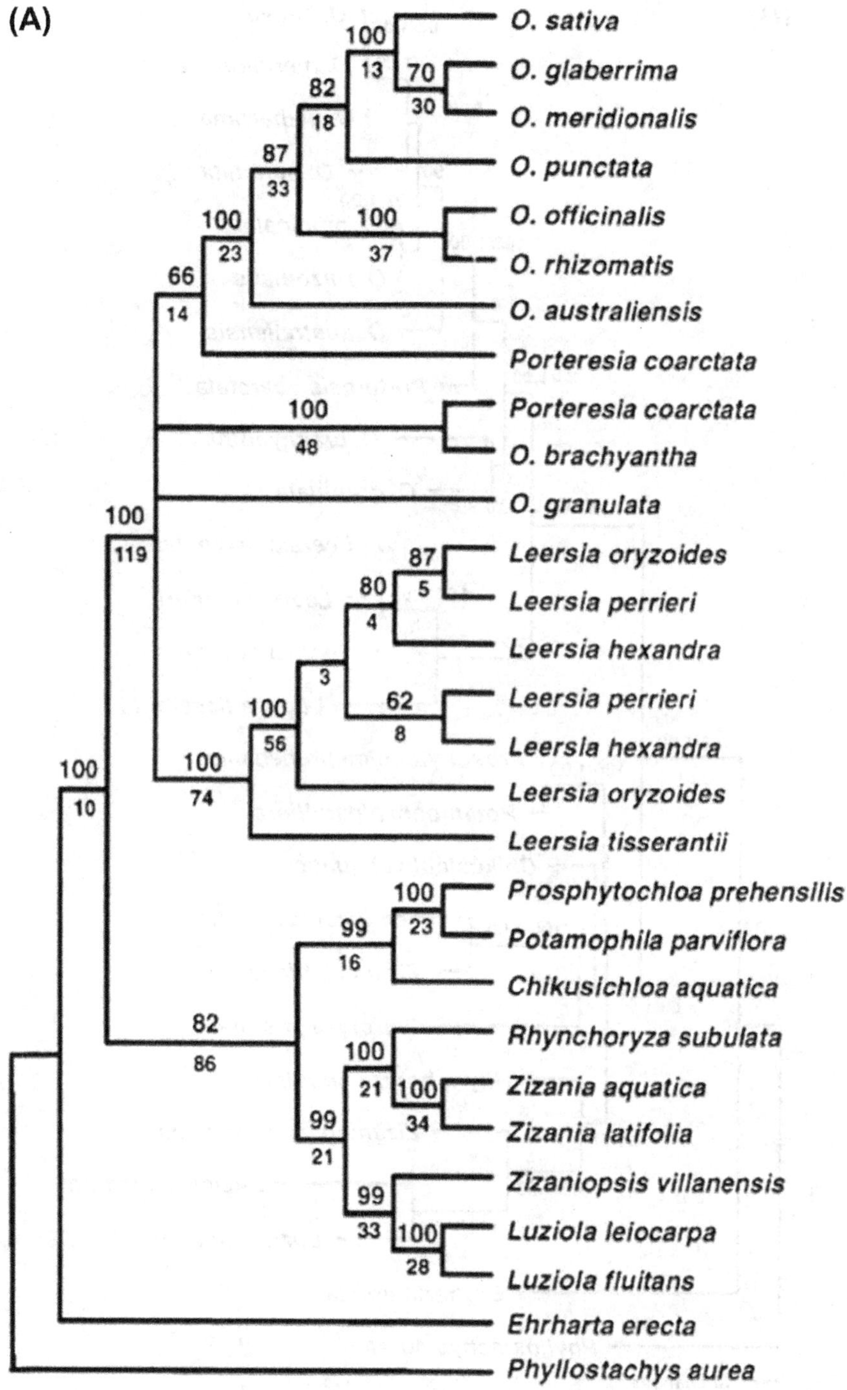

Figure 3.5 ***Phylogeny of the Tribe Oryzeae Based on Chloroplast, Mitochondrial, and Nuclear Genomes DNA Sequences.*** A. A single most parsimonious tree (MPT) based on the combined data set of chloroplast matK and trnL intron and mitochondrial nad1 intron 2 (tree length = 654, consistency index = 0.806, retention index = 0.852). Bayesian phylogenetic analysis generates a similar topology. Numbers near branches are bootstrap percentages followed by Bayesian posterior probabilities.
Reproduced from Guo and Ge (2005) with permission.

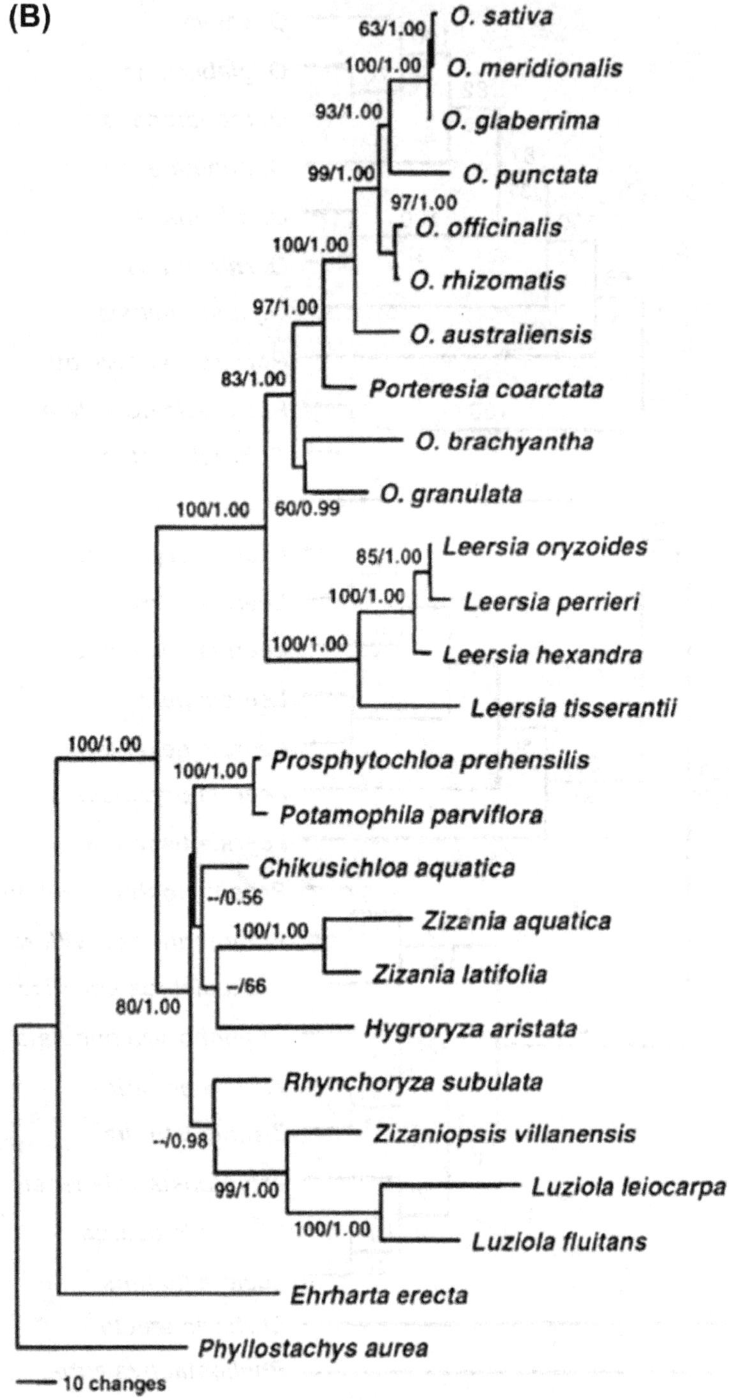

Figure 3.5 Cont'd B. The strict consensus of two equally most parsimonious trees of (MPT) from the combined Adh2 and GPA 1 data set after excluding Hygroryza aristata (tree length = 1529, consistency index = 0.661, and retention index = 0.751). Bayesian phylogenetic analysis generates a similar topology. Numbers above branches indicate bootstrap values above 50% and those below branches indicate nucleotide substitutions.

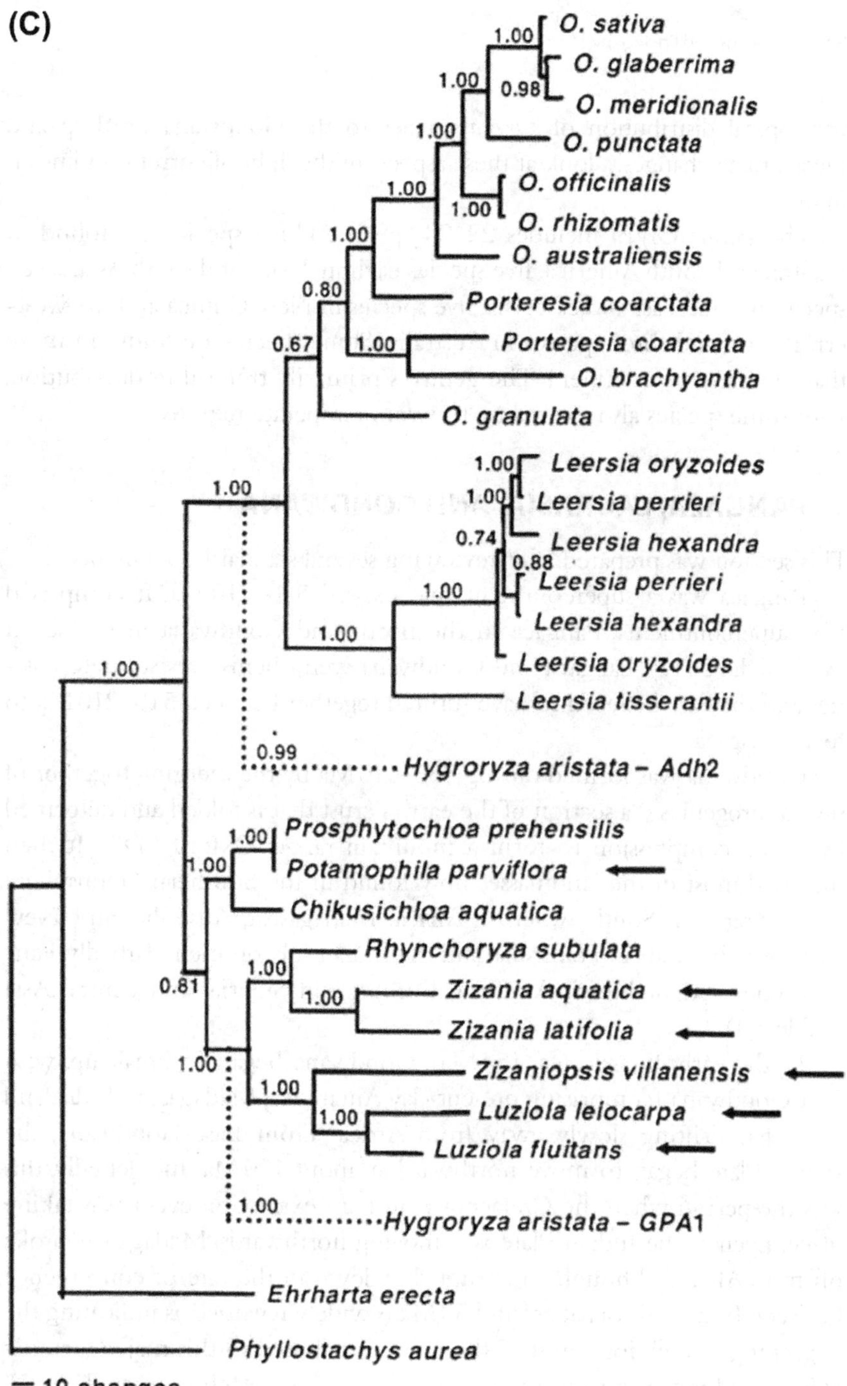

Figure 3.5 Cont'd C. Phylogeny of Oryzeae obtained from the combined Adh2 and GPA1 sequences by Bayesian inference under the TrN + G model after Hygroryza aristata was excluded. Numbers above branches are Bayesian posterior probabilities. The dashed lines indicate the position of Hygroryza aristata on the Adh2 and GPAI trees, respectively. Arrows indicate the species with unisexual florets.

pantropical distribution of *Oryza* species to the Gondwana breakup and metaclimate changes, a look at these aspects in the light of current findings is in order.

The genus *Oryza* includes 22–24 species. Three species are found in Central and South America, five species each in Africa and south Asia, seven species in southeast and east Asia, five species in New Guinea and the western Pacific, and three species in Australia. Some species are found in more than one region/continent. The genus is primarily tropical in distribution, with some species also extending to warm temperate regions.

7. PANGAEA, LAURASIA, AND GONDWANA

This section was prepared after reviewing several standard documents.

Pangaea was a supercontinent that existed 510–210 Ma. It comprised two supercontinents, Pangaea in the north, and Gondwana in the south (Wikipedia 2011). Pangaea and Gondwana were themselves separate entities and they are believed to have sutured together between 510–210 Ma to form Pangaea.

Gondwana was formed during 550–510 Ma by the merging together of several orogenies ("a section of the earth's crust that is folded and deformed by lateral compression to form a mountain range; Oxford ED"). It then included most of the landmasses now found in the Southern Hemisphere (i.e., Antarctica, South America, Africa, Madagascar, Australia, and New Zealand), the Arabian Peninsula, and the Indian subcontinent. Broadly, Pangaea consisted of North America, Europe, and Siberia, and central Asia (Table 3.1).

In the early Jurassic (ca. 184 Ma), Gondwana began to break up, with East Gondwana (comprising present-day Antarctica, Madagascar, India, and Australia) drifting slowly away from Africa. From East Gondwana, the Indian Plate began to move northward at about 120 Ma. Incidentally, this was the period when the Cretaceous–Tertiary extinction event was taking place. Even as the Indian Plate was moving northwards, Madagascar broke off from Africa. Although an abrupt slowdown in the rate of convergence between India and Asia at around 55 Ma is widely regarded as indicating the beginning of collision, most of the effects attributed to this major tectonic episode did not occur until more than 20 My later (Atchinson et al. 2007). These authors have indicated that the actual collision began around the Eocene/Oligocene boundary (~34 Ma).

Then, Australia began to separate from Antarctica ca. 80 Ma, and New Zealand did so around 85–13 Ma. The Australia–New Guinea landmass separated completely from the remainder of Gondwana between 55 and 45 Ma. The separation of Australia resulted in the opening of the southern ocean. By ca. 15 Ma, New Guinea had collided with the southwestern part of the Pacific Plate, resulting in the formation of the New Guinea highlands.

Elsewhere, South America began to gradually drift westwards from Africa ca. 130 Ma. The separation of South America from west Antartica – at perhaps ca. 30 Ma – caused major climate changes in the Southern Hemisphere because of the opening of the southern ocean.

In their classical study on angiosperm biogeography in relation to past continental movements, Raven and Axelrod (1974) have, among other findings, outlined the following points that are of relevance to the distribution of *Oryza* species. Direct migration last possible between West Gondwana (Africa + South America) and Australasia for warm temperate and subtropical plants (and animals), 100 ± 10 Ma; Direct migration between Africa and Madagascar, and probably also India, last possible about 100 Ma (possibly more recent; but evidence ambiguous); Overland direct migration last possible between South America and Africa nearly 100 Ma; Last direct connection between Africa and Eurasia prior to the Miocene ~ 63 Ma; More or less direct migration possible between South America and Australia via Antarctica until approximately ~38 Ma; Re-establishment of direct connection between Africa and Eurasia, 17 Ma; More or less direct migration first possible between Asia and Australasia, via New Guinea, ~15 Ma.

From the foregoing account, it is evident that the pantropical distribution of *Oryza* species in various continents cannot be connected to past tectonic movements or the Gondwanan breakup. Thus, the propositions of Chang (1976a,b,c), Second (1985a,b), and Khush (1997) do not seem reasonable in this respect.

8. PHYLOGENETIC TREES

"A phylogenetic tree is a diagram that portrays the hypothesized genealogical ties and sequences of historical ancestor/descendent relationships linking individual organisms, populations, or taxa. When species are considered, they are represented as line segments and points of branching correspond to subsequent speciation events. When possible, the lineage is presented in relation to a geological timescale" (King et al. 2006).

Since Charles Darwin drew the first evolutionary tree in 1837, they have become an important element in the study of evolutionary biology. In the genus *Oryza*, the first two phylogenetic trees were apparently created by Morishima and Oka (1960). They studied the pattern of interspecific variation in the species, based on the quantified degree of resemblance among them. They also used a system of factor analysis using a matrix of interspecific correlations. The procedure was developed by Michener and Sokal (1957) and Sokal (1958). Morishima and Oka (1960) employed 16 species (31 accessions) and 42 characters. The results are presented in Fig. 4.1. It will be discussed in Chapter 4 along with some of the molecular phylogeny studies done on *Oryza* interrelationships. Morishima and Oka's is the only study published on *Oryza* that used Michener and Sokal's method.

Several authors have pointed out the invaluable contribution of phylogenetic studies to determining relationships among and within different taxonomic entities. However there are pitfalls and discrepancies – as reflected by divergent results – when authors use different methods and programs for analyses and interpretations (Soltis et al. 1992, 1998, Doyle and Gaut 2000, for instance). This can be partially overcome by carrying out combined analyses of multiple data sets from various authors, as has been done by the Grass Phylogeny Working Group (2001) and the Angiosperm Phylogeny Group (latest AGP III in 2009).

Broadly, four methods are used in constructing phylogeny trees: the distance (neighbor joining (NJ) method, the unweighted pairing groups method with arithmetic means (UPGMA), maximum parsimony, and maximum likelihood methods, and Bayesian methods. Doyle and Gaut (2000) and Soltis et al. (1992, 1998) have discussed both the pros and cons of using various methods. However, several authors seem to overlook them for various reasons and choose the method based on the availability of the software and/or the speed of getting the results (e.g., neighbor joining method). This will be evident when we examine in Chapter 4 the results obtained by various authors.

9. MOLECULAR CLOCKS

Traditionally, the phylogenetic age of different groups of plants was being estimated on the basis of fossils. These have also been included in the foregoing discussions. The most widely cited instance of this in Gramineae has been the observation of Stebbins (1981) that the herbivores and grasses coevolved during the lower to middle Miocene (57–36 My) strata. He further observed that the time of divergence of maize and rice was ca. 50 Ma. The fossil records, vegetative parts, and pollen, however, take back the age by ca. 15–20 Ma (cf. pp. 38–39).

More recently, molecular biologists have employed molecular clock methods to estimate the time of origin of most of the higher plant and animal higher taxa. We have presented this information in the preceding pages. Emile Zucherkandl and Linus Pauling were the first to introduce the concept of the biological clock. They observed that due to the constancy of mutation rates, the amount of difference between the DNA of two species is a function of the time since their evolutionary separation. The principle of the biological clock is based on the theory of molecular evolution. In this, most molecular mutations are dominated by the random fixation of neutral genes, and not by natural selection. Neutral genes are those whose effects on fitness are too small for natural selection to play a determining role in their fate (Bromham and Penny 2003).

There are of course several critiques of the biological clock theory (cf., Doyle and Gaut 2000, Soltis and Soltis 2000, 2003). Bromham and Penny (2003) have observed that their reliability depends on the accuracy with which genetic distance is estimated, and on the appropriateness of the calibration methods (see also Soltis and Soltis 2003). Though this fact is widely known, most authors, especially those working on higher plants, do not choose to take these precautions literally. This appears to be the case also in some of the studies reported here. But, in the absence of more reliable information from other sources, this method does provide useful information about the phylogenetic diversification timings of different strata of taxonomic entities.

CHAPTER 4

Oryza Species and Their Interrelationships

1. INTRODUCTION

We have already discussed (in Chapter 3) the systematic position of the genus *Oryza* in the grass family. It belongs to the family Poaceae, subfamily Ehrhartoideae (one of 12 subfamilies, 120 species), and tribe Oryzeae (one of three tribes, the other two being Ehrharteae and Phyllorachidae; 9–10 genera, 60–70 species). The genus *Oryza* includes the two cultivated species, *O. sativa* L. (Asian rice, universally cultivated) and *O. glaberrima* Steud. (African rice, confined in cultivation to tropical West Africa), and 17–20 wild species (Tables 4.1 and 4.2). Seeds of several wild *Oryza* species used to be consumed like rice in times of scarcity, up to about the 1960s, wherever they grew in large stands. Some of these are used also for performing rituals, mainly *O. rufipogon, O. longistaminata, O. barthii, O. punctata, O. grandiglumis,* and *O. ridleyi* (Tateoka 1962, Kumar 1988, Vaughan 1990, Vaughan et al. 2005, Ahuja and Ahuja 2010, Kasem 2012).

Oryza species have a pantropical distribution. Most species are adapted primarily to wetlands, or seasonal wetlands, and streams. They occur both in the open and in shade. The species and taxa that are most closely related to the cultivated species prefer open, lowland habitats. A few species do grow in uplands, generally in the shade, in secondary woods, or deciduous forests.

The genus *Oryza* is considered to be a difficult one for classification, with various authors making attempts from time to time. Linneaus (1753) had described only one species, the universally cultivated Asian rice, *Oryza sativa*. The genus has been classified, or the species/taxa included in it have been enumerated, by about ten authors: Prodoehl (1922), Roschevicz (1931), Chevalier (1932), Chatterjee (1947), Sampath (1962), Tateoka (1963), Sharma and Shastry (1965), Vaughan (1994, 2003), and Clayton et al. (2010) (Table 4.1). In addition, Chang (1964) published the results of a poll of *Oryza* species, and Backer (1946) and Duistermat (1987) had revised the species and taxa occurring in southeast Asia and Australia.

Origins and Phylogeny of Rices
http://dx.doi.org/10.1016/B978-0-12-417177-0.00004-8

Table 4.1 Enumeration of the *Oryza* Species (Historical)

		Linnaeus (1753)	Baillion (1894)	Prodoehl (1922)	Roschevicz (1931)	Chevalier (1932)
1	*O. sativa (1753)*	+	+	+	+	+
2	*O. perennis (1794)*	–	–	–	–	–
3	*O. latifolia (1813)*	–	–	+	+	+
4	*O. subulata (1829)*	–	–	+	+	+
5	*O. minuta (1830)*	–	–	+	+	+
6	*O. coarctata (1832)*	–	–	+	+	+
7	*O. fatua (1839)*	–	–	–	–	+
8	*O. rufipogon (1851)*	–	–	–	–	–
9	*O. glaberrima (1853)*	–	–	+	+	+
10	*O. glumipatula (1853)*	–	–	–	–	–
11	*O. punctata (1853)*	–	–	+	+	+
12	*O. granulata (1891)*	–	–	+	+	+
13	*O. officinalis (1891)*	–	–	+	+	+
14	*O. meyeriana (1894)*	–	+	+	–	+
15	*O. ridleyi (1896)*	–	–	+	+	+
16	*O. stenosyrsus (1905)*	–	–	+	–	–
17	*O. barthii (1910)*	–	–	–	–	–
18	*O. brachyantha (1914)*	–	–	–	+	+
19	*O. breviligulata (1914)*	–	–	–	+	+
20	*O. longistaminata (1914)*	–	–	–	+	+
21	*O. schlechteri (1914)*	–	–	+	+	+
22	*O. australiensis (1915)*	–	–	–	+	+
23	*O. dewildemanii (1920)*	–	–	–	+	+
24	*O. abromeitiana (1922)*	–	–	+	+	–
25	*O. grandiglumis (1922)*	–	–	+	+	+
26	*O. mezii (1922)*	–	–	+	–	–
27	*O. schweinfurthiana (1922)*	–	–	+	+	+
28	*O. perrieri (1927)*	–	–	–	–	+
29	*O. eichingeri (1930)*	–	–	–	–	–
30	*O. stapfii (1931)*	–	–	+	+	+
31	*O. tisseranti (1932)*	–	–	+	–	+
32	*O. alta (1936)*	–	–	–	–	–
33	*O. longiglumis (1953)*	–	–	–	–	–
34	*O. angustifolia (1951)*	–	–	–	–	–
35	*O. ubanghensis (1951)*	–	–	–	–	–
36	*O. malampuzhensis (1958)*	–	–	–	–	–
36	*O. nivara (1965)*	–	–	–	–	–
37	*O. meridionalis (1981)*	–	–	–	–	–
38	*O. rhizomatis (1989)*	–	–	–	–	–
39	*O. neocaledonica (1994)*	–	–	–	–	–

Chatterjee (1948)	Sampath (1962)	Tateoka (1963–1965)	Ghose et al. (1965)	Chang (1985)	Lu (1999)	Sharma (2003)	Vaughan (2003)	Clayton et al. (2010)
+	+	+	+	+	+	+	+	+
+	+	−	−	−	−	−	−	−
+	+	+	+	+	+	+	+	+
+	+	−	−	−	−	−	−	−
+	+	+	+	+	+	+	+	+
+	+	−	−	−	−	−	−	−
−	−	−	−	−	−	−	−	−
+	+	+	+	+	+	+	+	+
+	+	+	+	+	+	+	+	+
−	−	−	−	+	+	+	−	−
+	+	+	+	+	+	+	+	+
+	+	−	+	+	+	+	+	−
+	+	+	+	+	+	+	+	+
+	+	+	+	+	+	+	+	+
+	+	+	+	+	+	+	+	+
−	−	−	−	−	−	−	−	−
−	+	+	+	+	+	+	+	+
+	+	+	+	+	+	+	+	+
+	+	+	−	−	−	−	−	−
−	+	−	+	+	+	+	+	+
+	+	+	+	+	+	+	+	+
+	+	+	+	+	+	+	+	+
−	−	−	−	−	−	−	−	−
−	−	−	−	−	−	−	−	−
+	+	+	+	+	+	+	+	+
−	−	−	−	−	−	−	−	−
−	−	−	−	−	−	+	−	−
+	+	+	−	−	−	−	−	−
+	+	+	+	+	+	+	+	+
+	−	−	−	−	−	−	−	−
+	+	+	−	−	−	−	−	−
+	−	+	+	+	+	+	+	−
−	−	+	+	+	+	+	+	+
−	+	+	−	−	−	−	−	−
−	+	−	−	−	−	−	−	−
−	+	−	+	−	−	+	−	−
−	−	−	+	+	+	+	+	−
−	−	−	−	+	+	+	+	−
−	−	−	−	−	+	+	+	−
−	−	−	−	−	+	−	−	+

Table 4.2 List of *Oryza* Species

S No.	Tateoka (1963)	Duistermat (1989)	Vaughan (1994, 2003)	Clayton *et al.* (2010) & WCPS (2011)	Consensual taxa (2013)
1	*alta*	NS[3]	*alta*	*latifolia*	*O. latifolia ssp. alta*
2	*angustifolia*[1]	NS[3]	NS[3]	*Leersia nematostachya*	*Leersia nematostachya*
3	*australienensis*	*australienensis*	*australienensis*	*australienensis*	*O. australienensis*
4	*barthii*	NS[3]	*longistaminata*	*longistaminata*	*O.longistaminata*
5	*brachyantha*	NS[3]	*brachyantha*	*brachyantha*	*O.brachyantha*
6	*breviligulata*	NS[3]	*barthii*	*barthii*	*O.barthii*
7	*coarctata*[4]	NS[3]	*Porteresia coarctata*	*Porteresia coarctata*	*Porteresia coarctata*
8	*eichingeri*	NS[3]	*eichingeri*	*eichingeri*	*O.eichingeri*
9	*glaberrima*	NS[3]	*glaberrima*	*glaberrima*	*O.glaberrima*
10	*grandiglumis*	NS[3]	*grandiglumis*	*grandiglumis*	*O.grandiglumis*
11	*latifolia*	NS[3]	*latifolia*	*latifolia*	*O.latifolia ssp. latifolia*
12	*longiglumis*	*longiglumis*	*longiglumis*	*longiglumis*	*O.longiglumis*
13a	*meyeriana ssp. meyeriana*	*meyeriana*	*meyeriana*	*meyeriana*	*O.meyeriana ssp. meyeriana*
13b	*meyeriana ssp. abromeitiona*	NS[3]	NS[3]	*meyeriana*	*O. meyeriana*
13c	*meyeriana ssp. granulata*	*meyeriana*	*granulata*	*meyeriana*	*O. meyeriana spp. granulata*
14	*minuta*	*minuta*	*minuta*	*minuta*	*O. minuta*
15a	*officinalis ssp.officinalis*	*officinalis*	*officinalis*	*officinalis*	*O. officinalis ssp. officinalis*
15b	*O. officinalis ssp. malampuzhaensis*	NS[3]	*officinalis*	*officinalis*	*O. officinalis ssp. malampuzhensis*
16	*perrieri*	NS[3]	NS[3]	*Leersia perrieri*	*Leersia perrieri*

17	*punctata*	NS[3]	*punctata*	*punctata*	*O. punctata*
18	*ridleyi*	*ridleyi*	*ridleyi*	*ridleyi*	*O. ridleyi*
19a	*Rufipogon*	*rufipogon*	*rufipogon*	*rufipogon*	*O. rufipogon ssp. rufipogon*
19b	(*O. glumipatula*[2])	NS[3]	*rufipogon*	*rufipogon*	*O. rufipogon ssp. glumipatula*
20	*sativa*	*sativa*	*sativa*	*sativa*	*O. sativa*
21	*schlechteri*	*schlechteri*	*schlechteri*	*schlechteri*	*O. schlechteri*
22	*tisseranti*[1]	NS[3]	NS[3]	NS[3]	*Leersia tisseranti*
23	*Rufipogon proparte*	*sativa proparte*	*nivara*[5]	*rufipogon*	*Spontanea*
24	NS[3]	*meridionalis*[6]	*meridionalis*	*rufipogon*	*O. rufipogon ssp. meridionalis*
25	NS[3]	NS[3]	*rhizomatis*[7]	*eichingeri*	*O. eichingeri ssp. rhizomatis*
26	NS[3]	NS[3]	NS[3]	*neocaledonica*[8]	*O. meyeriana ssp. neocaledonica*

[1] *O. tisseranti*, *O.perrieri*, and *O.angustifolia* transferred to the genus *Leersia* (Launert 1965)
[2] *O.glumipatula* described by Steudel (1853)
[3] NS: not studied
[4] *O.coarctata* redesignated as *Porteresia coarctata* (Tateoka 1965)
[5] *O.nivara* described by Sharma and Shastry (1965).
[6] *O.meridionalis* described by Ng et al. (1981)
[7] *O.rhizomatis* described by Vaughan (1981)
[8] *O.neocaledonica* described by Morat (1994)

Nayar (1973) had listed all the species recognized by the first six of the above authors. The consensual list of species and taxa of the above six authors and those recognized by the remaining authors – Sharma and Shastry (1965), Vaughan (1994, 2003), Clayton et al. (2010) and Duistermat (1987) – are given in Tables 4.1 and 4.2. Table 4.1 gives a historical list of the various species and taxa recognized by all 13 scientists who have enumerated all the species in the genus. Table 4.2 gives the enumeration given by the taxonomists beginning with Tateoka (1963).

2. SUBGENERIC CLASSIFICATION

Some attempts have been made at subgeneric classification, beginning with Baillon in 1894. Though at that point, very few valid *Oryza* species had been described, he grouped them into four sections, as have all subsequent workers since. Next, Roschevicz (1931) grouped the 19 species that he recognized in the genus *Oryza* into four sections.

Section Sativa: *O. sativa, O. glaberrima, O. stapfii, O. breviligulata, O. longistaminata, O. australiensis, O. officinalis, O. punctata, O. minuta, O. schweinfurthiana, O. latifolia, O. gradiglumis* (12 spp.).

Section Granulata: *O. granulata, O. meyeriana* (2 spp.).

Section Coarctata: *O. ridleyi, O. schlechteri, O. brachyantha, O. coarctata* (4 spp.).

Section Rhynchoryza: *O. subulata* (1 sp.).

Total: 19 species

The following year, Chevalier (1932) modified Roschevicz's classification slightly as follows:

Section Euoryza: *O. sativa, O. perennis, O. glaberrima, O. stapfii, O. breviligulata, O. longistaminata, O. australiensis, O. officinalis, O. punctata, O. minuta, O. schweinfurthiana, O. latifolia, O. grandiglumis, O. brachyantha, O. perrieri, O. tisseranti* (16 spp.).

Section Padia: *O. granulata, O. meyeriana* (2 spp.)

Section Sclerophyllum: *O. ridleyi, O. schlechteri, O. coarctata* (3 spp.)

Section Rhynchoryza: *O. subulata* (1 sp.)

Total: 22 species

Sharma and Shastry (1965) made the most elaborate subgeneric classification of the genus. They divided the genus into three sections and eight series as given below.

Section Padia

Series Meyerianae: *O. abromeitiana, O. meyeriana, O. granulata*

Series Ridleyanae: *O. ridleyi, O. longiglumis*

Series Schlechteriane: *O. schlechteri*
Section Angustifolia
Series Perrierianae: *O. tisseranti, O. perrieri*
Series Brachyanthae: *O. angustifolia, O. brachyantha*
Section Oryza
Series Sativae: *O. barthii, O. cubensis, O. balunga, O. nivara, O. sativa, O. breviligulata, O. glaberrima*
Series Latifoliae: *O. latifolia, O. alta, O. grandiglumis, O. officinalis, O. punctata, O. eichingeri, O. malampuzhaensis, O. minuta*
Series Australiensis: *O. australiensis*
Total : 26 species

Sharma and Shastry's (1965) subgeneric classification is essentially an extension of the classification of Tateoka (1963). Lu (1999) then updated the above grouping. In the intervening period, three species were removed from the genus (*O. angustifolia, O. perrieri, O. tisseranti*) and three new species were added (*O. meridionalis, O. rhizomatis, O. neocaledonica*). This involved removing series Perrierianae from Section Angustifolia and making Brachyanthae into a monospecific series. There have been also some nomenclatural changes during this time.

The object of classification and taxonomy is to facilitate the identification of a material/taxon. Subgeneric classification is useful in large genera. In a small genus like *Oryza* (21–24 species), it has only a limited application, which is why the subgeneric classification is hardly used in the taxonomic literature on rice.

3. SPECIES COMPLEXES

The species complex concept was first introduced into *Oryza* by Tateoka (1962, 1963). It has since found wide, broad-based use in rice literature. Tateoka introduced the concept to deal with species whose delimitation has been the subject of diverse opinions. He initially identified those species that were distinctive and easily separated from others. These comprised *O. schlechteri, O. australiensis, O. coarctata, O. brachyantha, O. angustifolia, O. perrieri, and O. tisseranti*. The three last named species have since been transferred from *Oryza* to *Leersia* (Launert 1965), as we have noted before.

Tateoka (1972) then identified the following five species complexes and placed the remaining species in them:

(1) *O. latifolia* complex: *O. grandiglumis, O. alta, O. latifolia, O. officinalis, O. minuta, O. eichingeri, O. punctata*

(2) *O. sativa* complex: *O. sativa, O. rufipogon, O. barthii* (now *O. longistaminata*)

(3) *O. glaberrima* complex*: O. glaberrima, O. stafii, O. breviligulata (*now *O. barthii)*
(4) *O. ridleyi* complex*: O. ridleyi, O. longiglumis*
(5) *O. meyeriana complex: O. meyeriana* and *O. granulata*

Practically all scientists who have studied *Oryza* species since then have used this concept, but the species included by them in their respective species complexes have sometimes changed marginally. Vaughan (2003) introduced a variant of this. He termed them "gene pools". The gene pool concept was first introduced by Harlan and de Wet (1971). It is defined as "the total genetic information possessed by the reproductive members of a population of sexually reproducing organisms" (King et al. 2006). Vaughan (2003) then added certain "bylines" to the various rice gene pools to highlight the most outstanding features of each group: Gene pools on the boundary of *Oryza, O. schlechteri*; Gene pool of dryland tropical complex, *O. meyeriana* complex; Gene pool of swampy lowland tropical forests, *O. ridleyi* complex; and gene pool with diploid–tetraploid series, *O. officinalis* complex. However, Vaughan's bylines are not based on comparable attributes.

4. SPECIES IN THE GENUS

Many authors have observed that the genus *Oryza* is a difficult one from the taxonomic point of view: WCSP (2012) enumerated the 139 species described in the genus to date. There have been about 10 attempts to enumerate the species in the genus and with some authors (Sharma, Vaughan, for instance), changing their lists more than once (tables 4.1 and 4.2). Taking various points of view into consideration, Table 4.2 presents a "consensual" list of species and taxa in the genus. Eighteen species are included in this list out of a total of 44 listed since the genus was first constructed in 1753. Ten subspecies are also given.

Five species have been transferred out of the genus by various authors. Out of these five, Lu and Ge (2004) proposed to restore *Porteresia coarctata* as *O. coarctata*. This aspect will be discussed later. The other four species that have been removed from *Oryza* are: *O. subulata* as *Rhyncoryza subulata* (IRRI 1964), and *O. tisseranti, O. perrieri,* and *O. nematostachya* into *Leersia* (Launert 1965). Five other species have been "downgraded" to either subspecies or synonyms. There have been three additions to the genus *Oryza: O. inandamanica, O. ubhangensis,* and *O. jeyporensis*. The latter two have not been described scientifically. *O. indandaminica* has been suggested to be a subspecies of *O. meyeriana* (WCSP 2012; also, Khush et al. 1990). *O. inandamanica* was formerly described by Ellis (1994). Similar problems exist for some other species also, viz., *O. alta, O. grandiglumis,* and *O. punctata*. These will also be discussed later.

All the taxonomists who have studied the genus have unanimously identified the following 12 species as valid ones: *O. sativa, O. glaberrima, O. rufipogon, O. barthii, O. longistaminata, O. officinalis, O. minuta, O. meyeriana, O. latifolia, O. longiglumis, O. ridleyi,* and *O. schechteri*. There is a lack of unanimity with regard to the taxonomic status of the remaining species/taxa (Tables 4.1 and 4.2) and of those listed as their synonyms, subspecies, etc. They are: *O. alta, O. glumipatula, O. meyeriana, O. granulata, O. indandamanica, O. malampuzhaensis, O. neocalidonica, O. nivara, O. meridionalis,* and *O. rhizomatis*.

Most of the information given below has been taken from the following sources: Roschevicz (1931), Chevalier (1932), Chatterjee (1948), Tateoka (1962ab, 1963, 1965ab), Duistermaat (1987), Vaughan (1994, 2003), and Clayton et al. (2010).

i. *O. latifolia* Complex

A. *O. malampuzhaensis* (2n = 48, CCDD)

Krishnaswamy and Chandrasekharan (1967, 1968), who described the species, collected their material from two sites in the Animalai hills in southern India, growing along forest river streams. It is a perennial, rhizomatous, and vigorous herb attaining about 160 cm height. Its ligules are short, 3 mm wide, with a fringe of fine hair at the tip; axils of the inflorescence are slightly woody and pubescent at the origin of branches. Spikelets are broadly oblong, 5.0–6.0 mm long, 2.5–3.0 mm wide, and awned. The 1.3–3.6 mm long spikelets become black on maturity.

Tateoka (1962) studied the above two samples along with *O. minuta* (2n = 48, BBCC) and *O. officinalis* (2n = 24, CC). He observed that the hairy ligules found in *O. malampuzhensis* are not uncommon in *O. officinalis*. *O. malampuzhaensis* has short rhizomes, while in *O. officinalis*, they are usually semi-woody. They sometimes produce new buds which bend somewhat obliquely at the base. He felt that this difference was not remarkable, and short rhizomes, as are found in *O. malapuzhaensis,* are not completely missing in the collection of *O. officinalis*. He observed further that although the spikelet length in the former species is longer, it came within the range of length of the latter. Tateoka (1962) therefore proposed that *O. malampuzhensis* is best considered as a subspecies of *O. officinalis*.

Subsequently, Joseph and co-workers (Joseph et al. 1999, Thomas et al. 2001a,b,c) conducted an extensive survey of southern Western Ghats (ca. 150 km length) for *O. malampuzhensis* and collected 11 accessions. The populations were found as single plants in small scattered patches at 100–1200 m above mean sea level. They were growing in diverse situations, in marshy lands along the margins of streams in the forest interiors that were shaded, partially shaded, or in the open (Joseph et al. 1999).

The authors studied the 11 populations for 15 characters along with three samples of *O. officinalis*. The intrapopulation variability was very low, while the interpopulation variability was high. Discriminant analysis showed that the first two characters, (ligule length [0.6427] and spikelet width [0.2321]) accounted for maximum variability. The diploid and tetraploid taxa showed an overlapping of morphological characters (Joseph et al. 1999).

In a subsequent study, Joseph and co-workers (Thomas 2001a,b,c) studied the relationship between *O. malampuzhaensis* and *O. officinalis* using 10 populations of the former and five of the latter. They employed 63 morphological characters and 23 primers for RAPD analysis and obtained 262 RAPD markers. Pair-wise comparisons of the data showed 60% genetic distance between the two taxa. This was significantly higher ($P < 0.01$) than the intraspecific distances in the two species. Cluster and principal component analyses of genetic distance estimations based on RAPDs and morphological characters also clearly differentiated the two taxa (Table 4.4). The authors then concluded that the two taxa are distinct species. However, the authors did not appear to have studied any of the features that are distinct in the two taxa, *malampuzhaensis* and *officinalis*, as given by Krishnaswamy and Chandrasekharan (1958) and Tateoka (1962). Thomas et al. (2001a) have given the spikelet shape as ovate-slightly lanceolate (in *O. malampuzhaensis*) and ovoid-cylindrical (in *O. officinalis*), also, possessed hairiness and rhizomes.

More studies are needed to determine the correct identity of *O. malampuzhensis*.

B. *O. officinalis* (2n = 24; CC)

This species was first described by Wallich in 1828. It has a wide distribution in south, east, and southeast Asia: India (west coast, southeast and northeast regions, Himalayan foot hills, Sikkim, Myanmar), China (south and southeast regions, Hainan), and throughout southeast Asia up to Papua New Guinea. Plants are erect, tall (up to 150 cm), vigorous, and are rhizomatous perennials. It has some morphological similarities with *O. sativa, O. minuta, O. rhizomatis,* and *O. punctata*. It occurs in forests or forest edges, swampy areas and wetlands, including the banks of streams and water bodies.

Tateoka (1963) had separated *O. officinalis, O. minuta, O. punctata,* and *O. eichingeri* on the basis of panicle characters, ligule length and hairiness, and spikelet and awn size (Table 4.3).

Table 4.3 Comparison of *O. eichingeri, O. punctata, O. officinalis,* and *O. minuta**

	O. eichingeri	***O. punctata***	***O. officinalis***	***O. minuta***
Panicle	Contracted with ascending branches	Loose with spreading branches	Loose with spreading branches, lower half of lower branches naked	Small with ascending branches, spikelets almost from base
Ligule	Always glabrous, <3 mm long	Always glabrous 3–10 mm long	Glabrous or hairy, 1–5 mm long	Glabrous, 1–4 mm long
Spikelet	Narrowly elliptic, 4.5–5.8 mm long, <2 mm wide	>5.0 (usually 4–5 mm) x 2.0 mm wide	Elliptic, 3.5–5.5 mm	Elliptic or narrowly elliptic, 3.6–5.0 x <2.0 mm
Awns	1–2 cm long	1–2 cm or more long	Sometimes awnless	<2 mm long

*Data from Tateoka (1963, 1965)

In the most detailed taxonomic study of the *O. latifolia* complex carried out so far, Tateoka (1962) examined 73 specimens comprising 50 sheets from several well known herbaria of the world, the East African Herbarium, Nairobi, 23 living collections assembled during his survey, and also those maintained at the National Institute of Genetics, Japan. He found several characters overlapped among all the four species. Most specimens could be identified only by using a combination of characters. For instance, 27/34 samples of *O. officinalis* and *O. minuta* could be identified only by using five characters (cf. Table 4.3), plant height, length of lowest leaves, panicle length, the point at which the spikelets are formed, and spikelet length.

	Character	***O. minuta***	***O. officinalis***
1	Plant height	<100 cm	<150 cm
2	Lowest leaves	ca. 28 cm long	ca. 65 cm long
3	Panicle length	<17 cm	>40 cm
4	Point at which spikelets are formed in the primary panicle branches	Lower one-third naked	Lower half naked
5	Spikelet length	3.6–5.0 mm	3.5–5.5 mm

Table 4.4 Comparison of *O. rhizomatis, O. eichingeri,* and *O. punctata*

	O. rhizomatis		***O. eichingeri***			***O. punctata***	
	Vaughan (1990, 1994, 2003)	**Clayton et al. (2010)**	**Tateoka (1963,1965)**	**Vaughan (1994)**	**Tateoka (1963, 1965)**	**Vaughan (1994, 2003)**	**Clayton et al. (2010)**
Life history	Perennial , rhizomatous	Perennial, short rhizomes	Perennial, base slender, not spongy	Perennial, base slender, hard culms	Perennial, base wide, spongy	Annual or weakly perennial, spongy, thick	Perennial caespitose, decumbent, spongy
Habit, height	Erect to semi-erect, 1–3 m	Erect, 60 – 100 cm		Short <1 m height hard slender stem	erect		Erect, decumbent
Ligule	3–8 mm, truncate	1–3 mm, truncate, entire	3.5 hard yellowish never split	Glabrous, < 3.5 mm, non-splitting	4–10 mm soft whitish easily splitting	> 3 mm, soft	3–10 mm entire or lacerate truncate or obluse
Inflorescence (panicle)	20–28 cm, well exserted, 5–9 branches, 1–2 from lower nodes	Open lanceolate, 10–25 cm primary branches ascending angular	Branches spreading			Spreading panicle base with 2–5 branches	Open elliptic 15–35 cm long ascending branches

Spikelets	Shattering, mottled black, 5.3–7.2 (> 6 mm in 1994) x 1.9–2.4 x 1.2–1.7	4.5–5.8 x 1.5–1.8 mm narrowly elliptic or laterally compressed	4.5–5.8 mm x 6.2 mm, narrowly elliptic	4.5–6.0 mm long	5.0 x 2.0 mm	> 5.4 x > 1.9	4.9–6.2 x 1.9–2.6 mm, elliptic, laterally compressed
Anther	3.1 mm long						
Awns	Often absent, if present, < 1.5 cm	Awned, 5–20 mm long	10–20 mm very flexuous	Flexuous with fine bristles	Awn, straight or flexuous, 10–20 mm long	Straight, flexuous with rigid bristles > 2 cm	Awned, 20–75 mm long

The remaining seven specimens possessed intermediate characters. Further, as stated above, *O. minuta* is 2n = 48. It has only a very restricted distribution (in the Philippines and New Guinea), while *O. officinalis* is 2n = 24 and it is distributed very widely (in south, east, and southeast Asia). He did not provide any explanation for his inability to distinguish the remaining seven collections out of the total 34 collections.

Prior to Tateoka (1963), Roschevicz (1931) and Chatterjee (1948) had recognized the distinctiveness of both *O. officinalis* and *O. minuta*. The following five characters were used by Roschevicz (1931) to establish the identity of the two species

Character	*O. minuta*	*O. officinalis*
Plant height	50–100 cm	up to 150 cm
Panicle length	7–12 cm	up to 30 cm, lower half naked
Secondary branches	up to 8 cm long	up to 11 cm long
Ligule length	1.5 mm	2–3 mm
Spikelet size	1.5–3.5 x up to 2.0 mm	4.5 x 2.25–2.5 mm
Awn length	5–20 mm	8–15 mm

Roschevicz had pointed out also the morphological similarities of *O. officinalis* and *O. minuta* with *O. sativa*. He then suggested that both the species could have contributed to the evolution of Asian cultivated rice.

Chatterjee (1948) highlighted the small spikelet size (length × breadth, 3–4 × up to 2 mm) and awn length (equal to or slightly longer than the spikelets) of *O. minuta*. Vaughan (1994, 2003) suggested a possible affinity of *O. minuta* with *O. punctata*. He observed that *O. minuta* has two characteristics – basal secondary branches of panicles, not whorled, and small spikelet size (< 4.7 × < 2.0 mm) – by which it could be distinguished from *O. punctata*. Since the genome composition of *O. minuta* is BBCC, he wondered how it could have acquired the BB genome, as it is present in only the African species, *O. punctata*. The very limited and discrete distribution of *O. minuta* – in central Philippines and southwest New Guinea – is another of its enigmatic aspects. In contrast, *O. officinalis* (2n = 24; CC) has a wide distribution spanning south, east, and southeast Asia. The habitats of the two species are similar: on the edges of forests in both seasonal and perennial wetlands, under shade, and in the open. *O. officinalis* is vigorous, erect, and perennial. The basal branches of its panicles are whorled and the panicles occur at half or more distance from the base (Table 4.3).

From the above account we can see that the two species, *O. minuta* and *O. officinalis,* can be distinguished only by using a combination of characters. There are also difficulties in determining their relationship to each other.

C. *O. punctata and O. eichingeri*

a. *O. punctata* (2n = 24, 48, BB, BBCC)

O. punctata was described in 1855. It is an annual, erect (ca. 100 cm), vigorous plant that grows in a variety of ecological situations such as open grasslands, wetlands, water holes, along streams, and in water up to 1 m in depth. It is widely present in tropical Africa from Cote d'Ivorie to South Sudan, Madagascar, and Swaziland. It shows some morphological resemblance to *O. officinalis* and *O. sativa.*

b. *O. eichingeri* (2n = 24; CC)

In contrast to *O. punctata,* this species has only a limited, disjunct distribution. It is present in middle Africa (Cote d' Ivorie, Central African Republic, Democratic Republic of Congo, Rwanda, east Africa (Kenya, Uganda, Tanzania)), also Sri Lanka. It appears to prefer shaded habitats.

Various authors have used different criteria to distinguish *O. punctata* and *O. eichingeri.* They are based on length of spikelet, awn, and ligule, and features of the culm and sterile lemma. Tateoka (1965) employed ligule length, awn, sterile lemma characteristics, and nature of sheath (sheath spongy in *O. punctata*). Further, *O. punctata* is annual or only weakly perennial, while *O. eichingeri* is perennial (Table 4.4).

Tateoka (1965) has carried out the most extensive studies of the above two species so far. In addition to surveying and collecting live material (17 accessions), he also studied herbarium sheets from some leading herbaria (total: 50 specimens). Of the 23 live specimens that he studied, two were intermediate between the two species and the rest were distinguishable into two distinct species. Spikelet length and breadth varied continuously. Broadly, *O. eichingeri* possesses smaller spikelets and *O. punctata* larger ones. He stated that characteristics of ligule, culm, awn, panicle, and sterile lemma together could be used to distinguish the two species.

Vaughan (1994, 2003), who had also conducted extensive herbarium examinations and field surveys, employed all the above characters to separate the two species.

Thus, these two taxa, *O. punctata* and *O. eichingeri,* can be separated only by a combination of characters. Incidentally, one of the reasons for the discrete and sparse distribution of *O. eichingeri* may be because large areas of tropical Africa still remain to be surveyed. The position of *O. punctata* in Africa may be analogous to that of *O. officinalis* in south and southeast Asia: wide distribution, much variability, possible formation of tetraploids in the peripheries of their distribution, and so on.

D. *O. alta* (2n = 48, CCDD), *O. grandiglumis* (2n = 48, CCDD), and *O. latifolia* (2n = 48, CCDD)

All these three species are American in distribution. All are tetraploids (2n = 48) and possess CCDD genomes. They also have similar features, in broad terms. In addition, *O. latifolia* possesses some similarity with the widely occurring Asian diploid species, *O. officinalis*.

O. grandiglumis was described in 1922. It is easily distinguished from the others by the presence of two sterile glumes that are similar to each other and are as long as the fertile glumes (lemma and palea). Broadly, this species is distributed in the Amazonia.

O. latifolia was first described in 1913. It occurs widely in Central and South America, from south Mexico to north Argentina and the Caribbean. It has the widest distribution of the three species.

O. alta was described in 1936. It is distributed in the central and northern parts of South America. The distribution of the three species appears to be in a sort of parabolic circles, with *O. latifolia* outside, *O. grandiglumis* inside, and *O. alta* forming the middle cycle (cf., Tateoka's (1962) distribution map).

Chatterjee (1948) separated *O. alta* and *O. latifolia* on the basis of spikelet and awn lengths and sterile lemma shape, as: *O. alta*, spikelets 7.5–9.0 mm long, awns 2–3 cm long, sterile lemmas acuminate; *O. latifolia*, spikelets 5–6 mm long, awns 1–2 cm long, sterile lemma acute.

In his key to the genus, Tateoka (1963) separated *O. grandiglumis* from the other two species by its characteristic and unique sterile lemmas, which are similar to the fertile lemma in length, size, and texture. He then separated *O. alta* and *O. latifolia* by their spikelet length: 4–7 cm in *O. latifolia* and >7 cm in *alta*.

Tateoka (1962) conducted the most detailed taxonomic study of the three taxa, examining 152 herbarium sheets. The spikelet length of the three species presented a continuous variation: that of *O. latifolia* varied from 4.2 to 7.0 mm, *O. alta* from 7.2 to 9.0 mm, and of *O. grandiglumis* within the range of *O. alta*.

All the three species show very vigorous vegetative growth. *O. alta* grows to 3–4 m or even taller, and *O. latifolia* and *O. grandiglumis* are 1–2 m tall.

Chatterjee (1948) and Swallen (1955) had earlier used culm height to distinguish between *O. alta* and *O. latifolia*, but Tateoka (1962, 1963) found this character to be inconsistent in expression. Swallen (1936, 1943, 1955) and Tateoka (1962a) observed also that the spikelet length in *O. latifolia* appeared to increase as one moved from northern and central regions of its occurrence to the southern region in tropical South America. Vaughan (1994) separated *O. latifolia* and *O. alta* on the basis of leaf width (<5 cm in *O. latifolia*, >5 cm in *O. alta*) and spikelet length (<7 cm in former, >7 cm in latter). All these observations highlight the broad morphological similarities and continuities of the three taxa.

Hooker (1897), in "The Flora of British India", volume VII, apparently misnamed *O. officinalis* as *O. latifolia*. This error was noticed by Roschevicz (1931), who then mentioned the obvious differences in leaf width, ligule characters, spikelet length, regions of natural distribution, and other characteristic differences between the two taxa. Tateoka (1962a) studied this aspect in detail. He examined 184 herbarium sheets of the two species. He observed that while the two taxa could be easily distinguished in most cases, less than 5% of the sheets showed difficulties in identification with overlapping of characters.

Since *O. officinalis* is present in tropical Asia and *O. latifolia* in tropical America, the most feasible explanation for the occurrence of populations which possess characters of both species is that both species share the C genome, and/or they show parallel evolution under comparable ecological environments. In this context, Vaughan (2003) proposed that the African diploid species, *O. eichingeri* (CC), might have the same CC genome as *O. latifolia* (CCDD), but he did not indicate how this could have happened. As we have noted before (Table 3.1), the timing of the Gondwana breakup does not match with the Gramineae differentiation.

ii. *O. ridleyi* Complex

O. ridleyi and ***O. longiglumis*** (both 2n = 48, HHJJ) belong to the *O. ridleyi* complex of Tateoka (1962). Both are species of lowland tropical wetlands (Vaughan 2003). *O. ridleyi* was described by Hooker in 1897 (in "Flora of British India", volume 7) and *O. longiglumis* by Janson in 1953.

O. ridleyi is an erect to semi-erect herb and 1–2 m tall. It is found in forests and open spaces, growing in marshes and/or along stream banks. It occurs throughout southeast Asia, including in New Guinea, but excluding the Philippines. *O. longiglumis* has the same ecological and morphological features asY *O. ridleyi*. The notable difference between the two taxa is that *O. longiglumis* possesses sterile glumes that are as long as lemma-palea. Its distribution is restricted to a small region in southern New Guinea (Vaughan 2003). Both the species are perennial (Duistermaat 1987). *O. longiglumis* has been collected from only three sites in New Guinea. The spikelet dimensions of *O. longiglumis* are only about two-thirds of those of *O. ridleyi*.

iii. *O. meyeriana* Complex

O. meyeriana (2n = 24, GG) and ***O. granulata*** (2n = 24, GG) belong to this complex. *O. granulata* was first described in 1891 and *O. meyeriana* in 1894. Broadly, *O. meyeriana* is distributed widely in south and southeast Asia (western region) and *O. granulata* in southeast Asia (eastern region), excluding New Guinea. The distribution of the two taxa overlaps in Thailand and Malaysia. They belong to the vegetation of the dryland tropical forest (Duistermaat 1987, Vaughan 2003). They are the only *Oryza* species that occur in dryland and upland habitats away from wetlands (Vaughan 2003). Both are perennial in nature. *O. granulata* is about 80 cm tall and *O. meyeriana* about half of this.

Roschevicz (1931) had seperated the two species, *O. meyeriana* and *O. abromeitiana*, on the basis of just one character, spikelet shape: oblong-orbicular in *O. granulata* and linear in *O. meyeriana*. Tateoka (1962a, 1963) did the most detailed study of this complex, using 79 specimens. He found overlap in spikelet characters between the two taxa (*O. meyeriana* and *O. abromeitiana*) and also in some other characters. He felt that it was difficult to completely separate the two taxa and also *O. granulata*. He proposed that all the three taxa were conspecific and also that could be re-designated as subspecies of *O. meyeriana*, as ssp. *meyeriana*, ssp. *granulata*, and ssp. *abromeitiana*. He gave their distribution as ssp. *meyeriana*: Indonesia (Java, Borneo), Philippines; ssp. *granulata*: India (Assam, Sikkim, Bengal, Tamil Nadu), Sri Lanka, China (Yunnan, Hainan), Indonesia (Java, Sumatra), Indo-China, Thailand, Myanmar; and ssp. *abromeitiana*: Philippines, Moluccas. All subsequent workers, including Duistermaat (1987) and Clayton et al. (2010), have accepted the merger of the two taxa as *O. meyeriana*, except Vaughan (1994, 2003), who identified them as two distinct species. Duistermaat (1987) had separated the two taxa as two varieties:

var. *meyeriana*: spikelets oblong, 3.2–5.1 times as long as wide; caryopsis 4.1–7.3 mm long;

var. *granulata*: spikelets elliptic, 2.1–2.7 times as long as wide; caryopsis 3.4–4.1 mm long.

Vaughan (2003) distinguished the two species as:

O. granulata: spikelets oblong to elliptic-oblong, < 6.4 mm long;

O. meyeriana: spikelets narrowly oblong to lanceolate, > 6.4 mm long.

At the same time, the frequency distribution of spikelet dimensions given by Tateoka (1962) clearly indicates that is difficult to carry out even subspecific classification of *O. meyeriana* on the basis of spikelet dimensions. These observations support the proposal of identifying the three taxa as subspecies of *O. meyeriana*.

iv. *O. sativa* Complex (2n = 24, AA)

Tateoka (1962b) included three taxa in the Sativa complex – the universally cultivated Asian rice, *O. sativa, O. rufipogon*, and spontaneas/red rices/*O. nivara*.

Asian rice is cultivated throughout the world, on all the continents. *O. rufipogon*, as conceived here, is present in Asia (south, east, and southeast Asia), Australia (northern part), and tropical America.

The wild rices – the taxa and populations that occur throughout the tropics and subtropics of the world – have always been an enigma to the taxonomists (IRRI 2000, FAO 2007). They look very similar to the two cultivated rices, *O. glaberrima* in tropical Africa and *O. sativa* elsewhere in the tropics. They are ubiquitous in rice ecosystems throughout the world, as weeds in paddy fields, on field bunds, along waterways and in seasonal and perennial wetlands, usually present in greater frequencies where the most closely related species, *O. rufipogon*, also occurs. Elsewhere in the world, in rice growing areas, where the wild related species does not occur, such as the USA, Japan, Korea, south Europe, etc., a weedy form commonly called 'red rice' occurs. It is a weed of great magnitude and intensity, and is one of the most serious weeds of rice everywhere (FAO 1999, 2007). However, it is in south and southeast Asia and tropical west Africa, that–at least until the 1960s (i.e., before rice research work began to make quantum improvements and rapid urbanization began to cause large-scale habitat losses of wild rice)–this became an irresolvable problem. We shall look at this problem in south and southeast Asia, because it is here that three-fourths of the world's rice is produced and consumed, and where wild rices occur, or at least used to occur until very recently, in maximum profusion and in the closest proximity to the cultivated rices.

Very early workers, such as De Candolle (1886), Watt (1891), and Hooker (1897), recorded the common presence of wild rices in the Indian subcontinent (now termed south Asia). There have been several reports of their presence, including results of experimental studies, from different regions of Asia from India (Roy 1921, MP), Bhide (1925), Bhalerao (1928, 1930) (both Konkan, Maharashtra), Sampath and Rao (1951), Ramiah and Rao (1953) (both south and southeast India), Chatterjee (1947) (northeast India), Nayar (1958) (Orissa), Nair et al. (1964) (Konkan-Malabar), Sharma and Shastry

(1965a,b) (central India), etc.); from elsewhere, from Myanmar (Beale 1927), Sri Lanka (Chatterjee 1948, Senaratna 1956), Malayan Archipelago (Ridley 1925, Backer 1946, TC Katayama 1963a), Philippines (Capinpin and Pancho 1961, TC Katayama 1963), Thailand and Indo-China (Chevalier 1932, Porteres 1956, Oka 1964a) and northern Australia (Tateoka 1963), Americas (throughout the tropics [Hitchcock 1935, Swallen 1943, Oka 1961, 1988, Vaughan 2003], FAO 1999, IRRI 2000), and so on. All of them had also observed the common occurrence of hybrid populations possessing characteristics of both wild and cultivated rices. The extent of natural crossings between the two species – cultivated and wild rices – had been estimated to vary between zero and 44.0%, in various countries by using different techniques (Table 4.6).

Some early workers had attempted to classify the wild rices into formal (taxonomic) and informal groups (Watt 1892, Roy 1921), but this has now been accepted as impractical. This is because wild rices show a tremendous amount of variability, showing a complete range of characters in all possible combinations, including in life history traits. It is obvious from the writings and conclusions of Watt (1891) that he was overwhelmed by its diversity. Initially, he identified three *Oryza* species, *O. granulata, O. officinalis*, and *O. sativa*, as being present in India (most of present-day south Asia). Then, in *O. sativa*, he identified four varieties, var. *rufipogon*, var. *coarctata*, var. *benghalensis*, and var. *abuensis*. While var. *coarctata* is possibly what is now known as *Porteresia coarctata*, and var. *benghalensis* is possibly the perennial floating wild rice, *O. rufipogon*, the identity of the remaining two forms, var. *rufipogon* and var. *abuensis*, can only be conjectured from his descriptions, as being what are termed in the current literature as spontanea or fatua rices. Roy (1921) had classified the wild rices of central India into 24 groups. For African rices, Porteres (1945, 1956, 1959) gave an extensive classification of the wild and cultivated rices (both *O. glaberrima* and *O. sativa*) (Table 4.5).

Roschevicz (1931) had designated the cultivated annual rice as *O. sativa*. He then identified two taxa under it, forma *spontanea* and forma *aquatica*. From his description, it is clear that *f. spontanea* represented the annual spontanea forms and *f. aquatica* the perennial aquatic forms, now designated as *O. rufipogon*. He then went on to note that: "the species is subject to extreme variation in both size and shape as well as color of individual parts of the spikelet and of the entire plant ....it is a heterogeneous species, which when sufficient material is available will need to be split into a number of smaller units".

Roschevicz gave the distribution of forma *spontanea* as Indian subcontinent, southeast Asia, north Australia, and the entirety of tropical Africa, and

Table 4.5 Comparison of *O. meridionalis, O. nivara, f.spontanea,* and *O. rufipogon*

S No.	Characters	*O. meridionalis*		*O. nivara*		*O.sativa f. spontanca*	*O. rufipogon*	
		Ng et al. (1981)	Vaughan (1994, 2003)	Sharma and Shastry (1965a)	Vaughan (1994, 2003)	Roschevicz (1931)	Vaughan (1994)	Clayton et al. (2010)
1.	Life history, plant height	Annual herb, 1–2 m, with adventitious roots, no rhizomes	Annual, erect, 1–2 m, "Similar to *O. nivara*", annual, AA genome, Oceanean wild rice	Annual, 1 m, basal internodes spongy	Annual, <2 m	Annual, strong, 1.5 m	Perennial, no rhizomes, but perennial rootstock, adventitious roots, nodal tillering	Perennial with rhizomes, elongated culms 30–100 cm, spongy
2.	Habit	Semierect to erect	Erect	Semierect to decumbent	Semierect to decumbent, basal internodes spongy	Erect, geniculate	Tufted and scrambling, decumbent or floating	Geniculately ascending or decumbent
3.	Leaf	Linear, thin, 41–78 cm long, 1.3–1.6 cm wide	____	Lanceolate, 20–50 cm long, 6–9 mm wide, inconspicuous at base	____	Convolute, linear-laneolate, 50 x 1.5 cm	____	20–40 x 0.5–1.0 cm, surface scabrous

Continued

Table 4.5 Comparison of *O. meridionalis, O. nivara, f.spontanea,* and *O. rufipogon*—cont'd

S No.	Characters	***O. meridionalis***		***O. nivara***		***O.sativa f. spontanca***	***O. rufipogon***	
		Ng et al. (1981)	**Vaughan (1994, 2003)**	**Sharma and Shastry (1965a)**	**Vaughan (1994, 2003)**	**Roschevicz (1931)**	**Vaughan (1994)**	**Clayton et al. (2010)**
4.	Ligule	Acute, 9–21 mm long	____	Membranous 10–25 mm, acuminate, apex bipartite, scabrid	____	40 mm, membra-nous, acicular	____	Eciliate membrane, 12–17 mm, lacerate
5.	Inflores-cence (panicle)	Freely branch-ing, stiff erect branches tightly adpressed to main axis, primary branches bear 1–2 sec. branches with 1–3 spikelets each, panicle base well exserted	Compact, erect with ascending stiff panicle branches, tightly adpressed to main axis, secondary branches rare	Erect, 10–20 cm long, 40–100 spikelets, 5– 9 branches, pulvinate, 0–2 sec. branches with 2–3 spikets each	Half exserted, few primary and secondary branches, thick stout rachis	30 cm, com-pressed, branched 2–3 times, erect	Well exserted and spreading, few branches and sub-branches	Open, linear 10–20 cm long, primary branches appressed, simple linear

6.	Anthers	Short, 1.3–2.2 mm	____	3 mm	____	2.5–2.6 mm	>3 mm	5 mm
7.	Spikelets	Markedly shattering, slender, 7.4–8.3 x 2.1–2.3 x 1.4–1.6 mm	Narrow, shattering, 7.6–8.0 x 1.9–2.2 x 1.3–1.5 mm, long scabrous awn, 7.8–10.3 cm	Laterally compressed, 6–8 x 2.5–3.0 mm, cuneately oblong, often black at maturity	Shattering, 6–8.4 x 1.9–3.0 x 1.2–2.0 mm	Orbicular or oblique oblong, com-pressed 6–8(–12) x 4 mm	Slender, 7–9.3 x 1.9–2.6 x 1.6–1.9 mm, highly shattering, embryo 1–1.5 mm long	Appressed, solitary oblong, laterally, 8–9 x 2–2.5 com-pressed, 8–9 x 2.0–2.5 mm, Rachilla elongation stout, 0.5 mm
8.	Awn	Very long, 63–115 mm, very rough, brittle, thick	78–103 mm, scabrous	Robust, barbiculate, 6–11 cm, often red	Long, strong, 4–10 cm long	Strong, long	5–11 cm, up to 16 cm	0–6 mm

of forma *acquatica* as central India and southeast Asia. He gave a separate nomenclature for the African species.

Chatterjee (1948), in his revision of the genus *Oryza*, identified two taxa of wild rices, *O. perennis* and *O. sativa* var. *fatua*. In the former, he included the perennial floating wild rices of Asia and Americas (*O. glumipatula*), and in the latter, the annual weedy, wild rices with "deciduous" spikelets. These two taxa may somewhat correspond to the *O. sativa* f. *aquatica* and *O. sativa* f. *spontanea* of Roschevicz (1931). In his monograph on the grasses of Asia, Bor (1960) identified two species, *O. sativa* and *O. rufipogon*, based on the persistence/caducousness of the spikelets. He included in his *O. rufipogon* the complete range of wild rices. He observed that *O. rufipogon* hybridized freely with cultivated rices and it was often found as weed in rice fields and considered to be a noxious weed. Tateoka's (1963) revision of the genus *Oryza* also followed Bor's (1960) classification. He too included in *O. rufipogon* all the wild rices that occur in Asia, Australia, and the Americas, and which are known variously as *O. fatua, O. formosana, O. glumipatula, O. paraguayensis, O. perennis* var. *cubensis,* and all the four *O. sativa* varieties of Watt (1891).

In later work, Sharma and Shastry (1965) and Sharma (2003, 2010) split *O. rufipogon,* as conceived by Bor (1960) and Tateoka (1963), into two species, *O. rufipogon sensu stricto* and *O. nivara*, the former broadly corresponding to the floating perennial forms and the latter, the annual forms, but occurring in upland areas. Sharma gave the characteristics of *O. nivara* as: "annual, aquatic, stem semi-erect to decumbent, basal internodes spongy, ligule membranaceous, 1.0–2.5 cm long, acuminate, bipartite at apex; panicle 10–20 cm long bearing 40–100 spikelets, branches 5–9, alternate, 0–2 sub-branches per branch bearing 2–3 spikelets each; spikelets 6–8 mm long, 2.5–3.0 mm wide, cuneately oblong, often black at maturity, fertile lemma cymbiform, coriaceous, 5.5–7.0 mm long, awn robust, 6–11 cm long".

Sharm and coworkers stated that *O. nivara* hybridizes with *O. sativa* in nature and produces partially sterile hybrids, and also that plants similar to *O. sativa* but with a few introgressed features of *O. nivara* or *O. rufipogon* occur in the cultivated fields. They referred to them as "field spontanea". Sharma (2003, 2010) observed further that "the shedding of the spikelet is a highly misleading character for establishing the limits of various species of this group". The author however did not give here the taxonomic limits within the wild rices of *O. nivara, O. rufipogon sensu stricto,* and their "field spontanea". Sharma further elaborated that the panicles of *O. nivara* could be fully or partially exserted from the flag leaf sheath, the spikelets are not as long as in *O. rufipogon*, the anthers do not fill the spikelets completely, the awns are stiffer than in *O.*

rufipogon, and the kernels are often red. *O. nivara* grows in seasonal pools that dry up completely in summer. He observed further that the genetic barrier between the three species, *O. sativa, O. nivara,* and *O. rufipogon,* is not complete, and hence natural hybridization occurs in nature and the hybrids occupy the same habitats as of the recurring parents.

Numerous authors have tried to determine the nature of the wild rices in the past (Watt 1891, Roy 1921, Bhalero 1928, 1930,Ramiah and Ghosh 1951, Oka and Chang 1959, 1961, 1962, 1964b, Sampath and Rao 1961, Oka 1956a, 1988, Nayar 1958, 1973, Sampath and Govindaswamy 1958, Morishima et al. 1961, Morishima 1969, Chu and Oka 1970). They are characterized by heterozygosity and semi-sterility. Nayar (1958) found that 17/20 single plant progenies of annual forms segregated for various characters. In a hybrid between *O. sativa* and *O. rufipogon*, the F_1s showed dominance for wild characters and the F_2 population looked similar to the annual wild rices. The characters segregated with complementary, modifying, and duplicate modes of inheritance.

In the classification of *Oryza* species in Malaysia and Australia, Duistermaat (1987) recognized only two species, *O. sativa* and *O. rufipogon.* She identified the latter by its longer anther length (3.5–6.2 mm long), perennial nature, tufted or stoloniferous habit, culms decumbent and floating, and so on. She included in it "Sharma and Shastry's (1965) *O. rufipogon sensu stricto*". Duistermat indicated *O. nivara* as a synonym of *O. sativa,* along with some others like *O. formosana* and *O. glutinosa.* She noted specifically that the two paratypes given by Sharma and Shastry (Lowrie 1975, DD, W; Put 1979, BM, DD) "clearly belong to the present species" (i.e., *O. sativa*), but that she had not seen the holotype. "There is nothing in the description to say that another species might be included", she concluded, indicating that *O. nivara* is synonymous with *O. sativa.*

Vaughan (1994, 2003) on the other hand, accepted *O. nivara* as a valid species. His description of the species varies somewhat from that of Sharma and Shastry (1965) in certain details. He gave the following as the features of *O. nivara*: height <2 m, spikelets large (6.0–8.4 x 1.9–3.0 x 1.2–2.0 mm), awn strong, long (4–10 cm), grows sympatrically with *O. sativa* and *O. rufipogon,* usually grows in shallow water in seasonally dry open habitats, but found also in swampy areas, edges of ponds and tanks, besides streams, in and around rice fields.

Vaughan separated *O. sativa* and *O. nivara* by the persistence/deciduous nature of the spikelets, and gave *O. sativa* a somewhat overlapping spikelet size with *O. nivara* (*sativa,* 4–8.5 x 2–4 mm; *nivara,* 6–8.4 x 1.9–3.0 x 1.2–2.0 mm), strong seed dormancy and high spikelet fertility. Further, he gave the following characters as those of *O. rufipogon:* perennial, decumbent or floating, awns 5–11 cm, grains (7–9.3 x 1.9–2.6 x 1.6–1.9 mm), highly shattering, low seed set. Please note that the spikelet dimensions are overlapping, and also Sharma's (2010) observation that shedding of spikelet is a highly misleading character in establishing the limits of the various species of this group. Incidentally, the habitat description, variable panicle sterility etc. given by Vaughan are very tenuous and they are characteristic of a taxa of hybrid origin as described by several well-known workers (Mayr 1942, Heiser 1949, Anderson 1951, 1953, Stebbins 1950, Baker and Stebbins 1965, Harlan 1965, Harlan and de Wet 1965, and more recently, Reiseberg 1995, Ellstrand 2003, and so on). Sharma (2003 and earlier) has never given the species delimitation of *O. nivara* vis-à-vis what he terms as "field spontanea" and the weedy red rice.

It is obvious from the above that the features of *O. nivara* are intermediate between those of *O. sativa* and *O. rufipogon*. Nayar (1973) had earlier given a strong justification in support of this contention, citing established authors (such as Harlan and de Wet (1965), Baker (1965), Baker and Stebbins (1965), Zohary (1965), Grant (1967), Harlan (1970)) that this material is of hybrid origin. Recently, Clayton et al. (2010) has not recognized *O. nivara* as a valid taxon (also, WCSP 2012).

v. *O. glaberrima* Complex

The three taxa included in this complex are:

(i) The west African cultivated rice, *O. glaberrima*;
(ii) The annual, weedy, *O. barthii;*
(iii) The vigorous, strongly perennial, *O. longistaminata.*

Taxonomists are in agreement about the identity, taxonomic limits, and nomenclature of these three species.

A. O. glaberrima

This species was described by Steudel in 1955 from a specimen collected in 1845–1848 from Portuguese Guinea (now, Guinea Bissau) by a French naval official, Edelstan Jarden. Its cultivation, which was originally limited to tropical west Africa, has been now all but replaced by the Asian rice (*O. sativa*), and more recently, by the new NERICA (New Rices for Africa) rices, which are being released by the Africa Rice Center. NERICAs are derived from *O. sativa*–*O. glaberrima* crosses.

O. glaberrima is recognized by the glabrous lemma-palea (which are sometimes hispid), usually awnless spikelets, short ligules of the lower leaves

(< 6 mm long) with round or truncate tips, and stiff panicles having only few or no secondary branches.

Nayar (2010) has recently reviewed the history of African rice. Please see also Chapter 5 of this book.

B. O. longistaminata

O. longistaminata was described by Chevalier and Roehrich in 1914. Hutchinson and Dalziel (1936 *in* "Flora of West Tropical Africa") changed this name to *O. barthii*, which was retained by Tateoka (1963). Chatterjee (1948) then changed it to *O. perennis*. Clayton (1968, 1970) restored the name as *O. longistaminata* based on the Botanical Code. This name has since been accepted by all taxonomists.

O. longistaminata is a tall (> 2 m or more), sturdy, perennial plant with long, strong, underground rhizomes. Its synonyms include *O. dewildimanii, O. madagascariensis, O. perennis, O. barthii,* and *O. silvestris* (WCSP 2012). It occurs widely throughout tropical Africa, growing in swampy areas, on the edges of lakes and ponds, edges of rice fields, irrigation channels, etc. Once established at a location, it is difficult to eradicate it. Its ligules are long (15–45 mm), the inflorescence is large, open, and drooping, with narrow spikelets, 8.0 x 2.5 mm in size. The anthers are long and almost fill the spikelets. It is cross-pollinated and partially self-incompatible (Nayar 1958, 1967, Oka and Morishima 1967). It has the widest distribution in Africa of all the *Oryza* species. It sometimes occurs in extensive stands, which local people then harvest for food and use in rituals.

C. O. barthii

This is the third species in this complex. It is an annual with a very wide distribution throughout tropical Africa. Its synonyms include *O. mezii, O. stapfii,* and *O. breviligulata*. As its former name, *O. breviligulata* indicates, its ligules are short (< 13 mm long, 3–4 mm wide) (Roschevicz 1931). Its spikelets are large (7.8–11.0 mm), easily shattering, and the awns are usually long and strong (10–20 cm).

There has been much discussion of the identity of *O. stapfii*. It is now accepted as a synonym of *O. barthii,* following Tateoka (1962b). It was first described by Roschevicz (1931) and named in honor of Dr. Stapf (Kew; a well-known grass systematist). There have been suggestions that it may be considered either as a subspecies of *O. glaberrima* or as a miniaturized form of *O. stapfii* or is synonymous with it. Their distinctions were based on length of spikelets and sterile lemmas. Tateoka (1962b) studied 58 specimens and found that they showed continuous variations for all these characters. Tateoka (1962b, 1963) then concurred with the views of Hutchinson and Dalziel (1936, *in* Flora of West Tropical Africa) that as this taxon possessed features of a wild taxon, it is most appropriately treated as synonym of *O. barthii*. This proposal has since been accepted by all the rice systematists. Clayton et al. (2010, 2013) have also endorsed this viewpoint. Incidentally, Nayar (1973, 2012) has proposed that *O. barthii* is a hybrid derivative of the two cultivated rices, *O. glaberrima* and *O. sativa*.

vi. The Outlier Group

In this grouping, we have included a motley of six taxa that do not fall clearly within any of the above five species-complexes, or which have been described more recently. They are: *O. australiensis*, *O. brachyantha*, *O. coarctata*, *O. meridionalis*, *O. rhizomatis*, and *O. schlechteri*.

A. *O. australiensis (2n = 24; EE)*

This has been one of the "consensual species" included in the genus *Oryza* since Roschevicz's (1931) time. It was described by Domin in 1915. It is a strong, tall (1–2 m), perennial herb with an elongated scaly rhizome, ligule an eciliate membrane, 2–3 mm long, entire or split, spikelets ovate-oblong to lanceolate, 6.5–9 x 2.25–3.31 mm, peduncle axis minutely ciliate. It occurs in swamps and seasonal wetlands in low eucalyptus woodlands in northern Australia.

O. australiensis may be very broadly aligned with the *O. officinalis* complex.

B. *O. brachyantha (2n = 24, FF)*

This species was described by Chevalier and Roehrich in 1914. This is also one of the consensual species from the time of Roschevicz (1931). It is short (30–80 cm), annual or weakly perennial (Vaughan 1994), caespitose, with roots from lower nodes, ligule eciliate, membranaceous, 1–2 mm long, entire or lacerate, truncate; spikelets oblong, laterally compressed, 6.5–9.5 x 1.25–1.5 mm.

O. brachyantha occurs throughout tropical Africa. Its affinity with the other species or species complexes is uncertain.

C. *O. coarctata (2n = 48, HHKK)*

O. coarctata was described by Roxburgh in 1832, and has been accepted as a valid species by Roschevicz (1931), Chevalier (1932), Chatterjee (1948), and Bor (1960). Then, based on his studies, Tateoka (1963a,b, 1965) erected a new genus, *Porteresia*, and transferred *O. coarctata* to this genus. According to him, the attributes of this taxon which differ from those of other *Oryza* species, are the presence of oblong silica bodies in the leaf (vs. oryzoid in the rest) and large embryos almost half the size of the caryopsis (vs. one-fifth to one-third size in the other species).

Roschevicz (1931) had earlier observed that in this taxon, the lemma is 7–9 nerved and the palea five-nerved, while in all the *Oryza* species, lemma is five-nerved and the palea is three-nerved. He observed further that *O. coarctata* possesses certain other features that were different from those of other *Oryza* species. They are branched stalk (culms), hard leathery leaves, indistinct articulation below the upper empty glumes, and non-flattened spikelets. The separation of this taxon, *Porteresia coarctata,* has since been accepted by other taxonomists (Vaughan, 1993, 2003, Clayton et al. 2010). The species is adapted to saline soils and is found in the river estuaries of Pakistan, eastern India, Bangladesh, and Myanmar.

Later on, Lu and Ge (2005) argued that, based on the affinities shown in molecular biology studies, *Porteresia coarctata* should be restored to *Oryza*. We shall discuss this matter further later.

D. O. neocaledonica (2n = 24, GG?)

This is the latest species described in the genus *Oryza* (Morat et al. 1994). It has been found about one hectare area in New Caledonia Island, in the south-west Pacific Ocean (about 1,200 km east of Australia, south of 20°S latitude). It is a perennial plant with short rhizomes, 60–80 cm tall, auricles absent, ligule ciliate, membranaceous, 0.5–1 mm long, panicles with 7–40 spikelets, and spikelets elliptical or oblong, laterally compressed, 7–9 x 1.5–1.8 mm.

Vaughan (2003a,b), in his key to the genus *Oryza,* considered *O. neo-caledomica* as a population of *O. meyeriana*. Two other variant forms of *O. meyeriana*, according to Vaughan (2003a), are *O. indandamanica* from Rutland Islands (Andamans, India) and ssp. *tuberculata* from China (described by Wu et al. (1994)). Clayton et al. (2010, 2013) however have recognized *O. neocaledonica* as a valid *Oryza* species.

Vaughan et al. (2008) included *O. neocaledonica* in their *O. granulata* group. They had suggested that this taxon might be considered as a subspecies or variant of *O. granulata* or *O. meyeriana*. Since *O. granulata* has itself been proposed here as a subspecies of *O. meyeriana,* it is more rational to consider *O. neocaledonica* also as a subspecies/variety of *O. meyeriana* (Bautista et al. 2006). Further studies are needed to establish its taxonomic status and relationships.

E. *O. meridionalis* (2n = 24; AA)

This species was described by Ng et al. (1981). It is an annual plant found in northern Australia, generally in swampy areas. It is semi-erect, 1–2 m tall, with ligules acute, 9–21 mm long, spikelets shattering, slender, 7.4–8.3 x 2.1–2.3 x 1.4–1.6 mm size, and awns 63–115 mm long. It was observed first by Morishima (1969) and Chu et al. (1969). They had called it *O. perennis* then, obviously in keeping with the then valid name for the taxon now known as *O. rufipogon sensu lato.* These authors aligned this taxon with the wild and weedy wild rices ("red rice") that occur extensively in the tropics, including in northern Australia. Duistermaat (1987) observed that it could be considered as both an annual and a perennial. Vaughan (1993) stated that it is similar to *O. nivara,* but with compact panicles and long scabrous awns, 7.8–10.3 cm long. Clayton et al. (2010, 2013) treated it as a synonym of *O. rufipogon* (Table 4.5).

Numerous authors have pointed out the extreme variability found in the wild rices, especially in south and southeast Asia (Watt, 1892, Ray 1921, Roschevicz 1931, Nayar 1973, Oka 1988, etc.). They range all the way from the cultivated rice at one end to the floating perennial rice, now designated as *O. rufipogon,* at the other. The wide variability is the consequence of natural crossings that take place between the two species, *O. rufipogon* and *O. sativa,* wherever they occur sympatrically. And until about 50 years ago, this used to take place almost throughout the rice growing regions of south and southeast Asia, and southern east Asia. It is also generally accepted that the cultivated rice, *O. sativa,* is also strictly not an annual species, except possibly for a few landraces that are of very short duration.

O. meridionalis can also be taken to represent a more stabilized homogeneous variant form of an intermediate hybrid derivative of *O. sativa* and *O. rufipogon,* as has been suggested for *O. nivara* and *O. glumipatula.* Further, because of its long isolation, it also appears to have developed partial sterility barriers.

Ng et al. (1981) had observed that: "*O. meridionalis* is morphologically similar to *O. nivara,* but differs in the more compact panicle, due to their short secondary rachises, which are tightly adpressed to the main axis and in the stiff erect ascending panicle branches".

These workers further reported that the hybrids of *O. meridionalis* with *O. rufipogon, O. nivara,* and *O. sativa* were all male-sterile.

F. *O. rhizomatis* (2n = 24, CC)

This taxon was described by Vaughan (1990). Vaughan collected his material in northwestern and southeastern Sri Lanka from seasonally dry grasslands. It is a perennial, 1.2–1.5 m tall, and erect with adventitious roots and stout branched rhizomes, ligule truncate, 3–8 mm long, inflorescence with widely spreading branches, 20–28 cm long, 5–9 well exserted primary branches with 1–2 of unequal length from lower panicle nodes, spikelets shattering, mottled black at maturity, 5.3–7.7 × 1.9–2.4 × 1.2–1.7 mm, awns frequently absent, and when present, 1.5 cm long, palea tip elongated.

Vaughan (1990) observed that this taxon, while it belongs to the *O. officinalis* complex, differed from the other taxa in this complex in spikelet and panicle structure: from *O. eichingeri* and *O. minuta,* by its larger plant and spikelet size, larger and spreading panicle and rhizomes; from *O. officinalis,* by its larger spikelet, elongated palea tip and absence of whorl of primary branches from lowest panicle node; from *O. punctata* by its shorter awn, lack of whorled primary branches and rhizomes; from *O. latifolia* by the absence of whorled panicle branches, narrower leaves and presence of rhizomes; and from *O. australiensis* by its smaller spikelets, lack of flexuous awn, and fewer primary branches in the panicle.

Vaughan did not appear to have examined the Sri Lanka material studied by the earlier workers – by Trimen in 1889 as *O. sativa* var. *collina*, or Hooker's (1897) identification of this material as *O. officinalis* var. *collina* (though he had apparently misnamed it as *O. latifolia* var. *collina* [cf., Biswal and Sharma 1987; they have also given a history or sequence of the various names given to this material]), or that studied by Biswal and Sharma (1987). Instead, he identified one of the materials collected by him as the holotype. Earlier, Tateoka (1962) had Commented that: "all Ceylonese material of this complex (*O. officinalis*), however, is really identical with *O. eichingeri*. This species represents an example of disjunct distribution between Ceylon and east and central Africa".

Sharma and Shastry (1965) originally designated the Sri Lankan material as a new species, *O. collina*. Biswal and Sharma (1987) later retracted this, and said that it should be designated as a variety of *O. eichingeri*. They had found also that hybrids between this taxon and *O. eichingeri* gave 10–20% pollen and spikelet fertility. Clayton et al. (2010, 2013) has listed *O. rhizomatis* as a synonym of *O. eichingeri*.

Considering all of the above, it seems more appropriate from the taxonomic point of view to go by the conclusions of Biswal and Sharma (1987) and Clayton et al. (2010, 2013) that the Sri Lankan material named by Vaughan (1990) as *O. rhizomatis* is *O. eichingeri*. More studies will be needed to determine whether Vaughan's material needs a subspecific or varietal status. What may be required to be done is to first compare the type specimens of Trimen, Sharma and Shastry, and Vaughan, and then decide about the identity of this material.

G. *O. schlechteri* (2n = 48; HHKK)

This species was described by Pilger in 1914. Even though all the rice taxonomists had in the past accepted it as a valid species, no living material of this species was available until Vaughan and Sitch (1991) collected it again in 1990. So far, it has been collected from five locations, all in the island of New Guinea (Vaughan 1994). Incidentally, when the species was first reported by Pilger in 1914, it was said to also occur in Australia (Roschevicz 1931).

O. schlechteri is a short (30–90 cm height), stoloniferous, annual herb. The culm is erect, the panicle 3–6 cm long and spreading, spikelets 1.5–1.7 mm long, oblong, laterally compressed, and awnless, the ligule an eciliate membrane, 1–2 mm long (Clayton et al. 2010, 2013). Since the gross morphology of the species has a certain similarity with *Leersia* species, Naredo et al. (1993) undertook some detailed studies of the spikelet employing SEM (scanning electron microscopy), and they confirmed that the taxon does belong to *Oryza,* as delimited by Launert (1965) and it is distinct from the genus *Leersia.*

From the foregoing review, we may note that the following are the points of disagreement about the taxonomic status of various entities in the genus *Oryza.*

(i) *O. abromeitiana:* whether it should be a subspecies or synonym of *O. meyeriana;*
(ii) *O. alta:* whether it should have a species status or be recognized as *O. latifolia* ssp. *alta;*
(iii) *O. glumipatula*: whether it should be a species or a subspecies/synonym of *O. rufipogon*;
(iv) *O. granulata*: whether it should have species status or be recognized as a subspecies of *O. meyeriana*;
(v) *O. malampuzhensis*: whether it should be a species, or a subspecies of *O. officinalis*;
(vi) *O. meridionalis*: whether it should be a species or a subspecies or synonym of *O. rufipogon*;
(vii) *O. nivara*: whether it should be a species, a subspecies or synonym of *O. rufipogon* or *O. sativa*;
(viii) *O. rhizomatis*: whether it should be a species, or a subspecies/synonym of *O. eichingeri;*
(ix) *Porteresia coarctata:* whether it should be restored as a species of *Oryza*;

We shall now see how far biochemical and molecular biology studies have helped to resolve these issues.

5. EXPERIMENTAL STUDIES

i. Biochemical and Biometrical

Since the early part of the 20th century, some authors have used cytogenetic, morphometric, and isozyme methods to complement taxonomical studies for gaining a better understanding of species relationships in the genus *Oryza* (Table 4.6). Cytogenetical studies have been reviewed by Nayar (1973) and Kurata (2008).

Morishima and Oka (1960) employed morphometric methods and Second (1982, 1985b) employed isozyme analyses toward this end. Then, in the last 25 years, extensive studies have been carried out using an array of molecular biology techniques. While these techniques are considered as additional tools for getting better understanding genetic variation and phylogenetic relationships (Wang et al. 1992), votaries of molecular biology techniques have argued that molecular techniques are the most reliable methodology for phylogeny studies, as they are least affected by various environmental factors (Agrawal 2003, for instance).

Most of the studies in this area on rice have been one-off studies either to test the suitability of a particular method for phylogenetic studies, or to establish priority by the authors in the use of a particular technique. Often, if some individual results show deviations from expected results they are attributed to misidentification of the received germplasm material. Only a few authors have performed follow-up studies.

The following account will show that no molecular study done so far in rice has been able to provide satisfactory solutions to the problems in *Oryza* phylogeny discussed above. Even when the same material was used. different molecular techniques have given divergent results (Bautista et al. 2001, Cheng et al. 2002, Cranston et al. 2010, for instance). While classical taxonomy has provided the basic operating system for determining species relationships, and cytogenetic studies have given the framework or foundation, and biochemical and molecular studies are expected to contribute to developing better understanding of the inter-relationships in this genus (cf., Chapter 3). However, the studies conducted so far highlight the need to fine-tune the methodologies, both molecular and biometrical, to give better resolution of the results. We need to keep in mind that the morphological expression of a character is the consequence of the cumulative action of the

Table 4.6 Estimated Outcrossing Rates in Rices by Different Methods*

S. no.	Material Used	Country	Method Employed	No. of Populations Studied	Result (%)	Reference
1	*O. rufipogon*	Taiwan	Marker gene	1	30.7	Oka (1956)
2	*O. rufipogon*	Thailand	Marker gene	1	44.0	Oka and Chang (1961)
3	*O. rufipogon*	Thailand	Isozyme markers	1	50.6	Barbier (1987)
4	Intermediate	Thailand	Isozyme markers	1	55.9	Barbier (1987)
5	*O. rufipogon*	India	Variance ratio	1	37.4	Oka and Chang (1959)
6	*O. rufipogon*	Sri Lanka	Variance ratio	2	22.4, 26.5	Sakai and Narisse (1959)
7	Annual	India	Variance ratio	1	21.7	Oka and Chang (1959)
8	Annual	India	Variance ratio	3	16.6–33.9	Sakai and Narisse (1960)
9	Annual	India	Marker gene	1	7.9	Roy (1921)
10	Annual	Thailand	Isozyme markers	1	7.2	Barbier (1987)
11	Weedy	India	Variance ratio	2	17.3–20.6	Oka and Chang (1959)
12	*O. barthii*	Africa	Variance ratio	2	30.2–19.7	Morishima et al. (1963)
13	*O. sativa* (indica)	India	Marker gene	34	0–6.8	Butany (1957)
14	indica	Africa	Marker gene	2	0–1.1	Roberts et al. (1961)
15	indica	Taiwan	Marker gene	4	0.1–0.3	Oka (1988)
16	japonica	Taiwan	Marker gene	5	0.6–3.9	Oka (1988)
17	indica	Sri Lanka	Variance ratio	1	3.6	Sakai and Narise (1960)

*Modified from Oka (1988)

effects of a large number of nucleotides, base pairs, and so on. It is also the center-piece in taxonomy classification. Even if the environment may modify some of the quantitative characters, and sometimes even qualitative characters, the results need to be analyzed in their proper perspective. Specialists in the field have highlighted the strengths and weakness of various molecular phylogenetic methods (Soltis et al. 1992, 1998, Doyle and Gaut 2000, Soltis and Soltis 2003, for instance).

In the only study using biometrical methods for determining species relations in the genus *Oryza*, Morishima and Oka (1960) employed two methods, factor analysis and Sokal's weighted variable group method (cf., Sokal and Sneth 1963). A summary of the study together with the two dendrograms are given in Fig. 4.1A, B.

The authors studied 16 species with five accessions/species each, using 42 characters (20 taxonomic and 22 quantitative). They first calculated correlation coefficients of each species with all the other species, and then calculated two correlation matrices, one based on records of all 42 characters given in code numbers (correlation matrix I), and the second based on records of 22 characters (all quantitative in characters) in code numbers and the remaining 20 (all taxonomic characters) in standard deviation units (correlation matrix II). Then, they factor-analyzed the data using Sokal's weighted variable group ('averoid') method.

With the exception of 3–4 species, both the matrices showed similar groupings into clusters (A, C, D, and the rest), but between-species relationships within the groupings were different for the two methods, except in the case of just two instances. *O. brachyantha* and *O. ridleyi*, and *O. minuta* and *O. officinalis* however came within the same clade by both the methods. The position of *O. malampuzhensis* (their *O. malabarensis*) was very different by both the methods.

Second (1982, 1985c) used isozyme polymorphism extensively to study the diversity of the two cultivated rices and some of the wild AA-genome taxa. Unfortunately, it is difficult to make much sense of these studies because of the use of old nomenclature. His *O. rufipogon* now consists of several species such as *O. rufipogon sensu stricto*, *O. longistaminata*, *O. meridionalis* and annual spontaneas. The gene diversity index (heterozygosity) was highest in *O. sativa* (0.23), medium in wild *O. barthii* (his *breviligulata*) (0.14), and low in 'weedy' *O. barthii* (0.06), and *O. glaberrima* (0.03).

A couple of authors used certain biochemical constituents – notably protein fractions – to ascertain their utility in phylogenetic studies. Pental and Barnes (1985) used protein fraction and two repeated DNA markers to study the relationships of the two cultivated rice species with *O. perennis sensu lato*

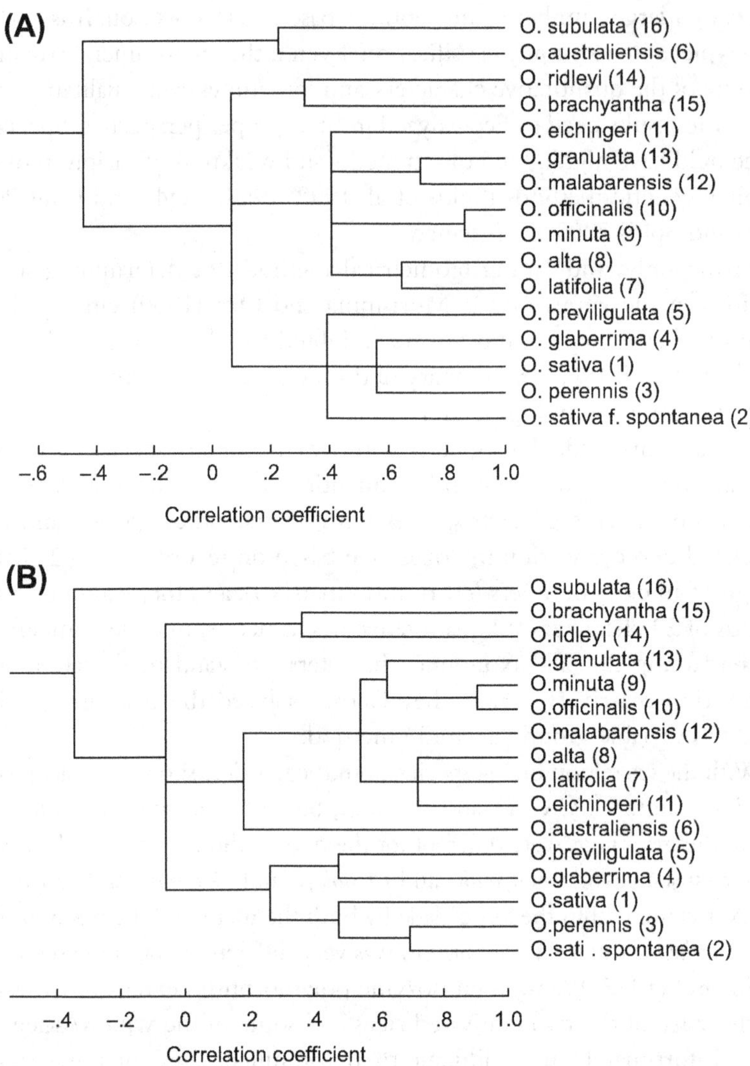

Figure 4.1 ***Species Relationship in* Oryza *Based on Biometrical Studies.*** A. Relationship obtained from correlation matrix I based on 42 characters. B. Relationships obtained from correlation matrix II based on 22 quantitative characters.
Note: *O. subulata* is now *Rhynchoryza subulata, O. malabarensis* is now *O. malampuzhensis, O. breviligulata* is now *O. barthii* and *O. sativa f. spontanea* is now *Spontanea rices / O. rufipogon pro parte / O. nivara pro parte*.
Reproduced from Morishima and Oka (1960) with permission.

(includes the present-day *O. rufipogon, O. meridionalis, O. longistaminata,* and *O. grandiglumis*), but the authors appear to have used only *O. rufipogon sensu stricto, O. grandiglumis,* and *O. australiensis*. All the taxa, except the South American *O. rufipogon, O. grandiglumis,* and *O. australiensis,* showed similar nuclear-encoded small subunit polypeptides and chloroplast-encoded large subunit polypeptides. The authors inferred that this was indicative of their close relationship. Their findings have no relevance to the already known information about their distribution and relationship (also Grover and Pental 1992).

Sarkar and Raina (1992) employed 55 accessions of 13 *Oryza* species to determine the pattern of seed-protein profile variation and their distribution among the species and genomes. They found that the banding patterns were not species-specific, but only genome-specific. The profile pattern of *O. brachyantha* (FF) was different from those of A and BC genome species.

These studies indicate that protein profile studies do not appear to given sufficient resolution for studying *Oryza* phylogeny.

ii. Molecular Biology Studies

Traditionally, relations among taxa within a taxonomic entity are determined on the basis of morphological and cytogenetical studies. There have, however, been situations where these two attributes have been unable to resolve the issues. The classification of angiosperms carried out by the Angiosperm Phylogeny Group (APG 2009) and of grasses by the Grass Phylogeny Working Group (GPWG 2001) are the best instances of these difficulties. But GPWG (2001) and APG (2009) have shown also that there are situations or instances which cannot be resolved even with the help of biochemical/ chiefly, protein) and molecular biology methods. They have however contributed substantially to sorting out many of the intractable problems. The two groups (GPWG, APG) had carried out combined parsimony analyses of several comprehensive studies using a range of molecular data sets (*ndhF, rbcL, rpoC2, phyB, ITS2,* and *waxy*, in GPWG 2001, for instance), chloroplast restriction site data, and morphological data to identify significant relationships that appear consistently in all the analyses of data sets (GPWG 2001).

Several authors have employed an array of molecular techniques mainly to resolve species and taxa relationships. They included nuclear restriction fragment length polymorphism (RFLP) (Wang et al. 1992, Bautista et al. 2001, Lu et al. 2002), random amplified polymorphic DNA (RAPD) (Ishii et al. 1996, Bautista et al. 2001, Ren et al. 2003), amplified fragment length polymorphism (AFLP) (Aggarwal et al. 1999, Park et al. 2003), simple sequence repeats (SSR) (Ishii et al. 2001, Ren et al. 2003), short interspersed elements (SINEs) and miniature inverted-repeat transposable element insertions (MITEs) (Mochizuki et al. 1993, Iwamoto et al. 1999, Cheng et al. 2002), and intron sequences (Zhu and Ge 2005). These references are only illustrative. No attempt will be made here to review the entire literature. The few chosen references do, however, illustrate that while the results of some studies have helped to clarify certain issues, none of the above techniques can be depended on to resolve all the problems satisfactorily under the conditions in which these studies have been done. The main limitation has been that practically all the studies have been one-off investigations, and that most authors have not carried out follow-up work, whenever and wherever they encountered problems or unresolved issues.

The study by Wang et al. (1992) may possibly be the first attempt in the genus *Oryza* to use a molecular technique to study its phylogeny (cf., also Causee et al. 1994). They studied 19 species/22 accessions/3–5 plants each *plus* the 71 accessions studied earlier by Second (1985c) and Dally and Second (199 with the same 0). They tested each sample of the first set of accessions with 15 single copy RFLP probes using single restriction enzyme, *EcoR1* to estimate within-population variation, and assayed 93 accessions with 25 probes that included 9/15 of the probes used earlier by Second. Their study may also be one of the most comprehensive done in *Oryza* (Fig. 4.2A,B).

The authors treated the hybridizing fragment detected by Southern analysis as a unit character and quatified the genetic distances both within and between accessions according to Nei (1987) using the software Hyper-RFLP (Cornell University). Then they constructed dendrograms using UPGMA with the same software. The authors conducted PCA analysis using DATA DESK (Figs. 4.2A, B).

The study gave 11.2 fragments/probe. Variations within the accessions were 17% and between accessions, 83%. They found that 78% of the accessions were monomorphic. The average genetic distance between species (0.077) was six times more than those within species. Their salient findings were as follows:

- **(i)** High within-species variation was observed in *O. australiensis, O. latifolia,* and *O. rufipogon;*
- **(ii)** *O. meridionalis* showed only low affinity with Asian *O. rufipogon;*
- **(iii)** *O. glumipatula* showed relatively closer relationships with *O. barthii* and *O. glaberrima;*
- **(iv)** *O. officinalis* from China and India fell into distinct groups;
- **(v)** *O. brachyantha, O. punctata, O. australiensis,* and *O meyeriana* formed different clusters;

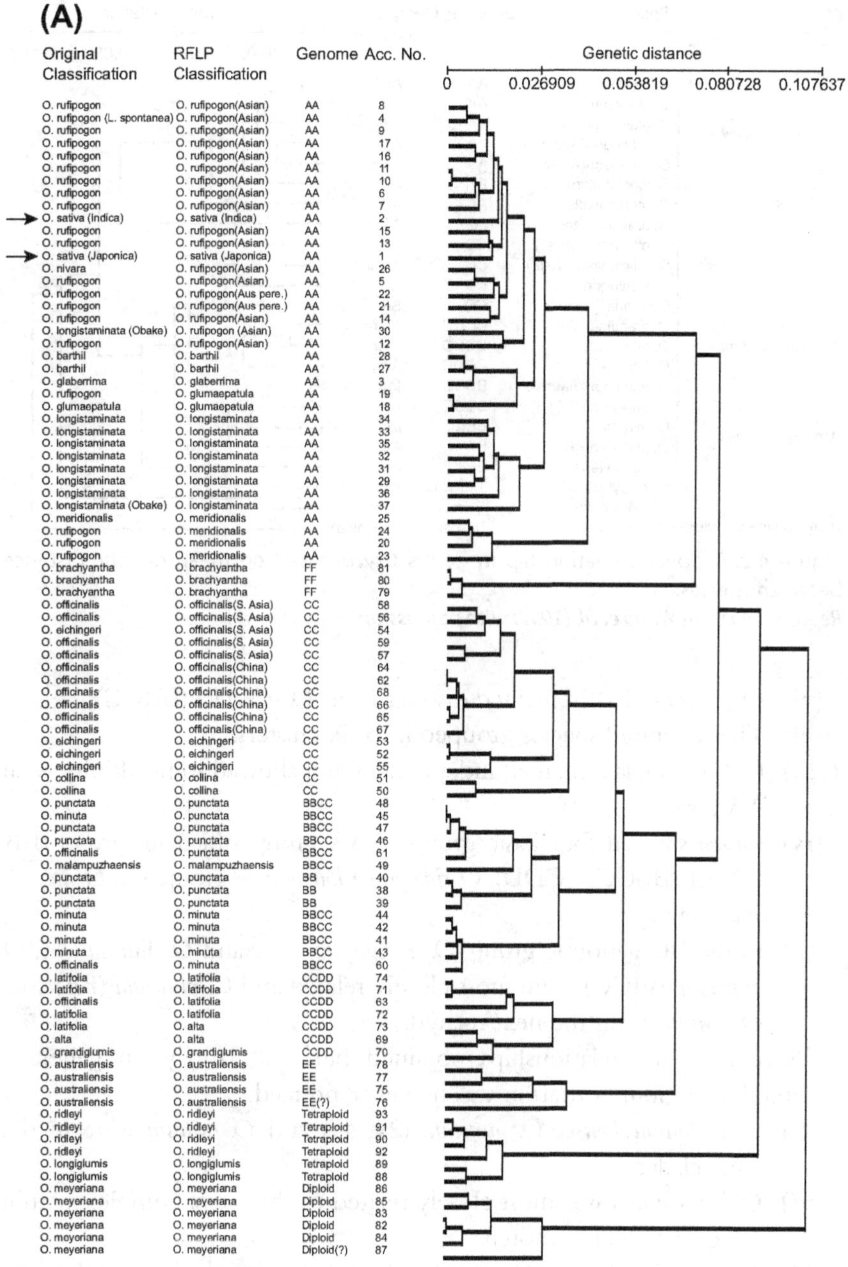

Figure 4.2 A. Species relationships in *Oryza* based on 95 accessions and 25 genomic clones of RFLP. Genetic distance is according to Nei (1987).
Reproduced from Wang et al. (1992) with permission.

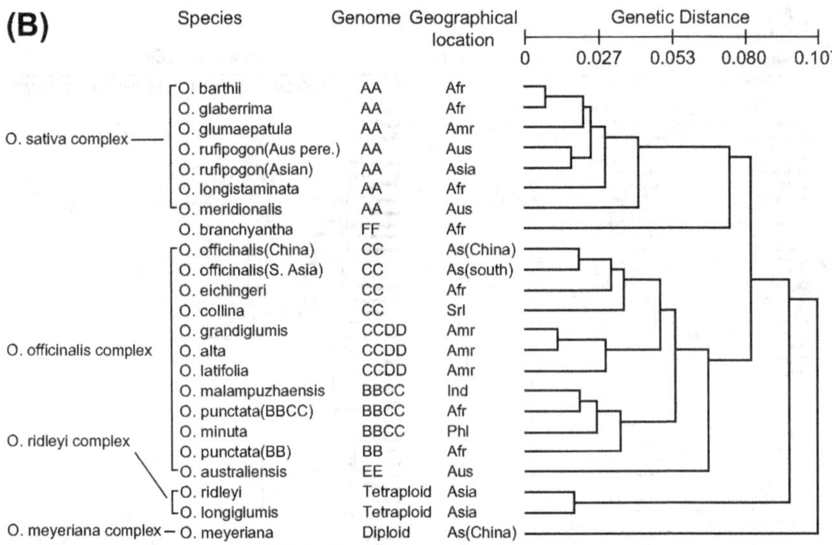

Figure 4.2 B. Species relationship in genus *Oryza* based on mean genetic distance between species.
Reproduced from Wang et al. (1992) with permission.

(vi) *O. punctata* (BB) remained distinct from *O. punctata* (BBCC);

(vii) The tetraploid species grouped into six clusters;

(viii) *O. brachyantha* remained independent, but showed some affinity with AA taxa;

(ix) *Oryza* showed four basic groups – AA group, officinalis group (BB, CC, EE, BBCC, CCDD), *O. ridleyi*/*O. longiglumis* group, and *O. meyeriana* group;

(x) In the BC genomic group, *O. malampuzhensis* and *O. minuta* (*sic.*, *O. punctata*, BBCC) were most closely related and O. *punctata* (BB) (*sic.*, *O. minuta*?) was the next related.

Several of the relationships obtained here fall largely on generally accepted lines. Some variation was however, noticed:

(i) *O. malampuzhensis, O. punctata* (2x, 4x) and *O. minuta* were in the same clade;

(ii) *O. brachyantha* was most closely related to the sativa complex within the genus, albeit remotely;

(iii) The relationship of *O. australiensis* with the officinalis complex was analogous to that of *O. brachyantha* with the sativa complex.

Ishii et al. (1996) used the RAPD technique to study the phylogenetic relationships in the A genome species of *Oryza*. They employed 14 decamer primers on 29 accessions consisting of eight *O. sativa*, six *O. glaberrima*, two

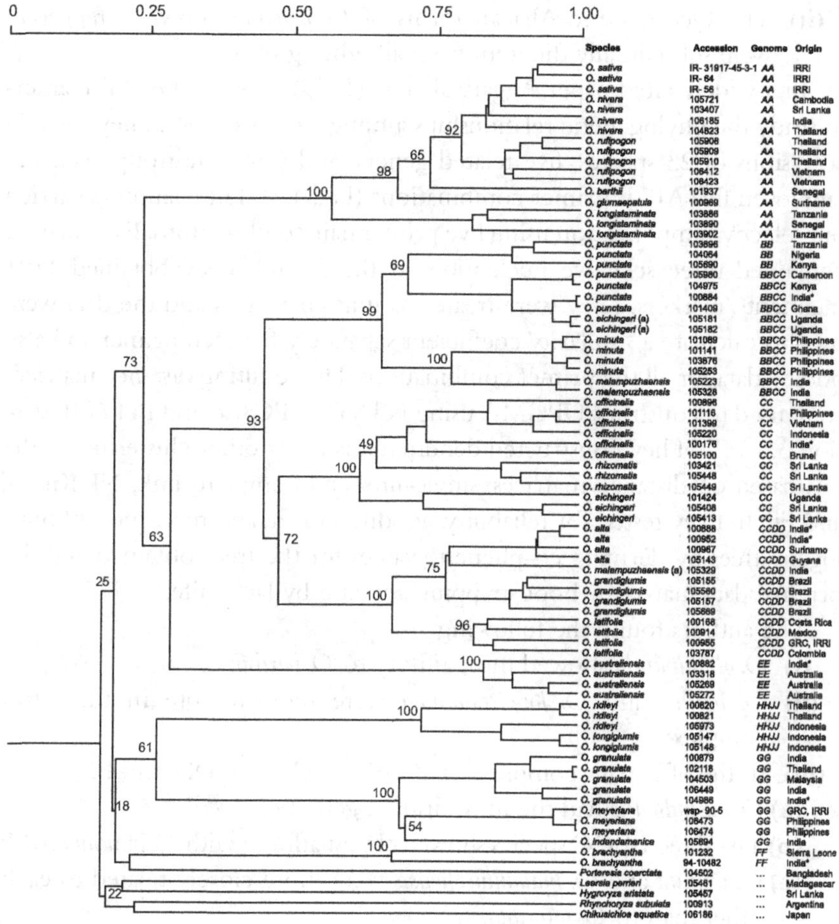

*=via Central Rice Research Institute, Cuttack, India

Figure 4.3 ***Species Relationships Based on AFLP Markers Obtained with Five Primer Pairs.*** Numbers shown at different nodes represent percentage confidence limits obtained in the bootstrap analysis. The scale shown above is the measure of genetic similarly calculated according to Nei and Li (1979). The species/accessions used, their genome(s), and origin are indicated on the right side of the panel. (Entries marked with (a) are those whose inferred genome(s) differ from the expected; see results.
Reproduced from Aggarwal et al. (1999) with permission.

O. breviligulata (now *O. barthii*), and 14 *O. perennis*. (Unfortunately, it is not possible to determine the composition of this grouping, as the authors have given only the continents of origin.) The authors concluded that:

(i) Most of the accessions formed clusters corresponding to their taxonomic groupings;

(ii) The Oceania and African forms of *O. perennis* (now, *O. rufipogon*) were substantially different from all other groups;

In a widely cited paper, Aggarwal et al. (1999) employed AFLP markers to study the phylogenetic relationships among *Oryza* species. They used 79 accessions of 23 species, five related genera, and three outgroup taxa, and employed five AFLP primer combinations (Fig. 4.3). These authors carried out selective amplification using five primer pair combinations. Each primer contained three selective nucleotides at the 3′ end. They obtained 1191 fragments/markers. They were treated as unit characters and the data were used to calculate a similarity coefficient separately for each primer and also pooled data for all the primer combinations. The resulting distance matrices were used to construct UPGMA using NTSYS – PC 1.8 and PHYLIP version 3.5 7C' . They constructed dendrograms using other clustering methods based on distance matrices, single-link and complete-link, NJ, Kitsch, and Fitch. They tested for reliability, goodness of fit, and robustness of phylogeny trees by deriving co-phenetic values for the trees obtained and the original data matrix by boot strapping and also by Jackknife.

The authors found the following:

(i) *O. glumipatula* showed most affinity to *O. barthii*;

(ii) *O. barthii* and *O. longistaminata* were most remote in the sativa complex;

(iii) In the officinalis complex – BB, BBCC, CC, CCDD, and EE:

- **(a)** *O. australiensis* had the most distinct genome;
- **(b)** The three BBCC species showed closest affinity with *O. punctata* (BB);
- **(c)** *O. minuta* and *O. malampuzhensis* were more closely related to each other than to *O. punctata*;
- **(d)** *O. punctata* was the nearest to *O. eichingeri*;

(iv) In the latifolia complex – GG, HHJJ:

- **(a)** *O. alta* and *O. grandiglumis* were closest to each other;
- **(b)** *O. granulata, O. meyeriana,* and *O. inandamanica* were well distinguished from each other;

(v) *O. ridley* and *O. longiglumis* were also well separated;

(vi) *O. brachyantha* was the most distinct;

(vii) Regarding phylogeny:

- **(a)** The study revealed eight monophyletic groups in *Oryza,* each corresponding to a different genomic construction with 95–100% confidence limits;
- **(b)** BB aligned with 55% genomic similarity; hence BBCC might have evolved by a recent hybridization event;

Adity APN-5-5

(c) Six of the monophyletic groups were derived from only two lineages, the officinalis group and the ridleyi–meyeriana group;

(d) All *Oryza* spp. have a common ancestry;

(e) Within the sativa group, the three geographical groups (Asian, African, American) are distinct. They appeared to have evolved independently when viewed both from their phyletic distribution and genetic distances;

(f) The American species were closer to the African species;

(g) The B, C, and E genomes evolved over a much longer time than the A genome;

(h) *O. officinalis* was the most diverged taxon of the three CC species.

The results are in broad agreement with the generally accepted phylogenetic relationships except for the divergence timings and some other variations. *O. malampuzhensis* (BBCC) showed a sister relationship with *O. minuta* (BBCC) in one part of the dendrogram. In another place, it (CCDD) showed a sister relationship with *O. alta* (CCDD).

Joshi et al. (2000) employed an ISSR technique to study the phylogeny of the genus. They used 19 species/42 accessions with 28 accessions representing all the nine genomes, ten cultivars (indica and japonica), one indica landrace, and three other genera close to *Oryza*. They screened 30 ISSRs and selected 11 polymorphic and informative patterns. They obtained/scored 302 bands from 11 patterns. Of these, 87 bands were present in either individual accessions, species, or genomes (Fig. 4.4).

Every amplified fragment obtained by using the ISSR primers was scored. This was then used to calculate similarity/genetic distances using a Dice coefficient (a statistic used to define similarity of two samples; sometimes called Sorenson--Dice coefficient), then the resultant distance matrix was used to construct an UPGMA phenogram using NTSYS-PC 1.8 and PHYLIP 3.5 7C′ to infer phylogenetic relationships. They performed bootstrapping using WINBOOT software (IRRI) to determine confidence limits.

The authors obtained the following results:

(i) The EE genome species *O. australiensis* was not part of the officinalis complex;

(ii) *O. punctata* (2n = 24, 48) was closest to *O. malampuzhensis*;

(iii) *O. alta* and *O. grandiglumis* were closer to each other than to *O. latifolia*;

(iv) The results for *O.granulata, O. australiensis, O. brachyantha*, and *O. ridleyi* showed variant results;

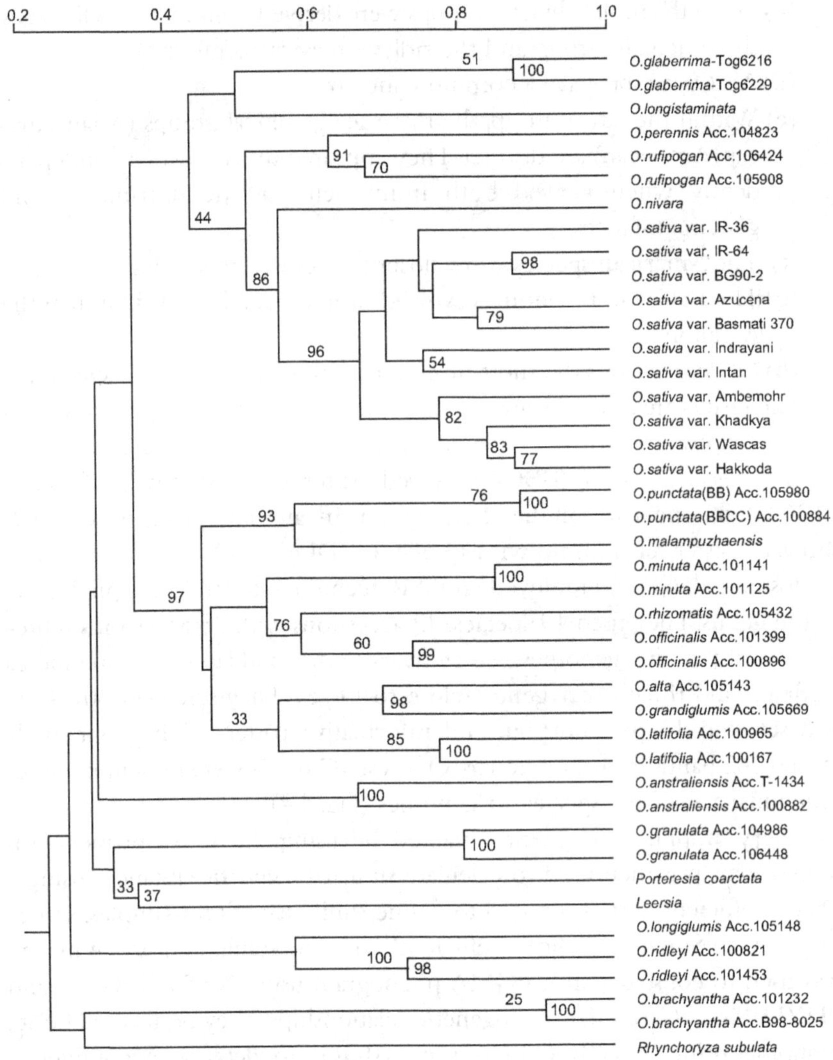

Figure 4.4 ***Relationship in the Genus* Oryza *Based on ISSR Polymorphism.***
The numbers at the forks indicate the confidence limits for the grouping of those species, which are to the right of that fork.
Reproduced from Joshi et al. (2001) with permission.

(v) *Porteresia, Rhyncoryza,* and *Leersia* were aligned with different groups;

(vi) *O. brachyantha* was the most diverged species, followed by *O. granulata* and *O. australiensis;*

(vii) *O. rhizomatis* aligned with *O. officinalis,* and together, they paired with *O. minuta;*

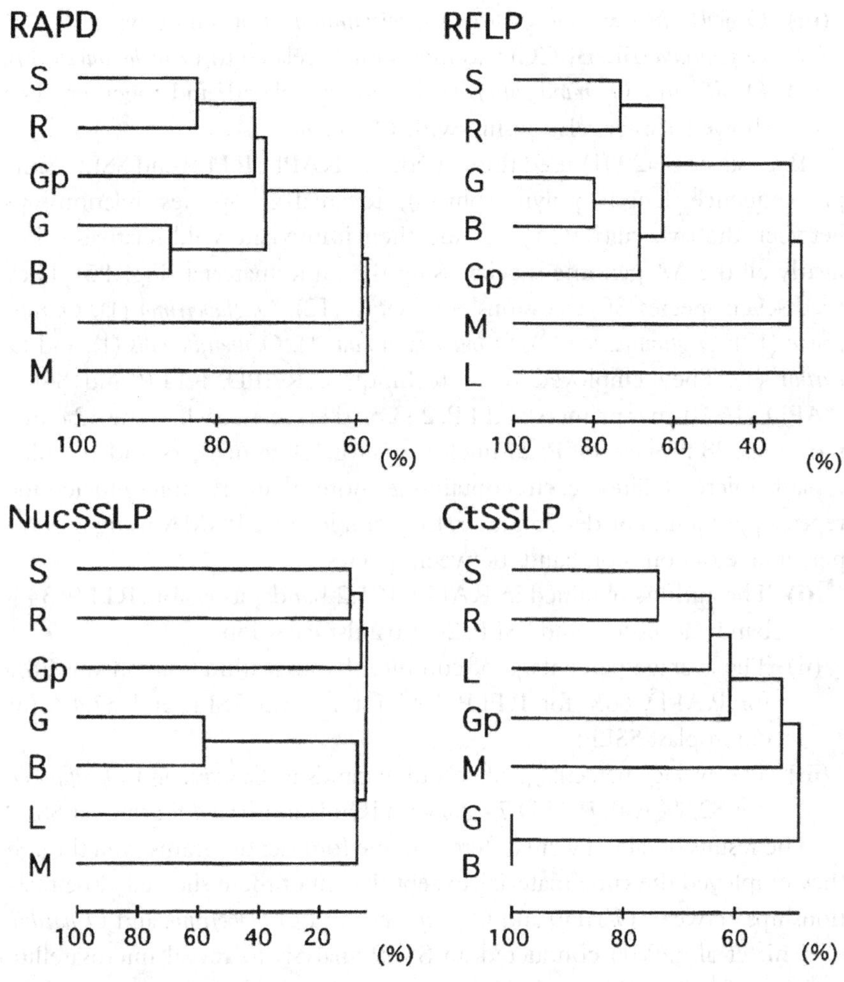

Figure 4.5 ***Genetic Relationship Among A-genome Species of Rice Constructed by RAPD, RFLP, Nuclear SSLP, Chloroplast SSLP Analyses.***
Clustering was carried out based on the average percentages of common bands among the six A-genome species, *O. sativa* (S). *O. glaberrima* (G), *O. rufipogon* (R), *O. glumipatula* (GP), *O. meridionalis* (M), and *O. barthii* (B).
Reproduced from Bautista et al. (2001) with permission.

(viii) *O. alta* aligned with *O. grandiglumis* at 98% confidence;
(ix) *O. ridleyi* aligned with *O. longiglumis.*

Several of the relationships seen in the dendrogram are at variance with generally accepted relationships. Some examples of this are:
(i) *O. brachyantha* was closest to *Rhyncoryza subulata*;
(ii) *Porteresia coarctata* was most closely related to *Leersia*, though remotely;

(iii) *O. glaberrima* was closest to *O. longistaminata*, but remotely;
(iv) *O. punctata* (BB, BBCC) was most closely related to *O. malampuzhensis*;
(v) *O. alta* and *O. grandiglumis* were closely related, and together, they showed a sister relationship with *O. latifolia*.

Bautista et al. (2001) used three methods, RAPD, RFLP, and SSLP (simple sequence length polymorphism), to analyze species relationships between the two cultivated rices and their immediate wild relatives – i.e., nearly all the AA genome species using the same material (Fig. 4.5). They used seven species/30 accessions – *O. sativa* (12), *O. glaberrima* (1), *O. rufipogon* (12), *O. glumipatula* (2), *O. longistaminata* (1), *O. meridionalis* (1), and *O. barthii* (1). They employed three techniques, RAPD, RFLP, and SSLP; RAPD: (16 10-mer primers); RFLP: 2–3 markers from each of the 12 chromosomes, 28 probes; SSLP: 24 nuclear microsatellite markers, and 10 chloroplast microsatellites, each containing more than 10 micronucleotide repeats per locus. For dendrograms, they employed UPGMA using average percentage of common bands between species.

(i) The authors obtained in RAPD: 101.2 bands/accession, RFLP: 34.5 bands/accession, and SSLP: 26.5 bands/accession;
(ii) The average percentage of common bands within a taxon was 78% for RAPD, 66% for RFLP, 10% for nuclear SSLP, and 67.4% for chloroplast SSLP;
(iii) The average percentage of common bands in *O. sativa* and *O. rufipogon* was 82.8% with RAPD, 74.0% with RFLP, and 10.4% for nuclear SSLP.

The results obtained were different in the four dendrograms even though they employed the same material, except that all of them showed close relationships between *O. sativa* and *O. rufipogon,* and *O. glaberrima* and *O. barthii.*

Ishii et al. (2001) conducted an SSLP analysis to reveal microsatellite variation (both nuclear and chloroplast) among the A genome species for studying their phylogenetic relationships. They used 29 accessions of cultivars (23 *O. sativa,* six *O. glaberrima*), and 30 accessions of wild species (12 *O. rufipogon,* five *O. glumipatula,* two *O. longistaminata,* six *O. meridionalis,* and five *O. barthii*). They employed 24 nuclear and 12 chloroplast microsatellite loci. Microsatellites amplified clearly in all the accessions giving an average of 18.4 satellites per locus. The PIC (polymorphism information content) values ranged from 0.85 to 0.94, with an average of 0.89. The highest average PIC values were obtained in *O. sativa* (0.79) and *O. rufipogon* (0.80). For chlorophyll microsatellites, the average PIC values were 2.9 alleles per locus and for nuclear microsatellites, 0.38. Both nuclear and chloroplast microsatellites showed parallel patterns of differentiation. *O. sativa* showed the highest similarity values to *O. rufipogon* and *O. glaberrima* to *O. barthii*. Among the three ecotypes of *O. sativa*, the average values observed between japonica and javanica were significantly higher (30.8 for nuclear and 70.8 for chloroplast microsatellites) than those between japonica and indica (12.8 for nuclear and 59.2 for chloroplast microsatellites) and javanica and indica (10.2 for nuclear and 64.2 for chloroplast microsatellites) varieties.

Cheng et al. (2002) employed *p-SINE1* members to study the interrelationships among the A genome species (Fig. 4.6A,B. They studied seven AA genome species (80 accessions) consisting of *O. glaberrima* (14), *O. barthii* (5), *O. glumipatula* (3), *O. meridionalis* (12), *O. rufipogon* (34), and *O. longistaminata* (12). They employed the retrotransposon, *p-SINE1* members. The presence/absence of *p-SINE 1* members at a particular locus was organized into a data matrix, and then they performed a Neighbor Joining analysis using PAUP 4.068 program.

The authors found the following (Fig. 4.6A,B):

(i) *O. longistaminata* and *O. meridionalis* were remotely related to each other and also to the other AA genome species;
(ii) *O. longistaminata* diverged earliest; *O. barthii* and *O. glumipatula*, the last;
(iii) All the strains of all the species showed a similar pattern, which indicated monophyly.

Park et al. (2003) measured genetic variations in the AA genome species using MITE-AFLP (miniature inverted-repeat transposable elements – AFLP). MITEs are the major transposable elements in *Oryza* species. The results obtained were on expected lines. *O. longistaminata* showed the highest variation among all the species. The clustering pattern of the species matched well with their geographical origins. The Australian taxon *O. meridionalis* was distant from the other species.

Ren et al. (2003) conducted a comparative study of RAPD and SSR markers to determine AA genome species relationships (Fig. 4.7). The authors used eight species/45 accessions – *O. sativa* (8), *O. glaberrima* (5), *O. rufipogon* (9), *O. nivara* (7), *O. glumipatula* (3), *O. longistaminata* (4), *O. barthii* (6), and *O. meridionalis* (3).

The authors used two molecular systems, RAPD and SSR. For RAPD, they used 27 primers and performed a data matrix analysis using NTSYS PC software and the Jaccard coefficient as a relative measure of similarity. For SSR analysis, they used 29 primer pairs, quantified the data using Nei (1978) ratios of shared DNA fragments, and determined similarity coefficient between accessions. Then, they created dendrograms from 181 RAPD

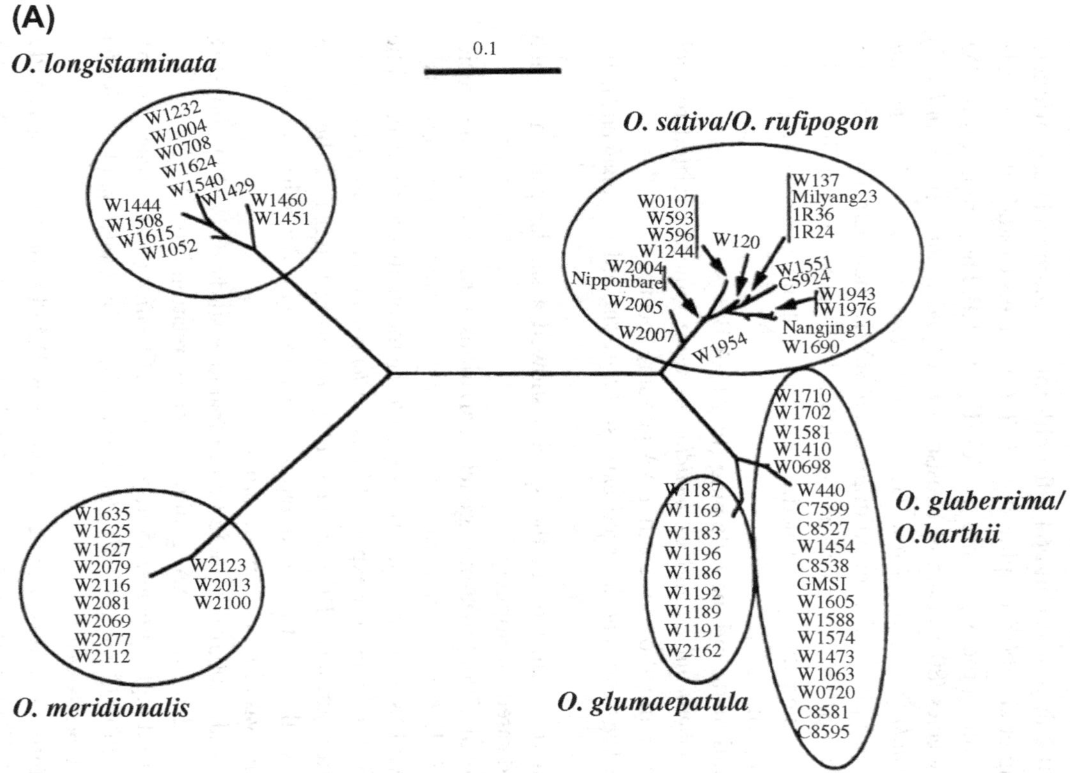

Figure 4.6 **A** Phylogenetic relationship among 72 strains having AA-genome species constructed by the NJ method based on the pattern of the presence or absence of *p–SINE1* members and three other transposable elements. All strains were clustered into five groups, corresponding to *O. longistaminata, O. meridionalis, O. rufipogon, O. sativa, O. glumipatula, and O. barthii / O. glaberrima*. The strains of *O. rufipogon (or O. barthii)* are not distinguished from those of *O. sativa* (or *O. glaberrima*) in the tree. The scale bar equals a distance of 0.1.

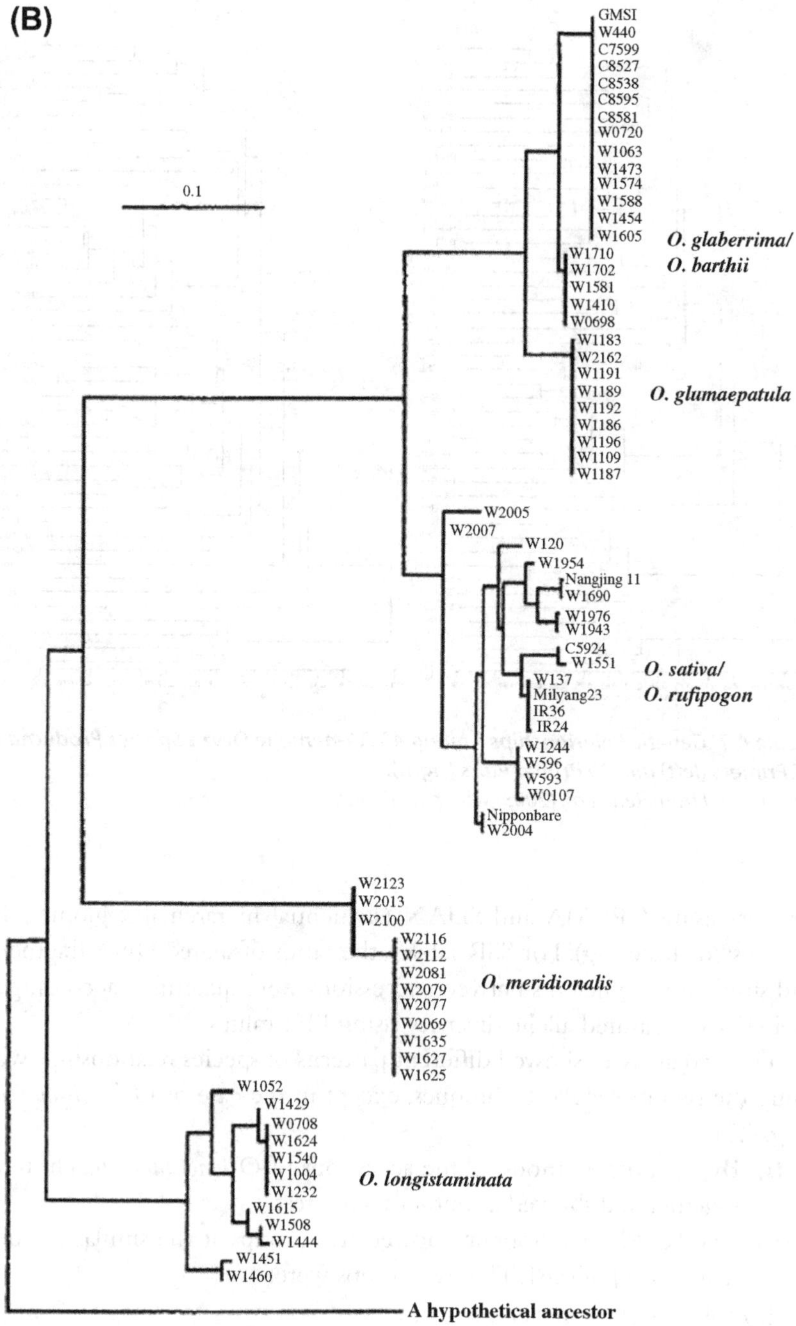

Figure 4.6 **B** Phylogenetic relationship among *Oryza* species having AA-genome and a hypothetical ancestor. The hypothetical ancestor is defined as the one without all insertions of *p-SINE1* and other transposable elements. The scale bar equals a distance of 0.1. *Reproduced from Cheng et al. (2002) with permission.*

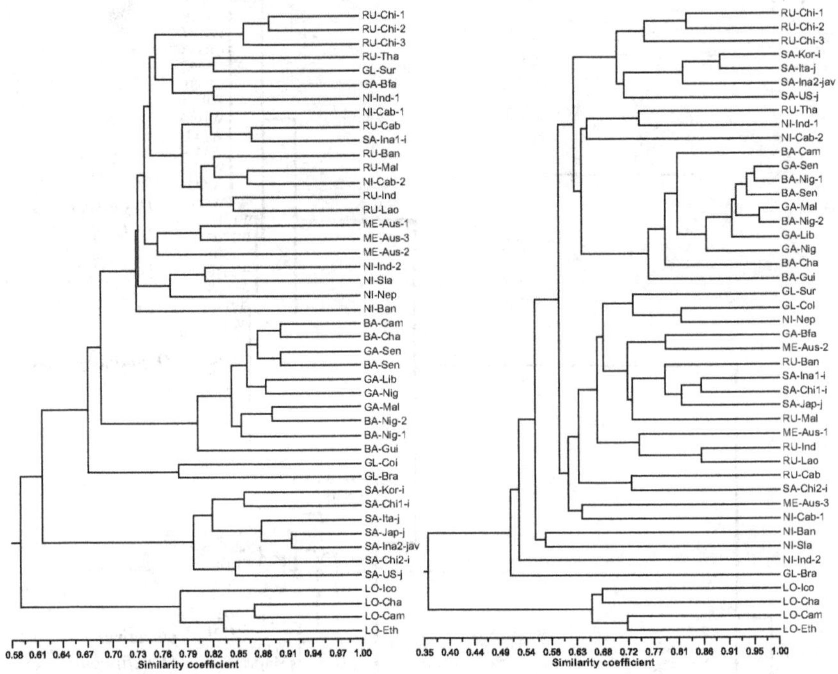

Figure 4.7 ***Genetic Relationships Among 45 AA-genome* Oryza *Species Produced by 27 Primers (left) and 29 Primer Pairs (right).***
Reproduced from Ren et al. (2003) with permission.

markers using UPGMA and SHAN (sequential hierarchial agglomerative and nested clustering). For SSR alleles, the ratios of shared DNA fragments and similarity coefficients between accessions were quantified according to Nei (1978) measured allelic diversity using PIC values.

The two analyses showed different patterns of species relationships with using the two molecular techniques, except in the case of *O. longistaminata* (Fig. 4.7).

(i) By both the methods, all the accessions of *O. longistaminata* clustered together and showed as outlier to the rest.

(ii) The RAPD dendrogram showed five groups at the similarity coefficient level of 0.61. The five groups were:

(a) All *O. longistaminata*;

(b) All *O. sativa*;

(c) 2/3 of *O. glumipatula*, with the third accession pairing with *O. rufipogon* –Thailand;

(d) 4/5 O. *glaberrima*, with the fifth *O. glaberrima* pairing with *O. nivara* – India;

(e) A large group consisting of *O. rufipogon*, *O. nivara*, and *O. sativa*, with all *O. meridionalis* embedded in it, and one each *O. glaberrima* and *O. glumipatula* scattered in it.

(iii) The SSR dendrogram seemed to show higher resolution, but other than *O. longistaminata* and *O. barthii–O. glaberrima*, all the other species showed a "musical chair distribution pattern", with the species/accessions scattered into ten groups.

Lu et al. (2002) studied the genetic diversity and relationships of *O. sativa* with its three most closely related taxa, *O. rufipogon*, *O. nivara*, and *rufipogon*. They used 58 accessions and employed 30 RFLP probes. The cluster analyses of the accessions did not show any grouping based on their species classification, instead those from the same geographic region tended to cluster together. They nevertheless concluded that the four taxa showed a "remarkably close relationship", and called for a reassessment of their taxonomic status. They observed that geographic isolation appeared to have played a "significant role" in species differentiation within *Oryza*. The authors did not mention the manner in which they separated *O. spontanea* and *O. nivara*.

Nishikawa et al. (2005) employed 12 SSR regions along with flanking regions on 21 species/50 accessions to study their interrelationships, i.e., nearly the entire genus (six diploid genomes and three different allopolyploid genomes) (Fig. 4.8).

The authors performed PCR with DNA polymerase, using 12 SSR regions (seven mitochondrial + five chloroplast) along with flanking regions. The sequences of *O. sativa* cv. Nipponbare were used as the standard. They obtained 58 SSRs in the mitochondrial genome. Then they carried out a phylogenetic analysis based on DNA sequences. They obtained a sequence matrix by multiple alignment using Clustal X, version 1.81, performed a phylogenetic analysis using PAUP version 4.0b 10, and developed the most parsimonious tree using a heuristic search involving 100 replications of random addition sequences and tree bisection reconnection (TBR) branch swapping. They developed both unrooted and rooted trees using the outgroup method (cf. Felsenstein 1985) and did Bootstrap to assess tree reliability.

The authors observed the following (Fig. 4.8).

(i) The sativa complex was monophyletic;

(ii) *O. glumipatula* consisted of two groups and they showed paraphyletic relationships;

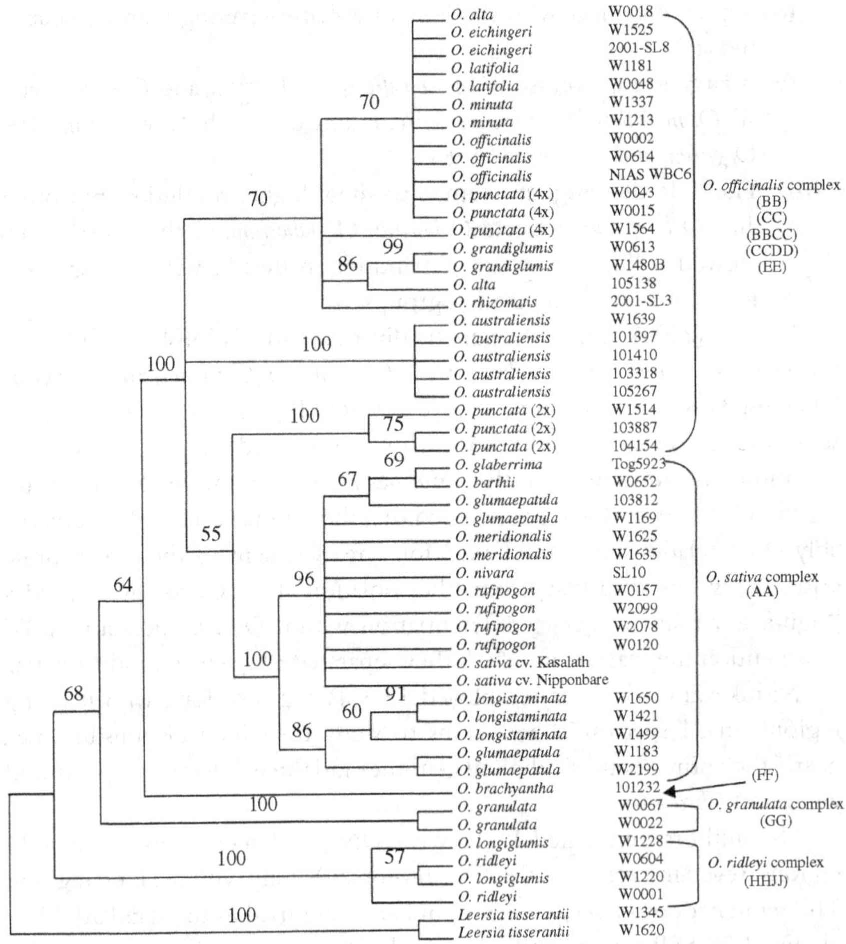

Figure 4.8 Phylogenetic tree of the genus *Oryza* based on 12 organelle SSRs and their flanking sequences, using the index = 0.6657, retention index = 0.9043, rescaled consistency maximum parsimony method. The tree is a strict consensus of the index = 0.6020. The numbers above the nodes represent bootstrap of 4,934 most parsimonious trees. Tree length = 356, consistency values expressed as percentage of 1,000 bootstrap replications.
Reproduced from Nishikawa et al. (2005) with permission.

(iii) The AA genome taxa in Latin America consisted of two groups – one indigenous, and the second introduced;

(iv) In the officinalis group:

(a) *O. australiensis* showed paraphyletic position in relation to *O. punctata* and the other species;

(b) The BB and EE species were well diverged;
(c) The CC genome was maternally inherited in *O. minuta, O. punctata,* and *O. latifolia*. This result contradicts the earlier observations of Dally and Second (1990) and Kanno and Hirai (1990);
(d) The CC genome species showed paraphyletic relationship with *O. punctata* (BB);

(v) In the latifolia complex:
(a) *O. alta, O. latifolia,* and *O. grandiglumis* were closely related;
(b) *O. latifolia* had a different chloroplast genome from those of the other two species;

(vi) *O. brachyantha* and *O. granulata* formed separate clades;
(vii) *O. longiglumis* and *O. ridley* diverged only recently;
(viii) *O. brachyantha* was embedded between the sativa–officinalis complex and *O. granulata*;
(ix) *O. punctata* (2x) was more closely related to the sativa complex than to officinalis complex, which included *O. punctata* (4x).

Zhu and Ge (2005) used intron sequences of four nuclear single copy genes (Intron/Gene names: *Adh 1-i3/Adh*1 [chromosome 11]; Os 1283/OsRFCD001283 [chromosome 1]; Os 17357/OsRFCD 017357 [chromosome 4]; Os9971/OsRFCD009971 [chromosome 2]) located on four different chromosomes and MITE (miniature inverted-repeat transposable elements) insertions to study the phylogenetic relationships among the AA-genome taxa. They used 37 accessions representing two cultivated and six wild species. The introns of the four genes present on the different chromosomes were sequenced and analyzed by maximum parsimony (MP) and Bayesian inference methods. All the species except *O. rufipogon* and *O. nivara* formed a monophyletic group. The Australian endemic, *O. meridionalis* was the earliest diverging lineage. The two ecotypes of *O. sativa*, indica and japonica, formed two monophyletic groups, which was suggestive of their polyphyletic origin. Based on the molecular clock method, the authors estimated that the AA-genome group diverged ca. 2.0 Ma and indica and japonica diverged ca. 0.4 Ma.

Zhu and Ge (2005) also studied the relationship between *O. rufipogon* (seven accessions) and *O. nivara* (five accessions). They chose the material to represent the full geographical range of distribution of the two taxa. The two taxa were scattered in two subclades within clade I (Fig. 4.8). They also mixed with *O. sativa*. This result suggested to the authors that the two taxa belonged to a single species, *O. rufipogon*, and hence, *O. nivara* could be considered as its ecotype. The study also indicated to the authors that the

African rice *O. glaberrima* originated from *O. barthii,* and that they were both related only remotely to *O. longistaminata.* The South American species, *O. glumipatula* showed closer affinity to *O. barthii/O. glaberrima* than to *O. rufipogon/O. sativa,* but one of the three accessions of *O. glumipatula* studied by the authors was seen nested among *O. rufipogon/O. sativa.* The authors attributed this phenomenon to introgression.

Rakshit et al. (2007) studied DNA polymorphism in 30 accessions of *O. sativa* (10 japonica, 8 indica), *O. rufipogon* (10), *O. barthii* (1), and *O. glaberrima* (2). They examined 126 kb of DNA sequence spanning 22 loci and one region distributed in four chromosomes (1, 2, 3, and 4). They concluded that indica and japonica were independently domesticated from *O. rufipogon,* and, while indica contained substantial genetic diversity ($\Pi = 0.0024$), the nucleotide diversity in japonica was extremely low ($\Pi = 0.0001$).

In the foregoing pages, we have summarized the phylogenetic studies carried out in the genus *Oryza* by 15 authors using biometrical, biochemical, and molecular biology methods. Four of these studies employed 2/4 methods to study the same experimental material. Even these studies showed different results. The differences were even greater in the other studies. This could be either because of the inadequate use of experimental material, or result of the inherent weaknesses of the experimental procedures used by the authors. One other factor that we may need to consider is the possibility of heterozygosity in the experimental material. This is because there is considerable indirect evidence for natural crossings taking place between related species throughout the genus. There is no other way to explain the overlapping nature of various taxonomical characters that we encounter in all the species complexes – say, between *O. officinalis* and *O. minuta,* between *O. punctata* and *O. eichingeri,* between *O. rufipogon* and *O. meridionalis,* between *O. granulata* and *O. meyeriana,* and so on. Hence, one needs to exercise utmost care when choosing experimental material, instead of "blindly" using whatever material is received from various germplasm repositories or collections.

After summarizing the taxonomical studies, we had listed nine taxa whose taxonomic status were uncertain on the basis of taxonomic studies. They were *Porteresia coarctata, O. meyeriana ssp. abromeitiana, O. granulata, O. malampuzhensis, O. glumipatula, O. nivara, O. meridionalis,* and *O. rhizomatis.* We shall examine their status in the light of the molecular biology studies.

Porteresia coarctata. Tateoka (1965) transferred *O. coarctata* to a new genus *Porteresia* created by him, and this change was accepted by all subsequent rice scientists. Then, Lu and Ge (2005) proposed that the taxon should be restored to *Oryza* based on the results of molecular studies of Aggarwal et al. (1999) using AFLP, Ge et al. (1999) using two nuclear analysis of ITS of nuclear rDNA, and Joshi et al. (2000) using ISSR. In Aggarwal et al.'s (1999) study, *P. coarctata* shows only a remote relationship with *O. brachyantha*. In Joshi et al's (2000) study, it shows a sister relationship with *Leersia*, and then together, they pair with *O. granulata*. Ge et al.'s (1999) results invite some attention. The two subsequent taxonomic treatments of the genus (Vaughan 2003, Clayton et al. 2010) have ignored Lu and Ge's (2005) proposition. This may however require further studies.

O. abromeitiana. No molecular biology studies appear to have been done on this taxon. Hence, it may continue to be considered as a synonym of *O. meyeriana*.

O. granulata. Morishima and Oka (1960) obtained different results about the status of this species in the two methods that they had used. The first showed a close relationship with *O. malampuzhensis*, and in the second, first *O. minuta* and *O. officinalis* paired with each other, and then together, they paired with *O. granulata*. In Aggarwal's (1999) study, *O. granulata* formed a single clade with *O. meyeriana*. Joshi et al. (2000) used only *O. granulata*, and no *O. meyeriana*. In this study, *O. granulata* showed itself as an outlier to all the species in the sativa and officinalis complex.

O. meyeriana. Wang et al. (1992) used only this taxon and no *O. granulata*. In this study, all its six accessions paired together, and together, it showed itself as an outlier to all the 18 other *Oryza* species studied by the authors. Thus, molecular biology studies have not given us any clear indication about the relationship between *O. granulata* and *O. meyeriana*.

O. malampuzhensis. At least four authors used this taxon in their studies. In Morishima and Oka's (1960) study, *O. malampuzhensis* paired with *O. granulata* in one study, and in the second study, it grouped with all six *O. officinalis* complex species. In Aggarwal et al.'s (1999) study, it first formed a single clade with *O. minuta*, and together, it constituted a megaclade with a second one that had been formed by *O. eichingeri* and *O. punctata*. Thus, this study hardly showed any relationship with *O. officinalis*. In the work of Joshi et al. (2000), it formed a clade with *O. punctata*, and then this clade showed itself as an outlier to all the other six *O. officinalis* complex species, which had formed into a large grouping. None of these studies indicated any close relationship with *O. officinalis*. Hence, under the present circumstance, it appears to be more prudent to maintain its subspecies identity. A detailed study to determine its relationship with all the other Asian and African

CCDD species, especially with *O. minuta* and *O. punctata* is called for. Both are tetraploids (2n = 48) incidentally.

O. glumipatula was studied by three authors. In the results of Wang et al. (1992), this species formed a single clade with *O. barthii/O. glaberrima,* and this clade was embedded in a larger clade that included all the *O. sativa* complex species. In Nishikawa et al.'s (2005) study, *O. glumipatula* (two accessions) paired with *O. longistaminata,* and elsewhere, it paired with *O. barthii/O. glaberrima* (as also seen in Wang et al. (1992)). The results here have also been inconclusive.

O. nivara was studied by three authors. In Aggarwal et al. (1999), the four accessions used in the study did not even form into a single clade, but they joined individually with three *O. sativa* accessions to form a clade. They ultimately formed a single grouping with all the other AA species. Joshi et al. (2000) used only one accession of *O. nivara*. In this study, the species formed a clade with a diverse collection of 11 *O. sativa* accessions. Ultimately, it was seen as one among a branch comprising all the AA species. In Nishikawa et al.'s (2005) work, *O. nivara* was seen to be embedded among all the other AA genome species.

In phylogenetic studies using a range of molecular biology techniques, one broad observation is that resolution up to the species complex level is adequate to some extent. However, within each complex, the interrelationships among species appeared to differ among studies. To a certain extent, if a comparable relationship is seen in two or more studies, we may take it as indicative of some affinity, and then we should subject it to a more rigorous and specific study, but as yet, this has not been for any species/taxon.

A possible relationship is shown between *O. glumipatula* and *O. glaberrima–O. barthii*. In a way, this is puzzling, because the former is indigenous to tropical America, and the latter two species are native to west Africa, so the relationship is difficult to explain. We cannot attribute it also to the Gondwana breakup, because South America broke off from Africa about 130 Ma, and the grasses originated only much later (55–70 Ma). The second possibility is as a chance introduction during the slave trade. During the early decades of the slave trade in the 15th century, slaves are known to have carried African rice to South America (Porteres 1949b, 1955b, Carney 2001), but it did not appear to have become established there.

In addition, what about *O. perennis*, an Asian species, occurring in America (2n = 48), *O. malampuzhensis* occurring in southern Western Ghats in southwest India, and *O. minuta* in the Philippines and in New Guinea in just one location? *O. malampuzhensis* (2n = 48) is nowadays considered to be a synonym or subspecies of *O. officinalis* (2n = 24), but occurs widely and it is present in south, east, and southeast Asia. The entire question of the dispersal of *Oryza* species needs to be studied in detail. Several of the dates given in Vaughan et al. (2008) for the rise of angiosperms and their further differentiation/divergence appear to be at variance with other published results (cf. Table 3.1).

The interrelationship of the "outlier" taxa within the genus *Oryza* is thus disparate in different studies. This has been indicated above.

The main limitation of these studies has been that practically all of them are one-off studies done mainly to get priority. The other problem may relate to either inadequate robustness in the methodology or error in the identification of the material. Examples of these are found in the studies by Morishima and Oka (1960), Bautista et al. (2001), and Ren et al. (2003). If follow-up studies are difficult, a combined analysis of all the studies conducted so far may be possible.

The nature of hybrid sterility in *Oryza* among and between different taxa has been extensively studied by Chu and Oka and their colleagues (Chu et al. 1969, Oka 1988, 1990, Sano 1991, Koide et al. 2008). They pointed out that varying degrees of male sterility are commonly encountered in such intertaxa hybrids in *Oryza*. The genus *Oryza* maintains a certain level of racial and species integrity, even while allowing some degree of cross-fertilization. A characteristic feature of the species within this genus appears to be that only about half the species possess sufficient distinct features to make their identification easy. The remaining species show a varying extent of "miscegenation" of characters of one or more species or taxa. One instance is that of *O. officinalis*–*O. minuta*–*O. malampuzhensis* with *O. punctata*. Within *O. sativa*, we are aware of the japonica–indica–javanica conundrum (Glaszmann 1987, Garris et al. 2005) and *O. glaberrima* heterogeneity (Semon et al. 2005).

The genus *Oryza* appears to have adopted a strategy of "controlled" cross-fertilization to ensure the sustenance and survival of the genus The best instances of this are indica–japonica rices. This aspect will be discussed further in Chapter 6.

Broadly, in all the species complexes, the American forms show grosser morphology. This is followed by the Oceanean and African forms, with the Asian forms being relatively the smallest. The morphologically intermediate-looking forms such as *O. nivara* may also be spontaneas which display a certain degree of hybrid vigor. This is apparent from the dimensions given by Sharma and Shastry (1965) and Sharma (2003) for *O. nivara* for panicle, spikelet, and awn dimensions, and other characters.

CHAPTER 5

The Origin of African Rice

1. INTRODUCTION

Of the two cultivated rices, Asian rice (*Oryza sativa*) and African rice (*O. glaberrima*), African rice is accepted to have originated in west Africa.

Rice is not a major crop in Africa if the continent is considered as a whole, as was discussed in Chapter 1. It ranks only fifth in area under cultivation, after maize, sorghum, millets, and cassava, and occupies only 4.4% of the cropped area. It provides 8.0% of the calories consumed by the population, whereas in the world as a whole, rice contributes 21% of the calories consumed; the most of any commodity (FAOSTAT 2008, Tables 1.1 and 1.4). Rice production in Africa (2010) was 22.98 million tonnes (as rice paddy) – 3.3% of global production (FAOSTAT).

The position of rice in west Africa is, however, better. Of the five regions into which the continent is divided – eastern, middle, northern, southern, and western – the production and consumption of rice is highest in western Africa (Table 1.4). Here, it occupies 7.7% of the cultivated area and supplies 12.2% of the food energy of the population. Here, it ranks sixth in terms of area under cultivation (7.7%), below millets (21.0%), sorghum (19.0%), tuber crops (yams, taro, sweet potato) (17.7%), maize (11.5%), and cassava (9.4%) (Table 5.1). Rice, alongside yams, has a very special place in the social customs of several ethnic groups in western Africa.

Outside western Africa, the major rice consuming countries of the continent are Democratic Republic of Congo, Egypt, Madagascar, Mozambique, and Tanzania (cf. Table 1.5). Of all the five continents, the rates of increase in production and per capita supply of rice have been the highest in Africa. Africa currently imports almost 60% of its annual requirements of rice. Egypt and Guinea are the only two countries that produce more rice than they consume in Africa, and Central African Republic and Chad are nearly self-sufficient in rice (FAO 2007).

The annual per capita consumption of rice in west Africa (32.80 kg) is much higher than that of all the continent as a whole (19.60 kg), and also higher than that of all the other regions (range: 10.10–18.30 kg/head/year) (Table 1.4 and 5.1). In 7/18 countries in this region, the annual per capita consumption of rice is higher than the world's average (52.9 kg). The annual

Origins and Phylogeny of Rices
http://dx.doi.org/10.1016/B978-0-12-417177-0.00005-X

Table 5.1 Principal Crops and Food Balance Sheet of West Africa**

Continent/ Region/ Country	Arable Land 1000 ha	Cereals	Paddy	Maize	Millets	Sorghum	Roots & Tubers	Cassava	Yams	Total Vegetal in Calories*	Cereals	Rice*	Coarse Cereals	Starchy Roots
		Area (% Age)									Per Capita Supplies In Percent			
Africa	199,405	50.38	4.4	13.83	10.46	13.3	10.96	5.82	2.11	2437 2259	50.43	8.21	11.12	14.07
Western Africa	72,794	59.8	7.72	11.48	21	18.97	15.74	9.42	5.38	2578 2466	48.41	13.27	21.76	18.23
Benin	2,650	34.26	0.87	25.02	1.7	6.57	15.7	8.98	6.3	2574 2462	36.79	9.13	8.12	34.3
Burkina Faso	4,840	73.93	0.99	8.98	29.05	34.65	0.12	0.02	0.06	2516 2397	73.73	5.6	50.79	0.48
Chad	3,600	56.86	2.64	3.61	20	22.22	2.61	0.75	0.67	2147 2005	50.86	4.56	39.12	8.85
Cote'd Ivorie	3,300	43.06	12.52	25.15	1.82	2.91	27.18	9.09	9.39	2644 2529	41.19	20.99	1.29	23.49
Gambia	315	53.33	3.17	6.67	34.92	8.57	0.95	0.95	nil	2288 2148	52.75	16.61	23.47	0.96
Ghana	4,185	34.93	2.8	18.92	4.95	8.27	35.13	19.28	7.67	2680 2559	29.51	7.31	4.78	42.24
Guinea	1,100	70.73	47.73	8.18	1.18	0.73	24.45	22.73	0.36	2447 2359	42.75	30.57	4.25	17.61

Guinea Bissau	300	40.43	18.33	5	10	5	14.67	1	nil	2051 1915	59.19	39.98	7.12	9.46
Liberia	382	31.41	31.41	nil	nil	NA	21.47	19.63	0.52	1930 1877	38.65	24.47	0.01	25.03
Mali	4,660	65.3	8.71	7.83	26.82	21.46	0.24	0.04	0.08	2236 2020	72.05	22.85	36.27	1.34
Mauritania	488	33.4	3.48	1.23	0.82	27.66	0.61	nil	0.82	2786 2288	50.68	10.37	6.35	0.61
Niger	14,483	54.83	1.73	0.06	35.21	19.33	0.06	0.35	nil	2170 2060	71.38	7.37	60.78	1.57
Nigeria	30,500	70.16	11.58	14.64	20	23.2	26.11	11.45	9.99	2714 2627	46.17	10.46	24.36	18.83
Senegal	2,460	54.11	3.58	7.15	34.84	8.46	1.5	1.46	nil	2374 2161	60.45	31.97	14.7	2.27
Sierra Leone	570	44.39	36.84	1.75	1.75	3.51	15.08	12.98	nil	1943 1869	51.62	38.81	3.29	12.4
Togo	2,510	32.39	1.39	17.89	3.78	9.16	7.77	4.78	2.19	2343 2265	48.95	8.41	10.03	27.49

*Numerator: total; denominator: vegetal in kcal
***Source:* Primary data from FAOSTAT (2006); Downloaded 01/06/2009 Computations by the author

per capita consumption of rice in west Africa is comparable to that of several countries in southeast Asia; for instance, in Guinea (western Africa), it is 128.2 kg/annually and Madagascar (east Africa), 102.1 kg annually.

Western Africa comprises 18 countries (as per the Africa Rice Center). They are Benin, Burkina Faso, Cameroon, Cape Verde, Chad, Cote d' Ivorie, Gambia, Ghana, Guinea, Guinea-Bissau, Liberia, Mali, Mauritania, Niger, Nigeria, Senegal, Sierra Leone, and Togo (Fig. 5.2). The region extends from 2°–25°N latitude approximately and covers 7.43 Mkm^2 area. This is 24% of the area of Africa. Its population is 29% of the continent's. Western Africa is bounded on the west and south by the Atlantic Ocean, by tropical forests in the east, and the Sahara desert, extending across about 10 Mkm^2, in the north. The Sahara desert occupies one and half times the area of western Africa. It has had, and continues to have, an overwhelming influence on the climate and peoples of the entire northern half of Africa, especially in the sub-Saharan region that includes western Africa. Hence, we shall briefly look at the physiography, paleoclimate, and ecology of the region.

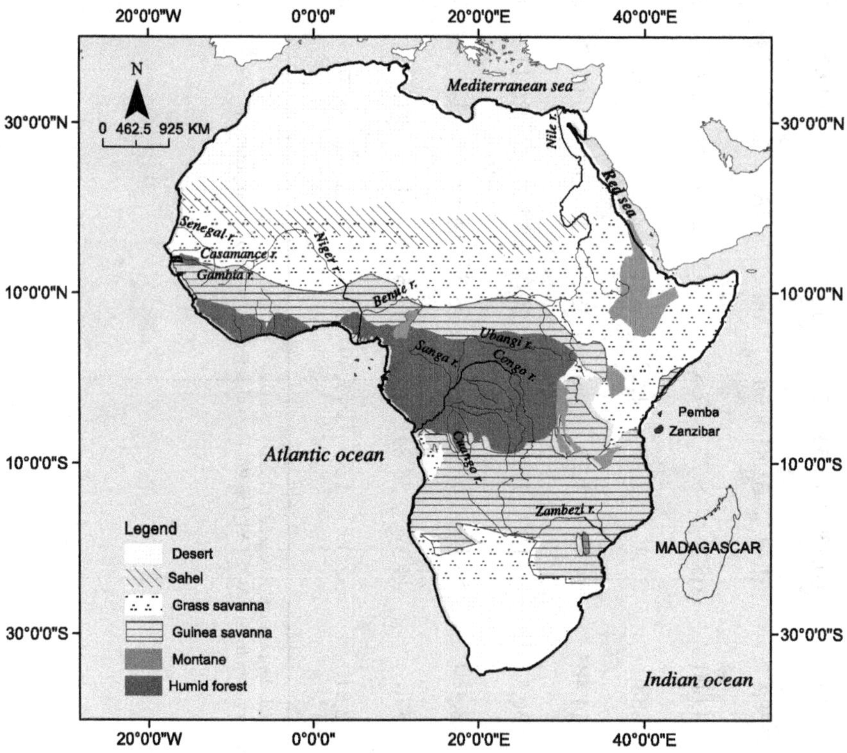

Figure 5.1 *Ecozones of Africa*

2. PHYSIOGRAPHY OF WEST AFRICA

Broadly, the entire western Africa is a flat land, less than 300 m in elevation. There are some undulations toward the north, the southwest (mostly upper Guinea), south (Cameroon), and the northeast (northeast Chad), which have some low mountains of 1,000–1,500 m elevation. Most of the major rivers of the region and their tributaries originate from the mountains of Guinea.

There is much variation in the amount of rainfall received in western Africa. It ranges all the way from 2,000 mm to 10–20 mm annually. Most of the southwest and southern coastal regions receive 1,600–2,000 mm annually (Fig. 5.1). Much of the coastline in the west and south consists of mangrove swamps. To its north lies the vast expanse of the west African flatlands, 1,200–1,400 km wide. It merges with the Sahara desert in the north (Fig. 5.1). The area can be divided into different zones that run broadly parallel from the west to east across the continent. The rainfall decreases progressively from the south to the north up to the Sahara desert. The desert may receive 10–100 mm annual precipitation, but very erratically.

3. ECOSYSTEMS OF WEST AFRICA

In the past, several authors have discussed the ecosystems of Africa, including Keay, Papadakis, Moorman and van Breeman, Chabrolin, Harries (1976), and Buddenhagen (1978). The dominant effect of the Sahara desert and its consequences on the waxing and waning of the various components of the ecosystem, on the biota, and on the lives and food habits of the people of the region during the Holocene period are outlined below.

The primary determinant of vegetation and agricultural crop pattern in this region is the length of the dry and wet seasons. The wet season may range from near zero to almost 12 months. The south and southwest regions contain moist forests (Figs. 5.1 and 5.2), and receive 9–12 months rainfall. To the north of this zone is the vast savanna region, and further to its north, the Sahara desert. Sandwiched between the savanna and the desert is the Sahel region, which receives 50–200 mm annual precipitation and supports some scrub and grass vegetation during the rains. Since prehistoric times, the extent and range of the Sahel and Sudan savanna (that lies to its south) have been shifting north and south in periodic climatic oscillations. The vast savanna region is broadly divided into three zones from south to north: the Guinea savanna, the grass savanna, and the Sudan savanna, with each zone receiving progressively less precipitation. Although the ecology of the

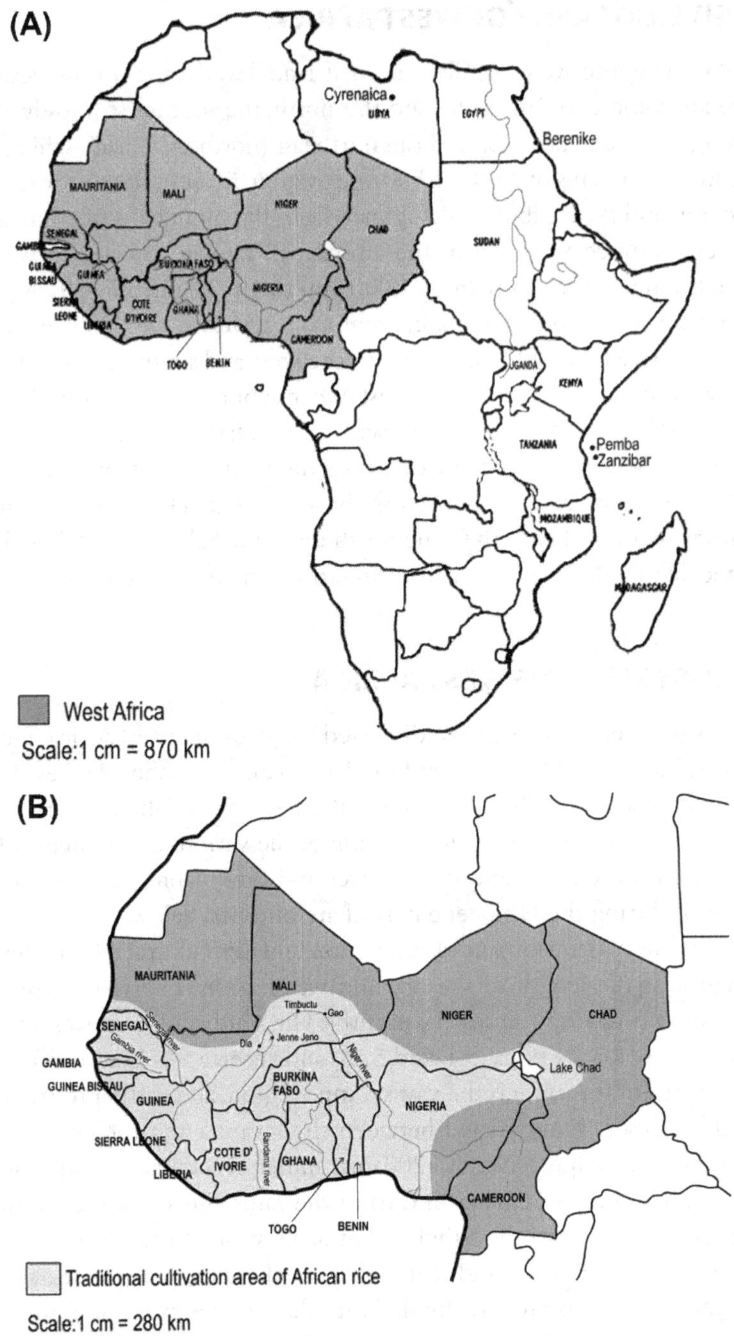

Figure 5.2 ***Africa Showing West Africa and Traditional Rice Growing Areas.*** (A) Africa showing west Africa. (B) West Africa showing traditional cultivated region of African rice.

Sahel and the northern savanna region is dominated by the hot, dry climate of the Sahara, its intensity is partially ameliorated by Lake Chad in northeastern Nigeria and the Niger river and its tributaries in south Mali.

This region is known collectively as the Inland Niger Delta (IND). Since prehistoric times – in fact, throughout the late Palaeocene and Holocene periods – it has witnessed some of the most intensive human activities, acting as the meeting point of trading caravans from the north (Mediterranean coast), northeast and east (south Sudan, Ethiopia) of the region, and as the refuge of nomadic pastoral people from the north during periods of intense aridity in the Sahara (Fig. 5.4). It has also been the home of some of the ancient trading and meeting centers such as Gao and Timbucktu, and also the seat of the ancient empires like Ghana, Mali, etc.

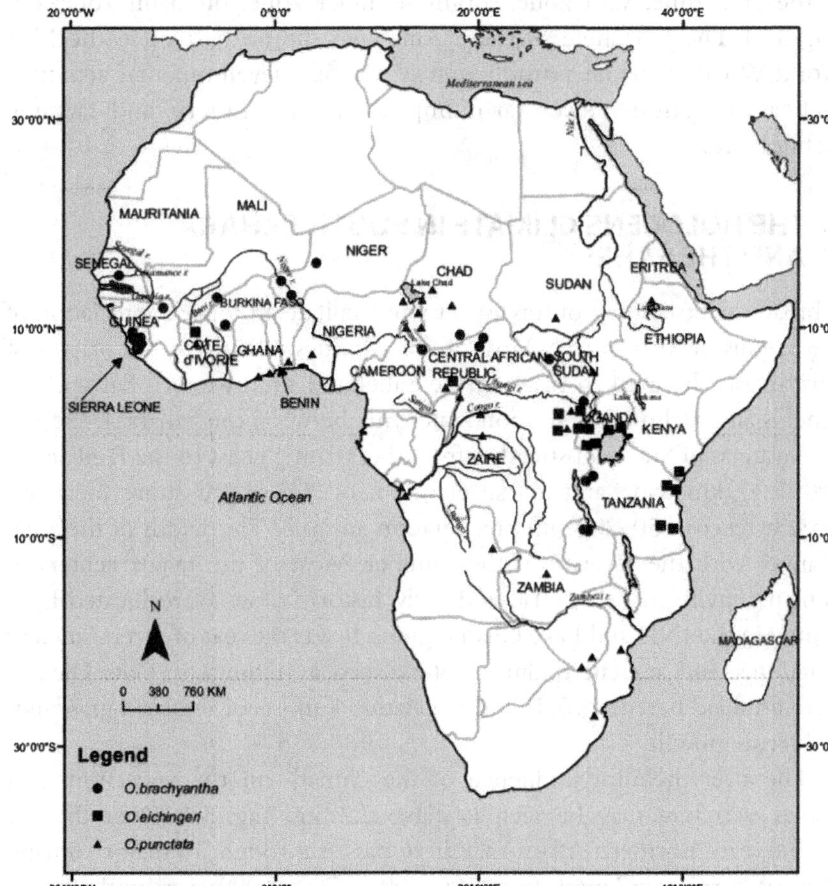

Figure 5.3 ***Africa Showing Distribution of Three Wild Rice Species, O. brachyantha, O. eichingeri, and O. punctata.***
Adapted from Vaughan (1994) with permission.

The extent and location of the Sahel and Sudan zones have waxed and waned, and shifted back and forth between the north and south from the late Paleolithic period (McIntosh and McIntosh 1993, Hassan 1997, 2002, Connah 2001,Vernet 2002 (also Philipson 2007) etc.) in consonance with the vagaries of climate in the Sahara and Sahel.

To the southwest of the IND zone lies the humid highlands of Guinea and the adjacent areas of Cote d' Ivorie and Burkina Faso. This region receives 400–1,000 mm rainfall annually. It is from here that all the rivers of the western region of west Africa originate – the Niger, Senegal, Gambia, and Volta rivers.

Traditionally, west Africa has been divided into different agricultural zones or ecosystems (Porteres 1955a, Johnston 1958, Harries 1976) such as the rice zone, yam zone, sorghum–millet zone, oil palm zone, etc. (Fig 5.1). The relevance of these zones has diminished since the First World War due to the introduction of various developmental activities, such as irrigation, newer farm implements, machinery, and farming technologies.

4. THE HOLOCENE CLIMATE IN SOUTH SAHARA AND THE SAHEL

A broad understanding of this aspect will facilitate a proper appreciation of the origins of African rice. Most of the activities relating to the origins of African rice have taken place in the Sahel and southwestern Sahara. The Sahel is an ecoclimatic transition zone lying between the Sahara desert and the savannas of Sudan. It stretches from the Atlantic coast to the Red Sea, is 400–1,000 km wide, and has an elevation of 200–400 m above mean sea level. It receives 50–200 mm precipitation annually. The width of the zone changes with the severity of the summer. Most of the major centers of human activity in prehistoric and early historic times were located here, mainly in the IND and Lake Chad regions. It was the seat of several ancient kingdoms, and ancient trading centers, such as Timbuktu, Gao, Djenne, were founded here (Fig. 5.4). The vegetation consists of semiarid grasslands and scrub growth.

The overwhelming influence of the climate on the vegetation and human activities may be seen in Table 5.2 and Fig. 5.1. Over the last 14,000 years, northern Africa may have passed through 30 major climatic changes, varying in length from 100–400 to 700–1,000 years, with smaller variations of 10–15 and 30–40 years (Nicolson 1976, Hassan 2002, Table 5.2). The chronology of various events is reported similarly in the accounts of

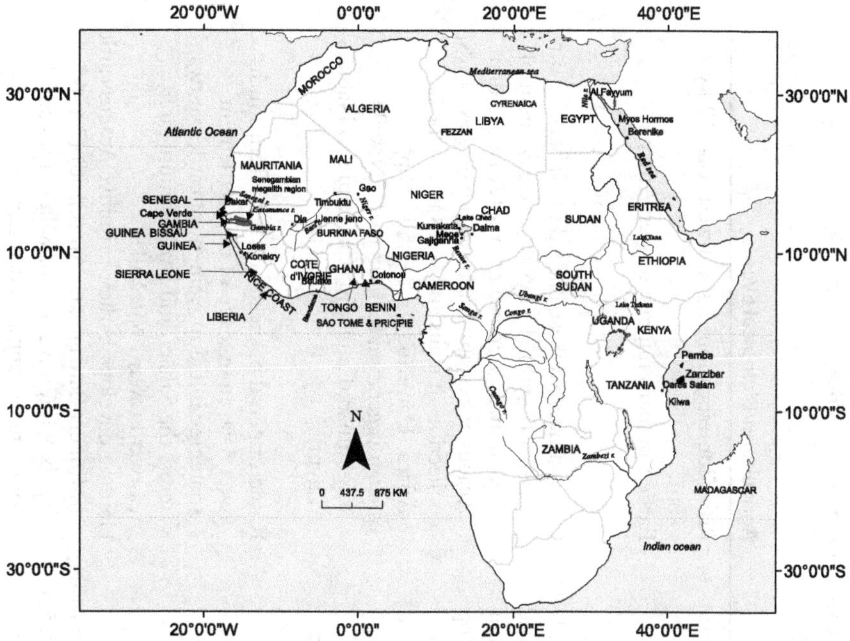

Figure 5.4 ***Africa, Showing the Places Mentioned in Chapter 5***

various authors (Hassan 2002a,b,c,d, Vernet 2002, MacEachern 2005, McIntosh 2006, etc.), but, as has been pointed out by Neumann (2005), there is no unanimity in the details. The Sahara has always displayed a mosaic of climatic events in its different regions. The several severe, abrupt, dry episodes appear to have stimulated a variety of anthropogenic responses that included demographic shuffle, dispersal, aggregation, adoption, or creation of a host of technological innovations, acceptance, and development of a variety of food producing modes, and a number of other sociological developments (Hassan 2002d). They could also have triggered the development of new variations in plants and animals, providing opportunities for early pastoralists to select forms that they would have found to be possessing more desirable attributes, as might have happened in the development of the African rice from its ancestral forms (Table 5.2).

There are some aspects of the prehistory of west Africa that are unique to the region. Firstly, the domestication and rearing of cattle preceded crop agriculture. Elsewhere in the world, farming had preceded cattle rearing (Marshall and Hildebrand 2002, Gifford-Gonzalez 2005). The earliest cattle remains discovered in Nabta (Egypt) date to 7,600 cal BP, while those in

Table 5.2 Chronology of Climate and Agricultural Events in Africa, with Special Reference to West Africa*

Calibrated ^{14}C Years, kyr cal BP	Uncalibrated ^{14}C Years, kyr BP	Climatic Events	Agricultural/Biological Events
16–13.4	13.3–11.5	After a long dry period during late Pleistocene, a first humid sequence followed by brief dry episode in north Africa (10.7 kyr bp) corresponding to Younger Dryas (11–10 kyr bp) of Europe	Plant domestication in southwest Asia (12 kyr bp); Earliest humans migrate to west Africa
–	10.0–7.5/7.2	Holocene humid period: lakes and ponds in Sahara cover large areas; Mega-Chad develops; but abrupt and severe spell of aridity in northeast Africa, ±8.1 kyr bp	Communities well established in Nabta and Bir Kiseiba (Egypt) (10 kyr bp); management of wild animals seen in northeast Africa (Adrar Acacus, 9 kyr bp); Intensive use of grass seeds, collection of fruits and seeds in northeast Africa (Uan Afuda, 8.9–7.5 kyr bp)
9.5–8.6	8.6–7.8	Wet conditions continue in the Sahel and Sahara; signs of increasing aridity in northeast Africa (8.1–7.9 kyr bp); deep walk-in wells	Sahelian vegetation seen up to 27°N; wild sorghum and millet in storage pits in Nabta (Egypt)
8.0–7.9	7.9–7.7	Abrupt, severe spell of aridity in Sahel and Sahara; succession of drought episodes	Rapid dispersal of sheep and goat from southwest Asia; emergence of new settlement strategies; initial movement of cattle westward across the Sahara, and a millennium later, eastward toward Nile Valley
7.0	6.5–4.2	Sustained drought for several centuries; long globally warm period with alternating humid (6.7, 4.5 kyr bp) and dry (6.4, 5.5, 4.8 kyr bp) peaks; but, overall a weak second Holocene humid period	Herding develops: earliest primitive Acacus cattle in Enneri Bardagu and Air (Chad) (7.4–6.5 kyr bp); intensive uses of wild grasses and sorghum in central Sudan region

*Based on Breunning and Neumann 2002, Hassan (2002b), Vernet 2002, and McIntosh (2006) west Africa have been dated to 5,990 cal BP (Delebe, Enneri Bardagne, both Chad) and 5,260 cal BP (Adrar Bous, Niger (Hassan 2002d).

Various reasons have been given for this, the most plausible of which appears to be that when people were living on the edge, it was much easier for transhumance with cattle than farming in times of sudden climatic stresses. Crop agriculture appears to have been taken up much later here – after 1,200 cal BC (Harlan 1992, van der Veen 1999, Breunig and Neumann 2002, Neumann 2003, 2005).

The most possible reason for this may be that the people of the region had several options to meet their food needs from natural sources, so that they might not have felt the need to initiate crop domestication earlier to ensure better food security. These resources consisted of a plentiful availability of game, wild grasses and sedges (including wild millet, wild sorghum, wild rice, and Bermuda grass), fish, and reared cattle. This can be inferred from the results documented by McIntosh, Neumann and co-workers (McIntosh 1995, Klee et al. 2000, for instance). Domesticating plants is a slow and time-consuming process requiring patient, systematic, and sustained efforts. For this, a sedentary and settled life is a prerequisite. Incidentally, Hassan (2002d) has stated that the decisive climatic factors for the appearance of cultivation in west Africa are clearly related to the arid conditions during the period from 2,200 to 1,300 cal BC.

5. THE HISTORY OF AFRICAN RICE

Nayar (2010) has recently summarized this subject.

i. Discovery and Description

When Linnaeus described the genus *Oryza* in 1753, he identified only one species, *O. sativa,* the Asian rice. It was Ernest von Steudel (1783–1856), who described the African rice, *O. glaberrima,* in 1855. He identified the specimen from material collected by Edelstan Jardin, a French navy commissioner, who had collected plant material from the tropical west African coast during 1845–1848. Jardin had originally labeled the material as *O. sativa.* It was known as *mole* in the vernacular (Porteres 1955). This specimen is now available in the National Museum of Natural History, Paris. Then, in a comprehensive publication (New Annales of the Navy and of the Colonies, July 1850 and May 1851), Edelstan Jardin stated that he had collected his sheet no. 9 (the one used by Steudel to describe *O. glaberrima*) from Loess Island (which is off the Conakry coast in present-day Guinea), on the southwestern Atlantic coast of west Africa (Fig. 5.4).

Then in 1875, while enumerating his plant collections of 1845–1848 (in *Bulletin de la Societe Linneene de Normandie* 2nd series, ix, pp. 247–339, 1875), Edelstan Jardin reaffirmed that his specimen (no. 9) had been collected from Loess Island, and that "the local people grow this 'small species (*O. glaberrima*) of *Oryza'* ('cultivent cette petit espece d' *Oryza*') under the name *mole*" (Porteres, 1955). It is unclear why Jardin described *O. glaberrima* as a small species, since it is no shorter than *O. sativa*. It could be that the epithet was used due to its rarity in the region vis-à-vis *O. sativa*.

According to Porteres (1955), *mole* is a generic name for rice in the Soussu dialect of the Mande linguistic group, and it denotes both the rice species. Rice is known also as *molo*, *moro*, and *mano*. Porteres also observed that the Guinea landrace, *saligboli,* which means 'open panicle and red caryopsis' is a semifloating type.

Prior to this, Leprieur, a navy pharmacist, had collected rice specimens from the Cape Verde peninsula near Dakar between 1824 and 1829, while exploring the vegetation of Senegal (Porteres 1955) (Fig. 5.4). In the National Museum of Natural History, there is a herbarium sheet of rice prepared by Leprieur in May 1826. It carries two panicles and is labeled as *Oryza sativa*. Porteres (*loc. cit.*) examined the sheet. He identified the right panicle as of *O. sativa* and the left as of *O. glaberrima*. He then declared that this was the oldest herbarium sheet of *O. glaberrima* and that it predated that of Jardin by 20 years.

The third specimens of *O. glaberrima* were collected by Augustus Chevalier in 1899 from upper Senegal–Niger along the Casamance River (Porteres, 1955).

From these records, two points become apparent. The first is the apparent rarity of *O. glaberrima* at least in the latter half of the 19th century. Otherwise, it is difficult to explain the considerable time lapse between the first three collections – the first in 1826, the second in 1845–1848, and the third in 1899, i.e., after gaps of 19–22 and 51–54 years. The second is the very close morphological similarities between the two rice species, as observed by Leprieur in 1826 and Jardin in 1845–1648. Otherwise, both of them could not have made identical misidentifications.

ii. Early History and Travel Accounts

Rice has a long history and tradition in west Africa. The coast stretching from the Casamance River (south Senegal) to Bandama river (Cote d'Ivorie) is known as the Rice Coast (Fig. 5.4). Here, especially in the western Senegal–Gambia region, rice forms part of many rituals and traditions among several linguistic groups. They have also developed efficient flooded rice production systems suited to the estuaries, levees and backwaters that occur widely in the southwest Rice Belt (Carney 2001, Linares 2002, Fields-Black 2008).

The first definitive account of introducing rice into Africa appears to be that of Alexander the Great in Egypt in 323 BCE, on his return journey from his expedition to India (Lu and Chang 1980). But rice did not become established then. This had to wait until the 7th century CE, when the Arabs introduced rice into Egypt. The Greek geographer and historian Strabo (63 BCE – 20 CE) observed rice growing in Augila (Aujila) oasis in Cyrenaica, Libya ca. 12 CE (Lewicki 1974). Cyrenaica has been one of the centers of trans-Saharan trade since prehistoric times (Figs. 2.3 and 5.4).

Incidentally, Garamantes (in Fezzan, Libya), has long been known as an ancient center of civilization, with rich agricultural traditions (900–400 cal BC). Pellig (2005) studied the state of agriculture in Garamantes and found that the agriculture there had borrowed crops from the east (Egypt), the south, and the west African Sahel. She found archaeological evidences of wheat, barley, sorghum, pearl millet, faba beans, peas, cotton, flax, and several fruit plants of southwestern Asia, together with a few from Africa, south of Garamantes, such as dates, watermelon, oil palm, pomegranate, etc. But she did not report rice in the excavated site.

The Islamic scholar, Al-Bakhri (1014–1094 CE) has provided the first indication of rice cultivation in west Africa in 1068 CE as growing along the River Niger (Carney 2001). There are in fact several accounts of Arab travelers to west Africa seeing rice throughout the region during the Middle Ages (Lewicki 1974). Lewicki also suggested that after pearl millet (*Pennisetum*), rice appeared to be the most common vegetable item in the food of the people in west Africa during the Middle Ages. However, none of the travelers have specified whether what they saw was wild or cultivated rice, and if it was cultivated rice, whether African or Asian. It has been known for a long time that large natural stands of wild rices occur widely in west Africa. These continue to exist, at least until about 50 years ago. Local populations used to harvest them for their use, for sale in the local markets, and for use in rituals. The extensive writings of Porteres, Oka, and Vaughan bear this out.

The four wild *Oryza* species that occur in west Africa are *O. barthii*, *O. longistaminata*, *O. brachyantha*, and *O. punctata*, in decreasing order of their range of distribution and frequency (Figs. 5.2, 5.3). *O. barthii* especially is very widespread throughout tropical Africa, and is often found in extensive stands, particularly in the IND (Inland Niger Delta) region.

In the 19th century, two prominent botanists, Barth (1857) and Schweinfurth (Maurizio 1926) also attested to their widespread occurrence in the region. Incidentally, two *Oryza* species have been named in their honor, *O. barthii* and *O. schweinfurthiana* (now, *O. punctata*).

Ibn Batuta (1304–1368/1369 CE) was a widely traveled nobleman from Tangier, Morocco. He was a religious scholar and inveterate traveler who spent 29 years traveling through southwest and south Asia, and north and west Africa (Gibb 1929). During 1852–1853, he traveled up to the then well-known Mali Empire (in present-day IND) and stayed there for several months. Although his primary interest appeared to be to chronicle the social life of the places he visited, he also recorded the foods consumed by the people. Ibn Battuta mentioned that the local people ate rice (Gibbs 1929), but his accounts suggested that millet was a more common food item.

Lewicki (1974) cited the travel accounts of several other Arab Muslim travelers to west Africa, some of whom traveled right up to the Senegal coast. They observed that rice was grown widely in the region. But, as also stated before, these travelers did not indicate the nature of the rice they observed – whether it was wild or cultivated rice, and if cultivated, African or Asian rice. Such travels came to an end in the 15th century with the ascendancy of European influence.

6. EAST AFRICA AND MADAGASCAR

When discussing the introduction of Asian rice into west Africa, some authors (Nayar 1973, 2010, Carpenter 1978, Cappers 2006) have suggested that this might have occurred through the east of Africa. *Periplus* records Arabian dhows and other boats traveling along the east coast (up to Rhapta, actual location not known, but thought to be near the present-day Dar es Salaam) for trade during Roman times (Fig. 5.4). *Periplus of the Erythrean Sea* is an anonymous account of the Roman trade that took place during the first century Common Era in the north western Indian Ocean (north Arabian Sea) (Schoff 1912, Casson 1989, McPherson 1993).

Sheriff (1981) has reviewed this aspect for UNESCO's History of Africa. He observed that only meager archaeological evidences have been obtained from this region so far. This means that we shall have to rely on ancient historical accounts (*Periplus*, Pliny, Ptolemy, Strabo, etc.) and numismatic evidences. The few coins obtained from the region have not also been studied systematically, so their context and interpretation cannot be determined with any accuracy. Most coins are from Ptolomaic Egypt, Imperial Rome, and Byzantium of the third century CE. Another collection was from

Mameluk and Ottoman Egypt of 13th century CE. Some Chinese (12th century CE) and Roman (1st–4th centuries CE) coins have also been obtained. These however indicate only that the region from where coins had originated had been visited by people of the region to which these coins belonged in prehistoric times, or vice versa (Sheriff 1981).

The above early historic accounts mention the active trade that was carried out with the Roman Europe and east Africa by Arab traders from southern Arabia. They traded in food items, silk, cotton, iron, and cloth from India, in exchange for timber, ivory, tortoise and other shells from east Africa. From accounts given elsewhere, it would seem that the food items included rice, but this is not definite as regards the east African coasts. There is also no evidence of any direct commercial intercourse between east Africa and India before the 7th century CE (Sheriff 1981). The *Periplus*, however, indicates the presence of Roman and Indian traders in the region from about the first century CE.

Madagascar is considered to be part of the African continent, but in many respects, it is distinct and unique, because it is an island. Madagascar (0.59 Mkm^2 area) is the fourth biggest island in the world. It lies more than 400 km from Mozambique in southeastern Africa and lies between about 12–25°S latitude (Fig. 5.4). Its present population is about 22 million. Because of its isolation, its flora and fauna display high levels of endemism.

A unique characteristic of Madagascar's population is that most of them are Austronesian in descent, and the language used in the island, Malagasy, belongs to the Austronesian linguistic family (Verin 1981). To determine the ancestry and structure of the population, Hurles et al. (2005) genetically analyzed both the paternal and maternal lineages of four ethnic groups of the Island and ten potential ancestral groups of Island Southeast Asia and the Pacific. They found that both the African and the Indonesian lineages contributed about equally on both the sides. Using available linguistic advances also, they inferred that most of the original populations had originated from south Borneo from along the Baristo river. In the UNESCO History of Africa, Verin (1981) has given the composition of the human population, based on physical morphology as: Indonesian mongoloid 37%, Negro-Oceanic 52%, Negro-African 2%, and European 9%. In a detailed genetic study, Tofanelli et al. (2009) was unable to pinpoint the homelands of the Austronesians, but they appeared to comprise populations from two vast areas, sub-Saharan Africa and southeast Asia. The habits, culture, and agriculture of Madagascar have been greatly influenced by oriental culture (house forms, terraced rice cultivation, ancestral worship, household items, musical instruments, most south and south east Asian crop plants etc.) (Verin 1981, Blench 2007, etc.). Nayar (2011) has recently summarized the present situation.

There is as yet no archaeological evidence to indicate the actual time of arrival in Madagascar of people from southeast Asia; so we must resort to ancient historical texts and evidence from other sources to estimate this, such as comparative linguistics, ethnology, physical anthropology, musicology, and oral traditions.

Lu and Chang (1980) quoted a rare Russian text which proposed that the first Asian rices were brought to Madagascar when the first settlers from Asia landed in the Tuliar region of southwest Madagascar in 1,000 BCE. According to them, the second major migration came from Indonesia in the early CE and the third from south India during the 10th–12th centuries CE. As a consequence, javanica rices were the first to be planted in the central plateau regions, and indica rices elsewhere on the island. At the same time, Verin (1981) (in UNESCO History of Africa) conjectured from linguistic and sociological records that the Indonesian migrations took place during the 5th to 12th centuries CE. More recent authors, including the well known philologist, Peter Blench (2007), have however proposed a date of the 3rd to 4th century CE, based on archaeo-zoogeographical evidence. It was known from late prehistoric period (*Periplus*, of Schoff 1912, Casson 1989) that it was possible to complete the voyages between Africa and south and southeast Asia within one year, if one took advantage of the southwest and northeast monsoon winds. A one-way trip could be completed in 6–7 weeks (Verin 1981, McPherson 1993, Parker 2008). Most authors have agreed, based on available evidences, that the outrigger sailing boats from southeast Asia used to stop over in Sri Lanka or on the Malabar coast en route to Africa. Some authors have also proposed that the last major immigration from Asia, which took place in the 12th century CE, included also people from India (Lu and Chang 1980).

The Asian plants introduced into Africa include banana (*Musa paradisiaca*), yam (*Dioscorea alata*), coconut (*Cocos nucifera*), sugarcane (*Saccharum officinarum*), taro (*Colocasia esculenta*), bread fruit (*Artocarpus altilis*), and jackfruit (*Artocarpus integrifolia*), besides of course, rice, *O. sativa* (Murdoch 1959, Ray 1993, and so on). The names of several Madagascan rice varieties are indicative of their origin in peninsular India (Porteres 1955) Incidentally, there are also some indirect evidences of direct southeast and south Asian influences in west Africa from the early centuries of the Common Era (Jones 1959, Murdoch 1959, Christie in Gray 1962, Hawley 2008). If there were strong cultural influences as the above authors have pointed out – as has there been with Madagascar – we can only rightly infer the Asian contacts would have brought with them botanicals also, including rice (also Rush 2008).

7. THE RICE ARCHAEOLOGY OF AFRICA

There are five archaeological accounts of rice in Africa – three from west Africa and two from eastern Egypt (Fig. 5.5).

i. Egypt

It has been determined that the rice samples obtained from the Roman ports of the Red Sea had been imported for the use of the Indian traders,

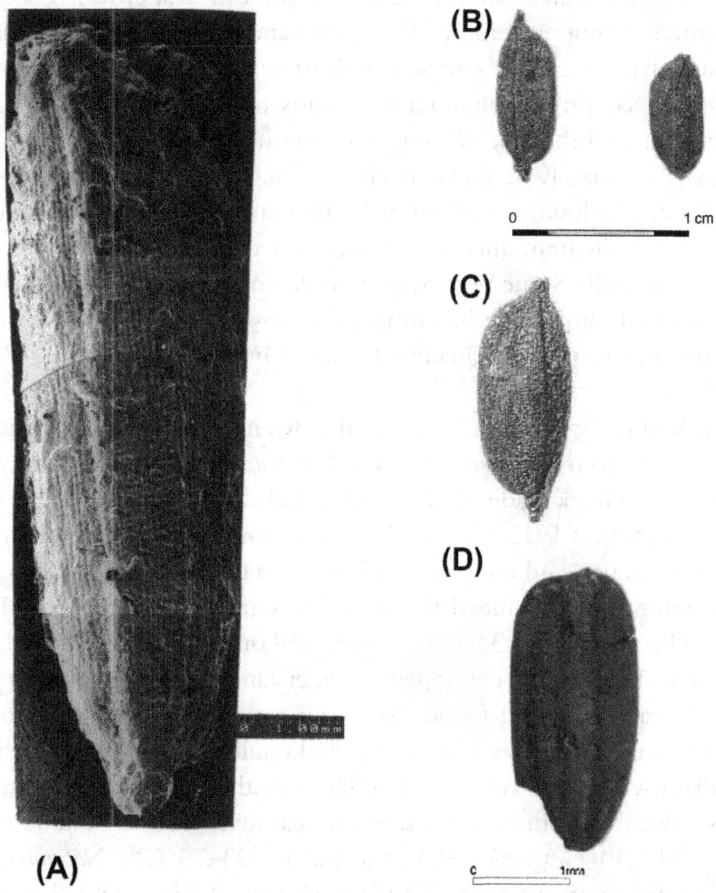

Figure 5.5 ***Rice Archaeological Specimens Obtained from Northern Africa.*** (A) Pottery sherd containing wild rice impressions from Gajiganna, northeast Nigeria(1100-1200 cal BC). Klee and Zach (1999). (B) Rice kernals imported from India 9early 1st century CE) from Myos Harmos, Egypt. van der Veen (2003). (C) Rice spikelets from south Asia recovered from berenike, Egypt(1-5th century CE.Cappers (2006). (D) Rice spikelets from Dia, IND, Mali (800-500 BCE). Murray (2006).

some of whom used to live in these ports for extended periods of time, a few even indefinitely. Along with rice, the excavation also revealed remnants of coconut husk and shell, substantial quantities of black pepper, etc. Black pepper was used extensively in the Roman empire for a variety of purposes – as an article of food, as a medicine for a range of illnesses, for certain rituals, and so on (van der Veen 2003, Cappers 2007, Sidebotham 2011).

Both samples of rice material obtained from Egypt are of the Roman period. At its height (ca. 27 BCE – 476 CE), the Roman Empire spread over an extensive area around the Mediterranean Sea, covering southern and central Europe, and most of west Asia and Egypt. It used to maintain extensive maritime trade contacts with east Africa (up to about present-day Dar es Salaam), throughout the coasts of the northern and eastern Arabian Sea, and the Bay of Bengal. Rome used to trade with south and even east Asia, mainly in spices, precious stones, textiles including silk, food items (grains including rice), with the Persian and Arab coasts in incense and essential oils (frankincense, myrrh), and with Africa in timber, ivory, and tortoise shells. Some ports were founded on the Red Sea coast for this trade during Roman rule in Europe, the most prominent of which were Berenike and Quseir al Quadim (earlier Myos Hormos) (23/24°N) in Egypt).

Much of the written information that we have about the trade and the times comes from the *Periplus Maris Erythracei*, which is an anonymous account of a Greek trader that was prepared during the first century CE (40–70 CE, Schoff 1912, Casson 1989). The *Periplus* gives detailed accounts of the trade routes and trade that took place in the northern Indian Ocean. The Erythrean Sea included the Red Sea, Gulf of Aden, and the Indian Ocean. The *Periplus* lists 34 items of botanical origin that were traded. Prior to this, Greek sailors had developed an understanding of the monsoon winds in the Ocean. They had found that a ship could easily voyage across the Indian Ocean and back within a year, and could travel between India and east Africa within 40 days (Casson 1989). Good accounts of Roman trade are also available from certain other classical texts such as Strabo's (64/63 BCE – 24 CE) *Geography*, Pliny, the Elder's (23–79 CE) *Natural History*, Claudeus Ptolemy's (90–168 CE) *Geography*, the *Peutinger Table* (last edition in fifth century CE), etc. In recent times, the seven seasons of archaeological excavations carried out at Berenike under the leadership of Steven Sidebotham (University of Delaware, USA) and his overview account of the same (Sidebotham 2011) have also yielded substantial amounts of information about Roman trade with the Indian Ocean littoral.

The Roman occupation of Myos Hormos has been dated to mid-first and second centuries CE (van der Veen 2003). At that time, the local grain crops consisted of hulled barley (*Hordeum vulgare*) and hard wheat (*Triticum durum*), together with several pulses, fruits, and nuts. The foreign foods consisted of rice, coconut, mango, and aubergine (egg plant, brinjal); all of Indian origin. "Rice was found in small quantities in both Roman (1st century CE) and medieval (11–15th century CE) deposits" (van der Veen 2003) (Fig 5.5).

The port of Berenike was founded in ca. 275 BCE and it was abandoned in the early 7th century CE, prior to the arrival of Arabs in Egypt. Roman activity in the Egyptian desert lasted from about 30 BCE to 638 CE. Rice was present "in reasonable quantities" during both early (1st–2nd centuries CE) and late (4th to early 6th centuries CE) (Cappers 1999, 2007, Sidebotham 2011) (Fig. 5.5).

ii. West Africa

There are three reports of rice in an archaeological context from west Africa. These are from Jenne (IND, Mali, 16°N), Gajigenna (Lake Chad basin, northeast Nigeria, 12°N), and Dia (IND, Mali, 16°N) (Fig. 5.5). IND, as we have seen, is a vast interior region of swamps and standing water, which cuts across the savanna grassland and scrub Sahel vegetation. Its deltaic hydrology is sustained by the River Niger and its tributary, Bani. They flood about 30,000 km^2 of land for 6–9 months of the year, as we have noted before. The region receives ca. 590 mm precipitation annually, mostly during July–September. Thus, the region is well north of the 750 mm isohyet border between the savanna and the Sahel. The annual temperature ranges between 14°C and 41°C (mean 28°C) (McIntosh and McIntosh 1995).

A. Jenne, Inland Niger Delta (IND), Mali

Jenne-Jeno was occupied during ca. 250 BC to 1400 AD (Fig. 5.2); one of the longest Iron Age sequences in west Africa. McIntosh and McIntosh (1995) have done the most extensive studies of the region. They divided their excavations into four phases. Phase I consisted of ca. 250 BCE to 50 CE. The pottery decoration they found suggested Saharan affinities. The inhabitants' subsistence was heavily reliant on wild resources (bovids, fish, water fowl) and domestic cattle. No domesticated plants were recovered. Other recovered items included stone beads and sandstone grinders. Incidentally, this appears to be consistent with the general phenomenon observed in all sub-Saharan excavation sites. Various authors have interpreted this as an over-reliance on wild grass seeds (Neumann 2003, 2005). Phase II covered a carbon date of ca. 50–400 CE. This layer revealed "a substantial sample of well preserved chaff of African rice, *O. glaberrima*" (McIntosh and McIntosh 1995). There is no further mention of this finding in any of their extensive later writings. Phase III (ca. 400–900 CE) was characterized by painted pottery of the period and (imported) copper, in addition to stone and iron. The subsistence pattern was broadly similar to that of Phase II. The fourth phase (ca. 900–1400 CE) was characterized by mud brick buildings, increased human activities, and societal customs.

In their detailed reports of the 1981 excavations (McIntosh 1995) and numerous other publications (McIntosh 2005, McIntosh and McIntosh 1980, 1984, 1988, 1993, etc.), the authors have dealt in detail with various sociological aspects, such as community structure, urbanism, pottery, metals, archaeofauna, including fish resources, etc. The botanical aspects were treated only briefly (McIntosh 1995). The results are summarized below.

"Almost every sample has some, and few samples are almost all, of the grass seed, *Brachiaria* (now, *Moorochloa*) *ramosa*, an annual grass encountered in marshy zones" (McIntosh 1995), as per the determinations done in RBG Kew. During McIntosh and coworkers' field visits, they found further that "seeds of *Panicum lactum, Echinochloa colona,* and *Oryza barthii*, besides *Brachiaria* (*Moorochloa*) *ramosa* were being collected on a large-scale along the middle Niger", and they were being sold in the open market and/or stored in underground silos. Incidentally, Porteres (1951, 1976) and Harlan (1989) also recorded that *Brachiaria* (*Moorochloa*) *subalbidum* and *B.* (*M.*) *deflexa* were cultivated in the region. Another very common material was Cyperaceae. Several authors working in Sahara and Sahel have observed that tubers of *Cyperus rotundata* (nutgrass, Bermuda grass) were being used very widely throughout the northern half of Africa, for food and other purposes. Elsewhere in the tropics throughout the world, nutgrass is one of the most serious weeds of agricultural lands.

Although McIntosh (1995) had initially stated that the rice samples were largely of *O. glaberrima,* the weedy wild rice, *O. barthii,* was, and still is, widely present in the region. According to the author, he could not make any further studies of the material because of the poor condition of the recovered material. Only a few specimens were well preserved and they were identified as those of *O. glaberrima*, "only on the basis of being awnless". But this assumption appears to be tenuous, as we have discussed earlier.

B. Kursakata, Gajigenne (Lake Chad Basin, Northeast Nigeria)

Scientists from the University of Frankfurt have conducted archaeobotanical excavations in this region since 1992 (Breunig et al. 1996). Prior to this, Connah (1981) had conducted some pioneering studies in this basin in Daima. One component of the University of Frankfurt study conducted at Furfuna (northeast Mali) exposed a dugout canoe (6,000 BC), which is the oldest recovered in Africa and one of the oldest anywhere. The Lake Chad basin comes within the old shoreline of the Bama Ridge, when the region was under water during the Mega-Chad of the early and middle Holecene Wet Phase. The largest extension of this lake was observed in 5,000 cal BCE, when Lake Chad had attained 40 times its present size (MacEachern 2005) and water levels were 40 m higher than now.

This area comes within Sahel, at latitude 12°N. It receives about 550 mm precipitation annually. Trees are rare in this region, and the open savanna is dominated by grasses interspersed with bushes and a few trees (Klee and Zach 1999).

Excavations have been carried out at three sites, Mege, Gagiganna, and Kursakata, which are within 30 km of each other. They have a typical Sahelian climate with a short rainy season. Gajiganna is the oldest of the three. It was occupied during 1,800 to 800 cal BCE (Breunig et al. 1996; Breunig and Neumann 2002). Habitation at Kursakata and Mege began after the occupation of Gajiganna ended – at around 800 cal BCE. As the charred plant remains were in poor state of preservation, Klee and Zach (1999), who had studied the site, used potsherd impressions to assess the food habits. All the sites showed the beginnings of village settlement and also the first evidence of animal domestication south of the Sahara (Breunig et al. 1996). Interestingly, the percentage of potsherds showing impressions steadily increased from nil to 90% from phase I (1,800–1,400 BCE) to their youngest phase IIc (1,000–800 BCE), i.e., during a 700 year period. The largest percentage of grass caryopsis belonged to Paniceae (predominantly, *Echinochloa pyramidalis*) followed by wild rice, *Oryza* sp. (Klee and Zach 1999). The domesticated *Pennisetum glaucum* (pearl or bulrush millet) appeared for the first time in phase IIb (1,200–1,000 cal BCE), and in the youngest phase IIc (800–1,000 BCE), pearl millet became the dominant component. The authors did not specify the actual *Oryza* sp. recovered by them. They have, however, stated that it was wild material.

Kursakata. The cultural layers of this settlement mound spanned 1,000 cal BCE – 100 cal CE. Pastoralism, cattle rearing, and fishing appeared to be the main economic activities. The transition from the late Stone Age to early Iron Age could be dated to the 1st millennium BCE. Contemporaneous with this was the appearance of sheep and goats (Breunig et al. 1996, Klee et al. 2000, Zach and Klee 2004). The excavations produced charred and uncharred seeds, fruits, and charcoal. The transition from wild to domesticated form of pearl millet was noticeable from 800 cal BCE. The entire excavations gave more than 6,000 seeds of 54 taxa, mostly carbonized seeds or fruits. They consisted of domesticated pearl millet and wild rice (407 samples of *O. barthii* and *O. longistaminata*), Paniceae (823 caryposis), some edible fruits and local vegetation. The *Oryza* material consisted of both culms and seeds (Fig. 5.5).

C. Dia (IND, Mali)

The third findings of rice in West Africa were those from Murray (2004, 2005) in Dia. This is located in upper MND (Middle Niger Delta), a few kilometers from the banks of the seasonal Diaka river, a tributary of the River Niger, and about 70 km northwest of Jenne (Fig. 5.4). The local vegetation consists of several leguminous shrubs and trees and grasses such as *Cenchrus biflorus*. The excavations were carried out during 1998–2002 (Bedaux et al. 2001), with the two locations yielding dates of 800/400 BCE–1,900 CE (Shoma site) and 500–1,900 CE (Mara site). Murray (2003, 2005) identified five horizons in the site: I (800/400 BCE – 0 CE), II (0–500 CE), III (500–1,000 CE), IV (1,000–1,600 CE), and V (1,600–1,980 CE). Out of the total 6,063 charred botanical items, as many as 1,376 items consisted of whole or fragmented rice grains. Of these, 24% of the specimens were recovered from Horizon I and 51% from Horizon IV. AMS dates of three grains gave 2,522 ± 47 BP, 2,485 ± 52 BP, and 2,520 ± 48 BP. AMS of another seven grains gave secure context through to the mid-1st millennium CE.

"Determination of the ancient grains as wild or domestic species was difficult", because "virtually all the grains were recovered naked" (Murray 2004, 2005). The author then tried to identify them by comparing length–thickness and length–width ratios of ancient grains (134) with those of modern grains (91 domesticated *O. glaberrima* and 68 *O. barthii* seeds) "Overall, when dimensions of the ancient and modern rices were compared, the majority of the ancient grains more closely resembled the modern domestic African rice than the wild grains".

Although the material suggests that the majority of archaeological rice from Dia is domesticated, two puzzling details have remained: "The first is that the ancient grains are consistently smaller than the modern grains, most notably in length. Second, statistical analyses of the ancient grains by horizon showed no significant difference in length, width, or thickness from the earliest to the latest levels, there was no significant change in size or shape

Table 5.3 Grain Dimensions of Ancient and Present-Day Rices from Dia (IND, Mali)*

	Ancient Grains charred (no.: 134)			
Dimensions (in mm and mm^3)	**Length**	**Width**	**Thickness**	**Volume**
Mean and sd	5.8 ± 0.5	1.8 ± 0.2	2.5 ± 0.2	25.8 ± 4.8
Range	4.6–6.8	1.2–2.5	1.8–3.0	14.4–41.6
	***O. glaberrima* charred (no.: 91)**			
Mean and sd	6.7 ± 0.8	1.5 ± 0.2	2.3 ± 0.3	22.5 ± 4.8
Range	4.9–8.7	0.9–1.8	1.8–3.0	9.7–33.6
	***O. barthii* dharred (no.: 68)**			
Mean and sd	4.4 ± 0.05	1.4 ± 0.3	2.0 ± 0.3	12.7 ± 4.5
Range	3.3–5.6	0.9–5.6	1.1–2.7	4.0–23.3

*Adapted from Murray (2007)

of African rice recovered from Dia over the course of more than 2,000 years" (Table 5.3).

The author explained this anomaly by stating that: "*O. glaberrima* and *O. barthii* show an evolutionary trend toward thicker and wider grains, which is especially marked in *O. glaberrima*".

The author then concluded that: "overall the lack of change through time supports the idea that rice at Dia was fully domesticated from the earliest occupation,and more likely, domestication occurred elsewhere and then spread to Dia around 800–400 BC" (Murray 2004, 2005).

The measurements of the charred grains recovered from Dia are: length, 3.3–5.6 mm (mean, 4.4 mm; s.d., 0.5); width, 0.9–2.0 mm (mean, 1.4 mm; s.d., 0.3); thickness, 1.1–2.7 mm (mean, 2.0 mm; s.d., 0.3); and volume 4.0–23.3 mm^3 (mean, 12.7 mm^3; s.d., 4.5) (Table 5.3).

In *Oryza,* it is inappropriate to use length–breadth–thickness ratios for discriminating between species (Nayar 2012). This author has suggested that the seeds recovered by Murray might have been of the wild species, *O. brachyantha,* which is known to be present in the region (cf., Vaughan 1991, Clayton et al. 2010). These aspects will be discussed further later.

8. RICE CULTURE IN WEST AFRICA

West Africa both produces and imports more rice than all other regions in Africa (Table 1.4). In this region, although rice occupies only 7.7% of the cultivated area and ranks only sixth in cropped area, it supplies 13.3% of the food energy.

Following a recommendation made by the IRRI symposium on Rice Genetics and Cytogenetics (1963), the National Institute of Japan (Misima, Japan) conducted two extensive field surveys in west Africa to collect *Oryza* germplasm (five scientists, 14 weeks in 1963 and 1977, 11 countries) (Oka and Chang 1964, Oka et al. 1978). Their two tour reports are two of the four most detailed accounts of rice cultivation practices in west Africa. They give some details about the distribution and ecology of wild *Oryza* species. The third is the publication, *Rice in Africa,* which is the outcome of a conference held in IITA Ibadan (Nigeria) during March 1977 (Buddenhagen and Persley 1978), and the fourth is Johnston (1958). Another invaluable reference is the proceedings of a meeting on African rice species organized by ORSTOM (1977) in Paris in January 1977.

The following paragraphs give a brief overview of the situation as compiled from the above publications. The physiography and weather of the region has been given already (Sections 2 and 4). They will not be repeated here.

i. Rice Ecosystems

Broadly, there are five rice ecosystems (Nayar 2010). They are mangrove swamp, lowland irrigated, upland irrigated, lowland rainfed, and inland deepwater regions (Table 5.4). The rainfed upland and lowlands regions constitute three-fourths of the area under rice cultivation in west Africa. These regions receive 1,000–2,000 mm rainfall annually.

The coastal region from the Casamance River (southern Senegal) to Bandama River (central Cote d'Ivorie) is known as the Rice Coast (Figs. 5.2–5.4). This region extends up to the Senegal River and includes the river banks, backwaters, and estuaries of all the rivers. It has had a long tradition of rice cultivation. Rice is cultivated along the river banks and mud flats after embankments and dykes have been constructed. The local ethnic groups used to make them with much ingenuity and sophistication employing just the simplest wooden implements. They have developed numerous rituals and customs around rice culture and uses. These have been discussed at length by some authors (Richard 1996, Carney 2001, Linares 2002, Fields-Black, 2008, etc.).

Rice is cultivated extensively also in the Inland Niger Delta (IND), the Lake Chad basin, and upper Guinea (Fig. 5.2). The IND is a vast expanse of flood plains and seasonal wetlands, covering more than 30,000 km^2 in area (Wetlands International 2013). It is criss-crossed by the Niger River and its tributaries and also the headwaters of the White Volta River. The entire area is

Table 5.4 Major Rice Ecosystems of West Africa*

SI No.	Type	Estmated Area (Mha)	Main Growing Areas	Sowing/Planting Method	Species Planted	Remarks
1	Inland deep water	0.5	IND (south Mali), Sokota valley (north Nigeria)	Broadcasting	*glaberrima*, partly *sativa*	Grows deep water varieties; red pericarp, early shedding; sowing April-May
2	Lowland rainfed	1.8	Throughout except in Sudan region	Broadcasting/ transplanting	*sativa*, partly *glaberrima*	Sown with advent of rains
3	Upland rainfed	1.8	Hill slopes of mainly coastal region	Broadcasting	*glaberrima* and *sativa*	Grows "early" varieties (<140 days); *sativa* in coastal region, *glaberrima* in inland region; sour with advent of rains
4	Lowland irrigated	0.6	South Mali, north Nigeria, Niger	Broadcasting/ transplanting	*sativa*	Grown in areas with assured water supply; sowing in June
5	Mangrove swamp	0.3	Guinea, Sierra Leone, east Nigeria, Gambia	Transplanting	*sativa*	Mostly recently introduced sativas; sowing in late June

**Source:* Nayar (2010) (Adapted from Buddenhagen (1977) and Ndjiondjop et al. (2008)

dotted with lakes. Even in the core part of IND, which lies north of 15°N, the climate is favorable for rice cultivation in spite of the region's proximity to the Sahara desert, because of the plentiful supply of water. Vast stands of two wild rice species, *O. barthii* and *O. longistaminata,* occur in this region, especially the former. Local people and farmers used to regularly collect their seeds for consumption and local trade. Populations of a third rice species, *O. brachyantha*, also occur in the region. Its seeds too, along with those of several other grass species, used to be collected and consumed during periods of food shortage (Fig. 5.3). This region has also been a meeting point for people and traders from all across the Sahara since prehistoric times. Such activity had reached its peak during the Middle Ages, during the high time of Arab traders and travelers (8th–15th centuries Common Era). Goods coming from different regions used to be traded or bartered here. Towns such as Gao and Timbuktu were then centers of great activity (Fig. 5.4). After the camel was introduced into north Africa from Egypt in the third century CE, it became the chief means of moving goods and of transport.

The area under rice cultivation in these inland deep water rice ecosystems is estimated to be 0.5 Mha (Tables 5.4, 5.5). The rice grown in this region used to be floating types. In keeping with the wide variations in the physiography and weather of the region due to the overwhelming influence of the Saharan desert, four systems of paddy cultivation are practiced here (Table 5.4). The most important characteristics of traditional rice cultivation in the region are:

Table 5.5 Characteristics of African and Asian Rices*

Character	*O. glaberrima*	*O. sativa*
Habit	Annual	Annual, but can be perennated
Ligule length	6–13 mm	14–45 mm
Spikelet morphology	Small tubercles on surface, generally glabrous	Rough with short thorny hairs
Seedling vigor	High	Normal
Tillering capacity	High	Normal
Drought resistance	High	Normal
Lodging	More	Less
Grain shattering	Easy	Normal
Yield potential	Lower	Higher
Flood resistance	High	Low, but variable
Primary branches	11.0–12.2	12.2–12.6
Secondary branches	15.2–24.0	37.5–41.0
Grain weight/panicle	75–150 g	250 g

**Source:* Various; cf. Nayar (2010, 2012)

(i) Rice cultivation is done in a very simplistic manner, employing minimum agricultural implements and farm labor. Cattle are not traditionally used for farm operations.

(ii) The crops have a comparative long period of growth (130–180 days).

(iii) The onset of the rains, and their distribution and duration have a decisive influence on crop production and yield.

(iv) *O. glaberrima* appears to show greater tolerance to both drought and flooding than *O. sativa.*

(v) Most farmers and farming systems employ mixtures of both African and Asian rices, even deliberately mixing the seeds of the two rices, before sowing. It is possible that this practice helps the crop to withstand the increased levels of uncertainty in rice farming in west Africa.

(vi) Both the rices are characterized by high levels of heterogeneity in the field.

(vi) Most African rices have a red pericarp, whereas most Asian rices grown in the region have a white pericarp.

(viii) Milling recovery of African rice is slightly lower than that of Asian rice. Breakage is also higher. The increased level of breakage appears to be due to the very dry weather and very low levels of atmospheric humidity. In some regions, rice is parboiled before milling. This reduces the breakage percentage and improves both the keeping quality and nutritional value.

(ix) Overall, African rice gives lower yields less than Asian rice, but it appears to show a greater resilience than Asian rice to withstand the uncertainties of seasonal weather fluctuations and has a better capacity to withstand competition from weeds.

(x) Overall, the incidence of pest and diseases is lower, but this may also be because the crops are raised under simple conditions and manuring is not usually practiced.

(xi) It is generally not easy to distinguish African and Asian rices in the paddy fields, especially during the vegetative phase, unless one looks closely at the ligule length.

(xii) Weeds are a major problem, especially in deep water and irrigated fields. The most persistent weeds are the two wild rice species, *O. barthii* and *O. longistaminata*. This is also true of irrigated paddy conditions.

(xiii) Natural hybrids of African and Asian rice and those with the wild rices are rare except to some extent between the African rice and *O. barthii*. See Table 5.4.

9. WILD RICES OF AFRICA

Five *Oryza* species occur in the region – *O. barthii, O. longistaminata, O. brachyantha, O. punctata,* and *O. eichingeri*. Of these, *O. barthii* and *O. longistaminata* are very common. They often occur in large stands, *O. brachyantha* occurs frequently, *O. punctata* sometimes, and *O. eichingeri* rarely (Fig. 5.2). *O. longistaminata* is more common in the inland regions than on the coast. It occurs generally in swamps and as weeds in paddy fields. It is semi-sterile, or sometimes completely sterile, in populations standing in stable habitats. It propagates vegetatively via well-developed rhizomes. Its natural hybrids with other species are hardly ever seen in the fields. *O. barthii* too appears to be present more common in the interior than in coastal regions. Natural hybrids between *O. barthii* and *O. glaberrima* are often observed. *O. stapfii* is now considered as a natural hybrid of *O. barthii* and *O. glaberrima*. Wetland rice populations have floating habit and have adventitious roots and branches from submerged nodes. Plants are vigorous and seed fertility is high. Where there are sizable populations, local people collect the seeds for consumption (Fig. 5.3).

10. CHARACTERISTICS OF AFRICAN RICE

The advantages of African rice over Asian rice are its greater resistance to various biotic and abiotic stresses, and a greater ability to compete with weeds. The other attributed advantages are better taste and its use by certain language groups in religious and social ceremonies (Carney 2001, Fields-Black 2008). Its disadvantages are that its grains are shed more easily, its resistance to milling, and lower yields (Oka and Chang 1965, African Rice Center, Annual Reports [www.warda.org], Dingkuhn et al. 1988, Sarla and Swamy 2005) (Table 5.5). The red pericarp and higher levels of breakage of grains during milling are indicated as negative attributes of African rice. In reality, the red pericarp rice possesses superior nutritional qualities and its use should be encouraged. The greater grain breakage is due to the prevailing very hot and dry atmospheric conditions in sub-Sahara during the harvest season. Porteres (1956, 1959) have incidentally recorded that varieties with white pericarp color are present also in the African rice. This may be the result of chance mutations taken place in African rice for white pericarp, or from natural crossings taken place with introduced white pericarped Asian rice varieties. Gross et al. (2010) have observed that white pericarp in African rice are apparently caused by unique mutations at the *Rc* locus, which controls pericarp control in Asian rice.

Various authors have pointed out that the rice fields in west Africa generally grow mixtures of *O. glaberrima* and *O. sativa*. The extensive surveys carried out by Oka and others (Oka and Chang 1963, Oka et al. 1978) showed that only 10 out of 37 fields contained pure stands of *O. glaberrima* (six fields) and *O. sativa* (four fields). The remaining 27 fields contained mixtures of the two rices in various proportions (Table 5.6). These authors also reported that most fields consisted of mixtures of 2–4 rice species (Tables 5.6 and 5.7). Of the 262 fields that they had examined, only 86 fields – i.e., less than one-third – were pure stands of a single species.

11. THE FALL OF AFRICAN RICE

After the First World War (1914–1918), the colonial powers who then occupied west Africa – mainly the French and the English – initiated some developmental projects in their respective colonies. One such program was the establishment of the Office du Niger in 1932 by the French in Segou, Mali. It envisaged the construction of a barrage across the river Niger to irrigate about 20,000 ha of this virtual desert region in Sahel to cultivate cotton, but as the cotton crop failed, rice cultivation was taken up instead. This – like all the other similar development projects taken up in the region in Senegal river basin, coastal Guinea, etc. – promoted only the cultivation of Asian rice, instead of African rice. This was a mortal blow to the African rice that led to its fall (Nayar 2010). All developmental projects which followed – including seed production and distribution programs – only used Asian rice cultivars (Buddenhagen and Persley 1977, Charbolin 1977).

This process was resumed after the Second World War, and this virtually rang the death knell for African rice. Presently, pure African rice is hardly grown commercially at all. It is, however, occasionally cultivated by some farmers by certain language groups for personal use, sentimental reasons, and use in certain ceremonies and rituals.

12. THE GENETIC TRANSFORMATION OF AFRICAN RICE

Rice research began in west Africa in the 1930s with the setting up of IRAT by the French in 1960 in Francophone Africa, and also rice research stations at Rokupr (Sierra Leone) and Baddegi (Nigeria) (Charbolin 1977) by the British. The work overwhelmingly emphasized Asian rice, mainly because it was soon realized that African rice was less responsive than Asian rice to various intensive agronomic practices. Its other innate deficiencies were also cited as reasons – easy grain shedding, difficulties in milling, easy lodging, etc. (Table 5.5).

Table 5.6 **Mixed Planting/Occurrence of *glaberrima–sativa* Rices in Different Types of Rice Cultivation (Number of Fields)***

Ecosystem	Only *O. glaberrima*	Mostly *O. glaberrima*	Half and Half	Only *O. sativa*	Mostly *O. sativa*	Total
Upland	5	4	2	–	2	13
Rainfed lowland	–	1	1	4	10	16
Deep water	1	5	1	–	1	8

**Source:* Oka and Chang 1963

Table 5.7 Sympatric Occurrence of *Oryza* Species in West Africa*

Species	*O. sativa*	*O. glaberrima*	*O. longistaminata*	*O. barthii*	*O. punctata*	3-Species Mixture	4-Species Mixture
No. of populations studied(Total: 262)	65	64	72	51	10	–	–
O. sativa	16	41	18	9	2	2	3
O. glaberrima	–	9	18	19	2	15	5
O. longistaminata	1	–	36	18	1	9	5
O. barthii	–	6	–	21	4	11	5
O. punctata	–	–	–	–	4	3	–
	–	–	–	–	–	–	–

**Source:* Oka and Chang 1963; Oka et al. 1977.

Rice research received a big fillip in west Africa in September 1969, when 11 countries in the region decided to set up the West African Rice Development Association (WARDA) in Monrovia, Liberia. It came into existence in 1971 (www.AfricaRice.org). In a way, this also led to a revival in the "fortunes" of African rice through its genetic transformation into a new avatar. WARDA has now been rechristened the African Rice Center (ARC/AfricaRice), after it was decided that its mandate should cover all the participating countries in Africa. At present, it covers the entire continent of Africa. ARC is now temporarily headquartered at Cotonou, Benin. Presently, Africa-Rice is one of the 15 research centers supported by the CGIAR. The ARC has assembled all the genetic variability present in west Africa. This consists of 1,130 accessions of *O. glaberrima* and 586 of *O. sativa*. In addition, the IRRI maintains 1,372 accessions of *O. glaberrima*, and ORSTOM France has 926 accessions (www.irri.org).

In the 1990s, WARDA embarked on a systematic and ambitious rice improvement program using *O. glaberrima*–*O. sativa* hybrids. The objective was to combine the favorable attributes of the two rice species. This led to the release of 18 new interspecific hybrid cultivars up to December 2005, suited to upland rainfed conditions. They were named NERICA 1–18. NERICA is an acronym of **Ne**w **Ri**ce for Afri**ca**. In 2008, it was estimated that NERICA rices were being grown over 200,000 ha. The large-scale release and cultivation of the NERICA rices have brought out both the strengths and weaknesses of the first generation cultivars (www.AfricaRice.org).

13. THE ORIGIN OF AFRICAN RICE

This aspect has been reviewed by Nayar (1973, 2010, 2012).

i. Ancestral Species

African rice, *O. glaberrima*, could have been either introduced from outside the region or continent, or evolved locally from one of the indigenous species. Since there are no reports of its introduction from elsewhere, it is generally assumed that African rice originated locally from one of the *Oryza* species present in the region. Five wild species and the Asian rice, *O. sativa*, occur in west Africa (Vaughan 1994, Clayton et al. 2011). The wild species are *O. barthii*, *O. brachyantha*, *O. eichingeri*, *O. longistaminata*, and *O. punctata*. Of these, the three species, *O. brachyantha*, *O. eichingeri*, and *O. punctata* have been ruled out for their remote relationships and their different genomic composition to African rice (*O. brachyantha* is FF, *O. eichingeri* is CC, and *O. punctata* is BB, BBCC, whereas African rice is AA) (Table 5.8).

Table 5.8 Comparative Taxonomy of *O. glaberrima* and Three Most Related Taxa

Character/Author	*O. glaberrima*	*O. sativa*	*O. barthii*	*O. stapfii*
Habit, Height				
Roschevicz (1931)	Annual, 0.5–1.0 m	Annual, up to 1.5 m	Annual, strong, up to 1 m	Annual, 0.30–0.80 m
Tateoka (1962, 1963)	Not given	Not given	Not given	Not given
Clayton et al. (2011)	Annual, 0.9–1.5 m	Annual, 0.45–1.80 m	Annual, cespitose, 0.60–1.20 m	Synonym of *O. barthii*
Leaf				
Roschevicz (1931)	20–40 x 0.5–1.0 cm	50 x 1.5 cm	50 x 0.80 cm	30 x 1.1 cm
Tateoka (1962, 1963)	Not given	Not given	Not given	Not given
Clayton et al. (2011)	20–30 x 1.0–1.5 cm	12–65 x 0.4 –1.8 cm	15–45 x 0.4–1.3 cm	Synonym of *O. barthii*
Auricles				
Roschevicz (1931)	Rounded/acicular hollows with two small falciform horn-like outgrowth ciliated along edge	Two falcate tough cilia along edge, fairly brittle up to 4 cm long	Small, listed	Two falciform horn-like with sparse, stiff ciolia along edge, in only lower leaves
Tateoka (1962, 1963)	Not given	Not given	Not given	Not given
Clayton et al. (2011)		Erect	Erect	Not given
Ligules				
Roschevicz (1931)	Orbicular, 3–4 mm long	Acicular, up to 4 cm membraceous, entire or often 2-loffed 1.5–5.0 cm, rarely shorter	Oblong, very short 3–4 mm	Oblong, up to 5 mm

Tateoka (1962, 1963)	Not given	1.5–5.0 cm, rarely shorter	0.6–1.0 cm, always round tip	Not synonym
Clayton et al. (2011)	Eciliate, membranaceons, 1.5–2.0 mm	Eciliate, membranaceous, 15–30 mm entire or lacerate acute	Eciliate membranaceous, 2–6 mm, brumcate, obluse	Synonym of *O. barthii*
Panicles				
Roschevicz (1931)	Up to 20 cm compressed with almost unbranched braches	Up to 30 cm, compressed, 2–3 times branched, spikelets from base or second third	Erect; compressed branching twice, rarely thrice	Up to 15 cm, compressed, not dense with almost no branching branches
Tateoka (1962, 1963)	Not described	Not described	Not described	Not described
Clayton et al. (2011)	15–25 cm, open, linear	20–50 cm, open, lancolate	20–35 cm, open, obovate	Synonym of *O. barthii*
Spikelets				
Roschevicz (1931)	Oblong-orbicular, 7–8 mm	Orbicular-oblong or narrow oblique 6–8(12) x 4 mm	Obliquily oblong, 10–11 x 3.0–3.5 mm	Oblique-oblong, 8–9 x 3.0 mm, strongly compressed
Tateoka (1962, 1963)	Not described	Mostly >7.0 mm	Not described	Synonym of *O. barthii*
Clayton et al. (2011)	Elliptic or oblong, 7–8 mm	Elliptic or oblong, 8–11 x 2.5–3.5 mm	Oblong, 7–11 x 2.5 – 3.4 mm	Synonym of *O. barthii*

Both the remaining two species, *O. barthii* and *O. longistaminata,* have a wide distribution throughout the middle of Africa. They are more common in tropical west Africa where they often occur in extensive stands. Local people have been collecting their seeds for millennia for consumption, use in rituals, and for sale in local markets. Incidentally, we have noted that since prehistoric times, people in south Sahara and the Sahel have used the seeds of a large number of grass species for direct consumption. Of the two rice species, *O. barthii* has a wider distribution (Vaughan 1994, Clayton et al. 2010). *O. longistaminata* is a tall, vigorous plant with strong rhizomes, characterized by semi-sterility and self-incompatibility (Nayar 1958, 1967, Oka and Morishima 1967). Cytogenetic and molecular biology studies confirm its remoteness from the remainder of the *Oryza* species. Consequently, it is generally agreed that *O. longistaminata* could not have been the immediate ancestor of the African rice, *O. glaberrima* (Table 5.8).

This leaves us with *O. barthii*, but we may also consider Asia rice, *O. sativa*. The general assumption about Asian rice vis-à-vis west Africa is that it was introduced in the early 16th century by the Portuguese colonizers, but there have been suggestions that it was introduced much earlier, in the 8th/12th centuries, or even earlier than that (Nayar 1973, 2010, Carpenter 1978).

A. O. barthii *as an Ancestral Species*

O. barthii is an annual wild species. It is a diploid (2n = 24) species possessing the AA genome. It has the widest distribution in Africa among all the *Oryza* species (Vaughan 1994, Clayton et al. 2010). In some regions, such as the IND and Lake Chad basin, it occurs in extensive stands. It occurs usually in seasonal wetlands, river and lake banks, and as weeds in paddy fields and field bunds (Oka and Chang 1964, Oka et al. 1978, early writings of Barth, Schweinfurth, Chevalier and Porteres, Vaughan, etc.). When it occurs as a weed, the farmers do not weed it out, both because it is not easy to distinguish it from cultivated rice in the vegetative stages of their growth and also because it yields as well as the rice crop. Its cooking and edible properties are also as good. *O. barthii* possesses bold grains, its seed fertility is good, and it yields as much or even more than the African rice on a single plant basis (Nayar 1973). It is a self-fertilizing species like the African rice (Table 5.9). Further, local people have been harvesting its grains since prehistoric times for food, use in rituals and for sale in the market, as we have seen before.

Table 5.9 Characters of *O. sativa, O. glaberrima, O. sativa x O.glaberrima*, and *O. barthii*

Characters	*sativa*†	*glaberrima*†	*sativa–glaberrima*†	*O. barthii*!‡
Ligule length (mm)	11.5–32.5	3.0–4.5	7.0–20.5	2.0–130
Spikelet (length x breadth) (mm)	7.3–8.5 x 2.8–3.6	8.1–8.5 x 3.3–3.6	8.7–9.2 x 3.0–3.3	7.0–11.0 x 2.5–3.5
Spikelet hairiness	Long to med long, dense to moderate	Trace	Long, dense	Glabrous or hispid
Panicle length (cm)	14.9–21.0	18.1–21.7	14.8–22.4	Not given
Plant height (cm)	62.8 (49.9–71.9)	66.8 (57.9–76.5)	71.8 (59.5–80.0)	Up to one meter
No. ear - bearing tillers	6.2 (4–9)	8.0 (5–11)	7.0 (5–11)	Not given

†*Source*: Nayar 2012

Roschevicz (1931) was possibly the first author to suggest *O. barthii* as the progenitor of African rice. Porteres (1945, 1950, 1956, 1962, 1976) and Morishima et al. (1962, 1963) have since supported this proposal. All subsequent rice scientists and other authors have unquestioningly repeated this proposition, possibly because of its simple and straightforward logic – the wide and sympatric distribution of the species, its close morphological similarity with African rice (Morishima et al. 1962, Chang 1976a,b, Harlan et al. 1976, Oka 1978, etc.), the frequent writings of Porteres and the absence of any other candidate species, and the assumption that Asian rice was a recent introduction into the region: in only the 16th century. *O. longistaminata* too occurs nearly as widely and frequently (Fig. 5.7). Its seeds too are harvested, but because of the prevalence of semi-sterility, the yield is less than that of *O. barthii. O. longistaminata* generally occupies more stable habitats than those of *O. barthii* (Oka and Chang 1963). It is genetically well isolated and is predominantly self-incompatible. Incidentally, the attributes of *O. barthii* are of a weedy species (cf., Baker and Stebbins 1965). This has been generally overlooked, mainly because of the absence of any study on this aspect.

a. Morphological Studies

The main differences in the morphological characters of the three rice species, *O. glaberrima, O. sativa,* and *O. barthii* and their hybrids, are in ligule length, panicle branching, spikelet hairiness, dimensions, and life history traits (Tables 5.8 and 5.9). Various authors (Porteres 1950, 1955, Morishima et al. 1962, Oka and Chang 1964, Oka et al. 1978 for instance) have also pointed out the prevalence of parallel variation for all the characters between *O. glaberrima* and *O. sativa*. *O. barthii* is similar to *O. glaberrima* in having short ligules, and it is nearly so in spikelet hairiness, and glabrousness, also. Incidentally, there are some crucial differences among the opinions of the five taxonomists on this (Tables 5.7 and 5.8). Overall, the spikelets appear to be almost glabrous or hispid in *glaberrima* vs. hard cilia or hispid in *O. barthii*. Its panicle branching is intermediate to that of *O. glaberrima* and *O. sativa*. *O. barthii* has bigger and bolder spikelets than both the cultivated species. All the five taxonomy treatments of the genus are in agreement on this. The only consistent morphological difference between *O. glaberrima* and *O. sativa* appears to be in ligule length: 1.5–13 mm in *O. glaberrima* vs. 15–50 mm in *O. sativa* (Tables 5.5, 5.8, and 5.9).

b. Genetic Studies

Hybrids of the three species – the two cultivated species and *O. barthii* – have been studied by several authors (see Nayar 1973 for a review of the early studies). In extensive crossings carried out among them (869 combinations/8,774 pollinations), Chu et al. (1969) obtained 51–62% success in intervarietal crosses with *O. sativa, O. glaberrima,* and *O. barthii*, 40% success in *O. sativa–O. glaberrima* crosses, 35% in *O. sativa–O. barthii* crosses, and 60% success in *O. glaberrima–O. barthii* crosses (Tables 5.8 and 5.9). While *O. sativa* hybridizes with moderate success with both *O. glaberrima* and *O. barthii*, the pollen sustainability in both the cross combinations is usually very low – less than 4%.

Hybrids between *O. glaberrima* and *O. barthii* too have been studied by several workers. The hybrids are highly fertile and F_1 meiosis is largely normal. The F_1 hybrids showed dominance of the wild parent for nine characters, heterosis in four characters, and dominance of the *O. glaberrima* character in just one case (Nayar 1958). The F_2 generation segregated in normal Mendelian fashion (Oka and Chang 1964). Natural populations of *O. barthii* contain plants that are intermediate in character between it and *O. glaberrima*. They are designated as *O. stapfii*. Most authors now consider *O. stapfii* as a synonym of *O. barthii* (WCSP Kew 2013). The seeds of *O. barthii* are bigger and bolder than those of *O. glaberrima* (Tables 5.6–5.9). The 100-grain weight and number of panicle-bearing tillers are also higher in *O. barthii*. These characteristics could be taken as indicative of hybrid vigor.

Hybrids between the two cultivated species show low pollen stainability, near-complete seed sterility, and low embryo sac fertility (Morinaga and Kuriyama 1957, Morishima et al. 1963). In F_1 hybrids, the ligule length is intermediate and spikelet hairiness is similar to that of *O.sativa*. The other characters of the hybrids are intermediate of the two parents. Thus, the morphology and breeding behavior of the hybrid progenies resemble that of *O. barthii* (Tables 5.8 and 5.9).

Sano and co-workers have studied the nature of hybrid sterility in *O. glaberrima* – *O. sativa* crosses in detail (Sano 1986). They are male-sterile and female-fertile. Four genes are responsible for the sterility, two of which are gamete eliminators. The other two are a pollen killer gene and a sporophytic sterility gene (Sano 1978, 1986, 1989, 1990). Although seemingly hybrid plants showing characters of hybrids between *O. glaberrima* and *O. sativa* are frequently encountered in the rice fields of west Africa, natural hybrids between the two cultivated species are stated to be rare in the field (Oka et al. 1978). This apparently contradictory situation is possibly because of two factors: the prevalence of extensive interface between the two cultivated species under diverse field conditions; and the prevailing permissiveness among the farmers about the presence of such hybrid plants in paddy fields.

Artificial hybrids between *O. sativa* and *O. barthii* have been studied by several workers (see Nayar 1973 for review). They are moderately fertile and meiosis is largely normal. The nature and extent of cytogenetic and genetic characteristics shown by these hybrids are no more extensive than those shown by interracial hybrids within *O. sativa*, such as between japonica and indica rices (Morishima et al. 1963, Chu et al. 1969, Nayar 1973).

A comparison of the yield components of *O. barthii* and *O. glaberrima* also brings out certain interesting features Tables 5.8 and 5.9). Spikelet dimensions and 100-grain weight are higher in *O. barthii*. The number of panicle-bearing tillers is variable, but overall, this may be also higher in *O. barthii*. Thus, *O. barthii* is seen to possess superior yield components and higher yield potential than *O. glaberrima*. The larger and bolder spikelets and higher plant vigor of *O. barthii* may be a manifestation of heterosis consequent on the taxon being the product of hybridization between *O. glaberrima* and *O. sativa*. The occurrence of *O. barthii* in disturbed habitats may reflect its weediness and hybridity. Incidentally, Anderson (1953), Stebbins (1950), Baker and Stebbins (1965), and several other authors (Rieseberg 1955) have dealt with this aspect in detail with examples from other genera (Table 5.10).

Table 5.10 Variability in Yield Components of *Oryza barthii* and *O. glaberrima**

Character and Authority	*O. barthii*	*O. glaberrima*
Spikelet size, mm		
Roschevicz (1931)	10.0–11.0 × 3.0–3.5	7.0–8.0 × 2.5–3.0
Sampath (1962)	10.0 (variable) × 3.0	8.0 × 3.5
Morishima et al. (1963)	8.8 × 2.9	8.4 × 3.4
Bardenas and Chang (1966)	9.2–11.0 × 3.1–3.4	7.4–9.0 × 2.9–3.6
100-grain weight, g		
Nayar (1958)	3.0	2.8
Sampath (1962)	3.0	3.2
Morishima et al. (1963)	2.4	2.3
Bardenas and Chang (1966)	2.3–3.4	1.6–2.8
Panicle-bearing tillers, number		
Bardenas and Chang (1966)	20–61	18–34
Number of spikelets per panicle		
Nayar (1958)	37	54
Sampath (1962)	73	125
Morishima et al. (1963)	42	90
Protein percentage		
Bardenas and Chang (1966)	13.4–17.3	9.5–13.8

*Reproduced from Nayar 2012 with permission.

For all the above reasons, it may be difficult to consider *O. barthii* as the ancestral species of the African rice. In fact, *O. barthii* may not even strictly fulfill the criteria of a valid species that is distinct from *O. glaberrima,* since it is neither morphologically discontinuous nor reproductively isolated from the latter. It does not have a habitat different from that of *O. glaberrima* except of course for the vast stands of *O. barthii* that occur in the region. A reproductive isolation, at least a partial one combined with geographic isolation, would have restricted gene exchange between the two taxa, which is a prerequisite for species formation (Stebbins 1950, Levin 2000, Coyne and Orr, 2004). In addition, morphological, genetic, cytogenetic, and ecological characteristics of *O. barthii* suggest that it could be a hybrid derivative of the two cultivated species, *O. sativa* and *O. glaberrima.* The parallelism of *O. barthii* with the Spontanea rices of south and southeast Asia appears to be obvious from its nature, distribution, ecology, and genetics.

B. **O. sativa** *as the Progenitor of African Rice*

a. Historical and Archaeological Aspects

The general assumption is that that Asian rice was introduced into west Africa by the Europeans after their arrival there in the mid-15th century. Alpern (2008) has even gone to the extent of stating that the first Asian rice was introduced into Sierra Leone only during 1574–1625 CE. We have, however, seen that much historical and archaeological evidence are available for the presence of Asian rice in the region from prehistoric to early historic times in northern (Egypt) and eastern Africa (Madagascar). While the Asian rice recovered from Berenike (southeast Egypt) is dated to the 1st century CE, the incontestable period when rice was brought from southeast/south Asia by the Indonesian migrants to Madagascar is 5th century CE (Verin 1981). Blench (2007) has proposed an even earlier date of 5th century BCE, albeit indirectly. There is much historical, biological, and linguistic evidence of people from southeast Asia traveling to and settling in Madagascar and east Africa from late prehistoric to early historical times.

In addition, there are also numerous accounts of Arab travelers and traders and also Berbers (the indigenous people of north Africa) traveling across the Sahara from prehistoric times, both from the north and northeast of Africa (Fig. 2.3). The historical evidence indicates that these travels may have started even as early as ca. 1,500 BCE and continued up to 16th century CE (Wikipedia 2013). Their purpose was business by barter and exchange. From west Africa, they carried gold, ivory, and slaves, mainly from the north and northeast of the region. The Berbers brought with them salt, cloth, beads, and metal goods from the north. The volume of trade increased substantially from the 8th century CE, after the Arabs and north Africans were converted to Islam. The introduction of camels into northern Africa in the 3rd century CE gave a great boost to trans-Saharan trade. A camel caravan used to consist of anything from 1,000 to 11,000 camels. This trade tapered off in the 16th century as Europeans became established in west Africa (Wikipedia 2013).

The extensive influence of external human contacts shown by the Ibo-Ukwa excavations in the southeast Nigeria forest belt (Insoll and Shaw 1977, Sutton 2001) might have opened up new possibilities for plant introduction into the region from Asia. Shah first excavated this site in 1959/60. The findings consisted of exquisitely made bronzes, ceramic pots, several thousand glass and carnelian beads, and human skeleton parts – indicating that this was a funerary site. There is now agreement on the date of the site as 7th century CE, the source of much of the material (pots and metals needed for bronze casting) within 100 km, and the route taken for introducing the beads (the Nilotic route south of the Sahara from the northeast). The source of the beads has been variously suggested to be Rome, the Middle East, and even the Orient. But, there is agreement that the beads originated from far outside the region (Sutton 2001).

It is against this background that Nayar (1973) suggested that Asian rice was possibly introduced into tropical west Africa before the 10th century CE, either from the north across the Sahara or the south (meaning Madagascar, as the country has been mentioned by the author in the context of Asian rice introduction). Carpenter (1978) too proposed that Asian rice might have been introduced into west Africa from east Africa across the tropical African forests, because of the extensive connections that the region (including Zanzibar and Pemba islands) has had with the Arabian, Iranian, and Indian coasts since prehistoric times, as documented in *Periplus* and other historical documents (Schoff 1912, Sheriff 1981) and the consequent immigration, diffusion, invention, and admixture that followed (Verin 1981, Kusumba and Kusumba 2006, and so on).

For the first time, Nayar (1973) suggested that *O. glaberrima* might have arisen from Asian rice, *O. sativa*. This proposition received partial support from Second (1986). No one else has commented on this proposal.

Second (1986) studied the relationship between *O. sativa, O. glaberrima,* and *O. barthii* (Second's *O. breviligulata*) at 40 loci using electrophoresis and multivariate analysis. The results showed an approximately equal genetic distance between *O. glaberrima* and the "ancestral" indica and japonica rices of *O. sativa,* and also that some weedy strains of *O. barthii* contained two alleles that are found frequently in *O. sativa* but are not present in either wild *O. barthii* or *O. glaberrima*. The author stated that this partially supported Nayar (1973)'s proposition regardng the hybrid origin of *O. barthii*.

b. Biological Aspects

The frequent, unpredictable, and often drastic climatic changes that have taken place in the Sudan and Sahel regions (Table 5.2) could have induced genetic mutations in *O. sativa,* or alternately, such regimes might have favored the expression of rare alleles leading to the formation of variant forms in nature. Some of these mutations of rare alleles might have been better adapted to the changing environment, and they would have been selected and retained by the farmers for further multiplication and cultivation. Such mutations would have been for characters that imparted increased resistance to one or more biotic and/or abiotic factors (such as increased tolerance to heat), better flood resistance and floating ability, rapid and greater tillering in the initial stages (this imparts improved competitiveness with weeds), and so on. These are the three physiological attributes present in *O. glaberrima* vis-à-vis *O. sativa* that give African rice an advantage in the prevailing traditional rice farming systems of the Sudan and Sahel regions.

The protection initially provided by farmers would have ensured the equivalent of the initial geographical isolation to the mutant forms that is necessary to provide initial protection from gene flow. This is required to maintain the genetic integrity of the selected forms. Those possessing less distinctive physiological and morphological changes would have remained unnoticed and unselected, or possibly, tolerated in the rice fields by the farmers. Rice farmers in west Africa are known to accept mixtures and heterogeneity in their rice fields. Carpenter (1978) had termed the "apparent newness and its primitive state" as the most notable feature of west African rice culture. The observations made by Oka and co-workers (Oka and Chang 1963, Oka et al. 1978) in their expedition reports lend further support to Carpenter's (1977) observation about the "incipient status" of *O. glaberrima*.

Farmers in west Africa accept mixed and heterogeneous populations in their rice fields. They also deliberately mix seeds of the two cultivated rices for planting (Oka and Chang 1964, Oka et al. 1978, Ghesquire et al. 1997, Semon et al. 2005). The latter two authors have stated that crossings do take place between the two species, but some early workers have noted noted that such natural crossings take place only rarely under field conditions, as there are strong barriers between the species. Incidentally, the multi-line concept that used to be advanced in the USA some decades ago to reduce the losses from rusts in wheat crops may be analogous to this prevailing age-old practice in west Africa. A novel form of the above might have then developed gradually into a neospecies, and it came to be designated as *O. glaberrima* in the course of time. The evolution of *O. glaberrima* from *O. sativa* may therefore be an instance of sympatric speciation. This aspect will be discussed in detail later.

There is much evidence supporting the proposition that *O. glaberrima* arose from *O. sativa*, including evidence from history, morphology, genetics, anthropology, and molecular biology. We shall discuss them below.

c. Morphological Evidence

The first piece of evidence is the repeated observations of Porteres (1945, 1950, 1956), Oka (1977 and other papers), and the descriptions given by various taxonomists (Roschevicz 1931, Chevalier 1932, Vaughan 1994, Clayton et al. 2000, etc.) that the two species show close morphological similarities (Table 5.8) and that it may only be possible to separate them on the basis of a combination of characters, such as short ligule length, glabrousness of spikelets and leaves, and reduced sub-branching of panicles. In addition, Porteres (1956) and others have pointed out that both the rice species show close parallelism in all their characteristics. Porteres has even identified *O. glaberrima* forms which are comparable to the indica and japonica ecotypes of *O. sativa* (Table 5.8).

The second piece of evidence is a report of an invalidly described species, *O. jeyporensis* collected from the Jeypore region (Orissa, southeast India) (Govindaswamy and Krishnamurthy 1958). The authors had observed that this taxon most closely resembled *O. glaberrima* morphologically. Sampath (1962), who examined the specimen, stated that it most closely resembled *O. stapfii*. WCSP (RBG Kew) (2011) has identified *O. jeyporensis* as a synonym of *O. rufipogon,* and *O. stapfii* of *O. glaberrima*. These opinions (Sampath (1962) and WCSP (2011)) are pointers to the close morphological similarity between the two cultivated species.

The third observation is the finding of *O. glaberrima*-like characters in the *O. sativa* germplasm collection from Assam (Seetharaman and Ghoral 1976). Both Jeypore (in southeast India) and Assam (northeastern India) are considered as centers of Asian rice diversity. Both teams (Govindaswamy and Krishnamurthy and Seetharaman and Ghosal), were engaged in rice germplasm survey and collection in the two regions.

d. Rarity of *O. glaberrima*

While preparing his *Species Plantarum* (1753), Linnaeus had included only one species in the genus *Oryza*, viz., *sativa*, the Asian rice. Then Steudel described African rice, *O. glaberrima* in 1858, after a lapse of 105 years. As stated before, Steudel used the material collected by Edelstan Jardin during 1845–1848, from the coast of Portuguese Guinea (now Guinea-Bissau), which Jardin had identified and labeled as *O. sativa*. According to Porteres (1955), Laprieur, a French pharmacist plant-collecting near Dakar (Senegal) during 1824–1829, was the first to collect *O. glaberrima*. One of his herbarium sheets available in the National Natural History Museum, Paris and labeled as *O. sativa* contains two specimens. One of this is *O. sativa* and the other is *O. glaberrima*. Porteres (1955) stated that this was thus the first collection of *O. glaberrima*. The third collection of *O. glaberrima* was made only after another 50 years. This was collected by Chevalier in 1901 from upper

Senegal-Niger, from the banks of the Casamance River. These events clearly point to the rarity of *O. glaberrima* and its close morphological similarity with *O. sativa*. Otherwise, both the earliest collectors of *O. glaberrima*, Jardin and Laprier, could not have misidentified their speciments as *O. sativa*.

The description of African rice as a new species in only 1858 is a comparatively late event to happen for a species that has been stated as the cultivated rice of west Africa since prehistoric times. Further, even this happened only more than 10 years after Jardin had deposited the herbarium sheet in the National Natural History Museum (NNHS), Paris under the name *O. sativa*. It is obvious that this situation could not have arisen if *O. glaberrima* had indeed been at least the dominant cultivated rice of west Africa since prehistoric times.

e. Cytogenetic Evidence

Numerous cytogenetic studies of *O. sativa–O. glaberrima* hybrids beginning with Ramanujam (1938) have shown the close affinity of the two cultivated rices (cf. Nayar 1973 for review). Hybrids between the two cultivated species are obtained without any difficulty. The rate of crossing success was no lower than that of the indica–japonica crosses of *O. sativa* (Morishima et al. 1962). In very extensive crossings, Chu et al. (1969) obtained 51% success in intervarietal crosses within *O. sativa*, 62% in *O. glaberrima*, and 58% success in *O. barthii* (Table 5.10). In interspecific crosses, the success rate was 42/39% in *O. sativa–O. glaberrima* crosses, 38/32% in *O. sativa–O. barthii* crosses and 59/62% within *O. glaberrima–O. barthii* crosses (the alternate figures are for reciprocal crosses). The F_1 hybrid plants of *O. sativa–O. glaberrima* grow normally and show hybrid vigor. They are usually male-sterile, but show some female-fertility also (Sano 1986, Heuer and Meizan 2003). The nature of this sterility is similar to, but of a higher magnitude, than that shown in inter-ecotype crosses within *O. sativa* (Hu 1960) (Table 5.11).

Meiosis in hybrids among the three species is nearly normal according to several reports (Nayar 1973 for review). The frequency and range of abnormalities observed in these interspecific crosses is no more than, or any different from, those found in inter-ecotype crosses of *O. sativa* (Chu et al. 1969). The cytology of the amphidiploids of *O. glaberrima–O. sativa* and *O. sativa–O. barthii* point out comparative affinities among them (Gopalakrishnan et al. 1964). While the F_1 hybrids were sterile, their amphidiploids gave 75% pollen stainability and 47% seed set in *O. glaberima–O. sativa* amphidiploids and 13% seed set in *O. sativa–O. barthii* amphidiploids.

Table 5.11 Pollen Fertility and Seed Setting (in percent) in Different Cross Combinations†

	O. sativa	*O. glaberrima*	*O. barthii*	*O. longistaminata*
O. sativa	74.7/1 – 99% 815/51%	– /42%	–/38%	–
O. glaberrima	0.4/0 – 3% 238/39%	95.8/43 – 99% 145/62%	–/59%	–
O. barthii	0.4 /0 – 4% 68/32%	95.6 /22 – 99% 219/62%	95.8/50 –/99% 79/58%	–
O. longista-minata	25.1/0 – 84% –/13%	29.3/0 – 49% –/4%	11.0/0 – 25% –/5%	63.3/25 – 85% –/3%

First row: Mean/range of pollen fertility. Second row: No. of cross combinations tested/seed set in per cent
†Adapted from Chu et al. (1969)

Ohmido and Fukui (1995) have characterized the chromosome complements of *O. glaberrima* and *O. sativa* using karyotype analysis, molecular cytology, and McFish imaging. The broad features of the prometaphase chromosomes were similar in both species. The chromosome length and arm ratios were identical in 9/12 chromosomes of the two rice species. Even the remaining three chromosomes showed only slight differences (1% significance): relative length in chromosome 1 (12.5 ± 0.8 vs. 13.6 ± 1.1) and arm ratios in chromosomes 4 and 5 (3.23 ± 0.24 vs. 2.93 ± 0.29 and 1.50 ± 0.13 vs. 1.24 ± 0.13). The faint unstable small condensation (FUSC) regions present in chromosomes 6 and 9 of *O. sativa* were seen also in *O. glaberrima*. Further, the three heavily condensed regions observed in chromosome 11 of *O. sativa* were present also in *O. glaberrima*. Similar was the case with the 5s locus recombinant DNA (rDNA) of chromosome 9 and *4s* rDNA locus in chromosome 11 (Ohmido and Fukui 1995). All the above observations indicate a close cytogenetic relationship between the two cultivated rices. This would be difficult – if not impossible – if the two rices had originated from different ancestral species (say, *O. barthii* and *O. rufipogon,* respectively) as is widely assumed, originated independently in two continents (west Africa and Asia), and evolved at two different points in time (3 millenia / 7 millenia BP), as some authors have stated.

f. Evidence from Molecular Biology

Additional evidence about the very close relationships between the two cultivated rice species can be deduced from molecular biology studies. Some of the studies cited here has been originally conducted with other objectives in view.

Lorieux et al. (1999) prepared the first *O. sativa–O. glaberrima* microsatellite based linkage map. It comprised 129 markers, representing 112 discrete loci in a total map length of 1923 cM. Its genome coverage was about the same as that of the two microsatellite maps of *O.sativa* published earlier by Akagi et al. (1996) and Chen et al. (1997), with which it was compared, except for some differences in chromosomes 4, 7, and 10. The authors obtained very good colinearity between the two species except for some insignificant markers and inversions.

Several studies have been carried out on the evolutionary relationship among the AA genome species using different techniques of molecular biology. For instance, Cheng et al. (2002) using *SINE* (short interspersed nuclear element) insertion analysis and Park et al. (2003) using *MITE-AFLP* (miniature inverted repeat transposable element–amplified fragment length polymorphism) found that *O. glaberrima* exhibited the least intraspecific variability of all the AA species. Further, these studies also showed the close sister-brother or parent–progeny relationships between the two rice species. Similar results have also been reported by numerous other authors.

A study of the recent sequence evolution of indica and japonica rices (Ma and Bennetzen 2004) has provided unintended support for the very close relationship between the two rice species. The authors had used *O. glaberrima* as the reference material. They employed 263 primer pairs and successfully PCR (polymerase chain reaction)–amplified 196 "single-band" products. They were then sequenced and BLAST (basic local alignment search tool)–searched. They found that as many as 168 of 196 (86%) of them were appropriately aligned to the indel regions of both indica and japonica *sativa,* which indicated that they were orthologous. *O. glaberrima* was equally distant from both indica and japonica. The authors then observed that their data "at the large-scale genomic sequence level strongly support the previous observations" on the prevalence of sister relationships between the two cultivated rice species. These observations lend further support to the proposition that *O. glaberrima* has a close relationship with *O. sativa.*

Another possibility is that *O. barthii* is an outgroup to both *O. glaberrima* and *O. sativa.* It is crossable with both the species. But, if *O. glaberrima* was under natural or artificial selection for sterility barriers with *O. sativa,* the observed sterility could only be a specific genetic incompatibility between the two species regardless of their phylogenetic relationships. The extensive studies on the nature of sterility between *O. glaberrima* and *O. sativa* may not support this latter proposition.

Semon et al. (2005) studied the prevalence of linkage disequilibrium (LD) in 198 accessions of *O. glaberrima* using 93 nuclear microsatellite markers, with *O. sativa* as the control. LD is the non-random distribution into the gametes of a population of the alleles of genes that reside on the same chromosome. The non-random distribution may result either from linkage of the loci in question, or because the loci interact with respect to their effects on fitness. This linkage may then generate disequilibrium (King et al. 2006).

The authors obtained abundant SSR (single sequence repeat) diversity (9.4 alleles per locus among 93 SSRs) with a mean PIC (polymorphic index content) value of 0.34, gene diversity (He) value of 0.27, and allele size range of 67–388 bp (base pairs). They detected significantly elevated levels of LD even among distantly located markers. The authors felt that this was not caused by physical linkage barriers, but by other factors, such as population structure. The sample (198 accessions) representing 12 countries in west Africa grouped themselves into five genetically distinct groups. Two of the subgroups clustered with *O. sativa* (indica and japonica), which indicated to the authors that some *O. glaberrima* accessions were mixtures. The other three subpopulations having specific combinations of phenotype traits appeared to reflect specific ecological adaptations. Of the 198 accessions, 88 (44%) were genetic admixtures with varying levels of ancestry shared among the five groups. Three of the groups shared 30–49% ancestry with indica-like subpopulations. Another 68 (34%) accessions shared at least 1% ancestry with *O. sativa*, and 11 accessions shared more than 52% ancestry with *O. sativa*. The authors attributed this apparently anomalous situation to widespread natural hybridization between the two rice species. The authors observed that the high LD levels in *O. glaberrima* "are primarily caused by population structure driven by introgression with *O. sativa* and the differentiation of ecotypes adapted to different growing environments in west Africa".

However, several other studies do not seem to support this assumption of widespread introgression.

They have shown that natural crossings between the two cultivated species take place only rarely in the field (Oka et al. 1978, Sano 1986, Nuijten et al. 2009). We have noted this before. The subpopulations or cultivars of *O. glaberrima* that showed various combinations and extent of phenotype traits of *O. sativa* could be explained as a consequence of mutations taking place in *O. sativa* favoring different ecological niches of cultivation. The admixture of "varying levels of ancestry" could therefore be a reflection of different stages in the evolution of *O. glaberrima* from its putative ancestor. Hence, this study may be taken to support the close biological "mother–daughter" relationship between the two species.

There have been a couple of other molecular biology studies aimed at determining the relationship between the cultivated rices. Ishii et al. (1993) studied chloroplast, mitochondrial and nuclear DNA differentiation, employing RFLP in eight *O. sativa* and two *O. glaberrima* accessions. Chloroplast DNA was digested with three (*Eco*R1, *Hind* III, and *Pst* I) and mitochondria DNA with five (*Bam* HI, *Hind* III, *Pst* I, *Pvu* II, and *Xho* I) restriction endonucleases. The authors found different degrees of nuclear genome variation in *O. sativa* and only a little variation in *O. glaberrima*. The average genetic distances between *O. sativa* and *O. glaberrima* were 0.477 (chloroplast DNA), 0.898 (mitochondrial DNA), and 0.961 (nuclear DNA) (see also, Zuccolo et al. 2010).

Sakai et al. (2011) have presented the genomic sequence of the African rice. Using this information, they compared its gene sequences with about 25% of that of *O. sativa* (japonica). The patterns of amino acid changes did not differ between the two species in terms of biochemical properties, *O. glaberrima* genes showed larger synonymous–non-synonymous substitution ratios, which indicated that *O. glaberrima* has undergone a "genome-wide relaxation of purifying selection". These studies also indicate a close genetic relationship and similarity between the two species (also, Lie et al. 2011).

The foregoing account leads us to conclude that the African rice, *O. glaberrima* arose from the Asian rice, *O. sativa*. Initially, a sterility barrier might have arisen in one or few plants of *O. sativa*, consequent on an abrupt, drastic environmental event such as flooding, sudden temperature variation, etc. The farmer would have selected the stray plants that survived such an environmental catastrophe, and nurtured and multiplied them in isolation. This would have ensured the initial protection needed for further developing isolating mechanisms under sympatric conditions. This could have then led to the strengthening of sterility barriers in the selected material, leading to the further development of a new population that is now designated as *O. glaberrima*. Such an event could have taken place under different environmental conditions such as sudden flooding, excessive weed competition, etc. and at different locations. We have prevously noted (Table 5.2) that sub-Sahara has been very prone to drastic climatic changes from early times. This may explain why there are not one or two single characters which can separate *O. glaberrima* from *O. sativa* morphologically.

However, under field conditions, sympatric speciation happens only rarely (Coyne and Orr 2004). The initial isolation and protection provided by the farmer might have been an enabling factor for the differentiation of *O. glaberrima* from *O. sativa*. We shall briefly discuss this aspect again later.

ii. Speciation in *O. glaberrima*

We have indicated that African rice, *O. glaberrima,* was formed from *O. sativa* by sympatric speciation.

Sympatric speciation is "an uncommon process by which populations inhabiting at least in part the same geographic range become reproductively isolated" (King et al. 2006). Mayr (1963) diagnosed the weaknesses of this process, and Futuyma and Mayer (1980) observed that sympatric speciation is caused by an initial restriction of gene flow caused by some biological features of the organism. Coyne and Orr (2004) observed that sympatric speciation could take place if the following four conditions are fulfilled:

(i) The two species must be largely or completely sympatric;
(ii) The species must be substantially isolated reproductively;
(iii) The sympatric taxa must be sister groups;
(iv) The biogeographic and evolutionary history of the two groups must make the existence of an allopatric phase very unlikely.

O. glaberrima fulfills all four of these conditions in west Africa with respect to *O. sativa.*

14. TIME AND PLACE OF ORIGIN

The French botanist, Roland Porteres, has been the only scientist to state anything about this aspect and to also study the biology of African rice. His studies spanned nearly four decades, and he covered almost all aspects of the diversity and origin of *O. glaberrima.* For instance, in *O. glaberrima,* he identified 13 botanical varieties, 41 botanical forms, and 1,500 landraces. He also mentioned about the very close parallelism in the diversity patterns of the two species, *O. glaberrima* and *O. sativa.* It is, however, only his work on the origin of African rice that has received any attention.

Porteres' (1962) proposals regarding the time and place of origin and on the ancestral species are quoted very widely in the literature, possibly because his was the only contribution on this topic. He proposed that African rice arose from the annual wild rice occurring in west Africa, *O. barthii* (his *O. breviligulata*), and that this happened 3,500 years ago in the inland Niger flood plains (here termed IND, Inland Niger Delta). He identified two secondary diversity centers for *O. glaberrima,* one in the upper highlands of Guinea and the other on either side of Gambia river, between the Casamance River in the south and the Sine River in the north (both in Senegal) (Fig. 5.4). He based his conclusions on Vavilov's theories regarding the centers of origin and diversity of crop plants. Porteres also advanced some linguistic evidences about the antiquity of the Senegambian megaliths, which he had linked the time of origin of African rice. The antiquity of the Senegambian megaliths, according to him, was 1,500 BCE. Porteres (1962) stated also that rice in the secondary center of diversity in Senegal was more highly evolved than the rices in IND. Senegal rices possess white caryopsis, non-floating character, non-shedding of spikelets, and 'supple' and finer panicles. Traditionally, *O. glaberrima* has a red pericarp, and most varieties have a floating habit. The situation in the Guinea highlands was somewhat intermediate between that in IND and the Senegal coast, he observed.

Porteres further proposed that Asian rice was first introduced into west Africa in two locations: between the River Cacheo and the River Casamance (both in Senegal), and between Conakry (Guinea) and Grand Bassa (Liberia). It was introduced into Madagascar in the 15th century and the names of several landraces of Madagascar were similar to south Indian rice landrace names, he stated.

Porteres (1962) further argued that the regions where the Mande language group names for rice (such as *malo, maro, mano*, etc.) were used indicated the antiquity of rice cultivation in the region, as the Mande people have lived there for a long time. Elsewhere, he stated that the local names for rice were derived from other non-African languages. Porteres postulated further that agriculture in Africa originated 3,500–2,800 years ago in three regions:

(i) in Egypt as an extension of Middle East agriculture (wheat, barley);
(ii) in Ethiopia (based on fonio (*Digitaria exilis*), tef (*Eragrostis lef*), and iburu (*Digitaria ibura*); and
(iii) in west Africa (based on cereals, mainly African rice and sorghum).

The paper that Porteres (1962) presented at the Third Conference on African History and Archaeology (1962) received much adverse comments in the conference itself (cf. Gray 1962). Tucker pointed out that it was not correct to make philological deductions without making comparative philological studies. Clark noted that the archaeological evidence did not support Porteres' contentions, since even 2,000 years ago, Neolithic people had occupied only a few parts of the sub-Sahara, and it was the hunter-gatherers who occupied most of the region. Wrigley proposed that agriculture, including several crop plants of west Africa, could have diffused from east Africa.

Porteres' (1962) postulation about the time of origin of African rice (3,500–2,800 BCE), linking it with the presumed age of Senegambian megaliths, has also attracted much criticism. He mentioned this first in 1950, and continued to reiterate it in his numerous publications up to his 1976 paper (Porteres 1976). Even at the Third Conference (cf. Gray 1962), Mauny observed the inappropriateness of dating the origin of rice with reference to the Senegambian megalithic structures, as according to Mauny, they had been erected by the Basseri group, who were non-rice eating people.

The Senegambian stone circles are now an UNESCO heritage site (Fig. 5.6). They consist of circles of variable diameter made up of 10–26 laterite stones that are up to 2.5 m in height and 10 tons each in weight. There are more than 1,000 such monuments found broadly in four groups located

Figure 5.6 ***Senegambian megaliths, two views (http://en.wikipedia.org/wiki/Senegambian stone circles. Accessed 12 December 2013.***

along about 300 km of the Gambia river and width of 100–150 km north–south. They are found (between the Gambia river in the south and Saloum river) in central Senegal and northern Gambia (Fig. 5.4). Several of these circles are built over human graves, but no one has been able to advance any link between the two. They have been dated to approximately the 3rd century BCE to the 16th century CE. "It presents a vast sacred landscape and reflects a prosperous, highly organized, lasting society", according to the UNESCO (www.unesco.com).

Some archaeologists have carbon-dated artifacts recovered from the site. They date to 750 + 110 CE from Wassu (Gambia) by Fagan (1969), and 200 BCE – 800 CE from Tiekene-Bossessoura by Posnansky and McIntosh (1976). Porteres' observations were only of a general nature and he had not related them to any specific experimental material or context of the Senegambian megaliths. To determine the age of the structure, Hill (1978) had studied pottery recovered from the megalith site. He observed that the Porteres' inferences about the pottery obtained from the region were far from correct, and so were his observations on the civilizational aspect of the Senegambian megaliths. Hill, who was very critical of Porteres' observations, concluded that "there is little beyond geographical propinquity to support it" (Porteres contention) (Hill 1978).

It seems that in the past rice biologists have either ignored all these contrary findings, or were oblivious of the above given archaeological developments, and continued – and still continue – to repeat the age of 3,500 BCE as indicate the time of origin of African rice (Harlan, Chang, Sweeny, Nesbit, NAS 1996, and so on). As early as four decades ago, Nayar (1973) had highlighted the criticisms surrounding Porteres' contentions about the time of domestication of African rice and those that he had faced in the Third Conference on African History and Archaeology (1962).

Of the three archaeological finds of rice from west Africa, one is of wild rice and the other two were claimed to be of cultivated rice, but the veracity of their identification is suspect, and has been questioned. Their antiquity also only takes the age of African rise to about 100 BCE at the earliest, and nowhere near the 1,500 BCE claimed by Porteres (1962). We have also discussed at length the antiquity of the Senegambian megaliths, which ranges only from the 3rd century BCE to 16th century CE, according to UNESCO. But no association has been established by anyone between the Senegambian megaliths and African rice origin. We have also discussed the linguistic and pottery evidences advanced by Porteres. These too are suspect, and their age is also nowhere near the dates (1,500 BCE) proposed by Porteres.

It is in this context that Nayar (1973) and Carpenter (1976) had proposed that Asian rice might have been introduced into west Africa during ca. 8th/11th centuries CE, and that African rice might have evolved from Asian rice as a consequence of mutational changes brought about by some drastic environmental events. Direct experimental evidence for this is not available, but the author (Nayar 1973, 2012) has advanced considerable evidence in support of this from various angles – archaeological, botanical, cytogenetical, and from molecular biology. Unless evidence to the contrary is advanced, we may accept this proposition, even if tentatively.

CHAPTER 6

The Origin of Asian Rice

1. INTRODUCTION

The Asian rice, *Oryza sativa* L., is the rice that is cultivated universally. It is the most important food crop of humankind. Even in tropical west Africa, the home of the African rice *O. glaberrima*. Asian rice has almost completely replaced it. Rice provides more energy to humans worldwide than any other crop plant, including wheat and maize. It is grown in 3.1% of the total agricultural land (covering 159.4 Mha area), and in 2010, its production was 696.3 Mt (FAOSTAT, downloaded 10.01.2012). The area under cultivation, production, and yield of rice is steadily increasing across the world (Tables 1.1 and 1.2).

2. PAST STUDIES

Because of its antiquity, especially in east, southeast, and south Asia, rice finds mention in most of the ancient scriptures and writings of the region. It also forms an integral part of the cultural traditions in most of the Asian societies. Readers can get partial accounts of this aspect from Kumar (1988) and from some of the chapters in Sharma (2010). Here, we shall not go further into this aspect.

Considerations on the origin of rice began in the latter half of the nineteenth century even before Mendel's laws were rediscovered in 1900 (de Condolle 1886, Watt 1892, etc.). The early propositions were based on oral traditions, linguistics, history, and field observations. They are summarized in Nayar (1973). Studies of the origins of crop plants were much stimulated and supported by the genetic experiments and field ecological studies conducted during the first half of the twentieth century.

The session on rice in the international symposium on crop plants origins in 1951 organized by the Indian Society for Genetics and Plant Breeding in Delhi provided the first collective opportunity to take stock of the work on the origin of rice up to that period. All of the three papers on rice presented in the Symposium (Chatterjee 1951, Ramiah and Ghose 1951, Sampath and Rao 1951) were followed by lively discussions (Anderson 1951, for instance). They helped to get a proper perspective about the then prevailing state of knowledge on the subject, and also identify gaps in knowledge.

Origins and Phylogeny of Rices
http://dx.doi.org/10.1016/B978-0-12-417177-0.00006-1

Ramiah and Ghose (1951) supported Watt's (1892) theory that the Asian rice originated in peninsular India. This was based on linguistic grounds and widespread presence of wild rices in the region. They supported also Watt's contention that Asian rice evolved from the annual wild rices, variously called fatua, spontanea, etc. The authors further suggested that the perennial wild rice species might have contributed to the development of some cultivated varieties. Chatterjee (1951) slightly modified his earlier stand (Chatterjee 1947, 1948), and felt that Asian rice might have originated anywhere in east or south Asia. He proposed a polyphyletic origin for the Asian rice with var. fatua (annual *rufipogon* as per his description) and *O. officinalis* as the putative ancestor.

Sampath and Rao (1951) suggested that the wild rices present throughout the tropics are the sole progenitor of both the Asian and African rices – which at that time were known as *O. rufipogon sensu lato* (these included what was then known as *balunga* in southeast India, now identified as *O. rufipogon*), *O. barthii* [since renamed *O. longistaminata*] of tropical Africa, and *O. cubensis* [*O. rufipogon*] in tropical America). This was because this taxon had the widest distribution of all the *Oryza* species, it had the same chromosome number as the two cultivated rice species (2n = 24), and it gave fertile hybrids with *O. sativa*. They further observed that some of the segregants from the perennial *rufipogon–sativa* crosses closely resembled the annual wild rices, which the authors termed spontanea rices. They stated that most spontaneas were heterozygous in nature and that "such a complex and artificial group as spontanea paddies are not likely to be a species ancestral to *O. sativa*."

Rice research received a tremendous boost when the International Rice Research Institute (IRRI) was established in Los Banos, The Philippines. The institute was dedicated on 7 February 1962, with initial funding from the Rockefeller and Ford Foundations. The IRRI arranged a symposium on rice genetics and cytogenetics in 1963, which brought together most of the scientists then working in the areas of rice genetics, cytogenetics, and taxonomy. The papers presented at the meeting and the discussions that followed went a long way toward clarifying the situation on various aspects of rice evolution and identify topics requiring further studies (IRRI 1964).

At the Symposium, Tateoka's (1964) presentation was noteworthy for clarifying several taxonomic problems regarding the *O. sativa* complex (the species and taxa most closely related to the two cultivated rices). He proposed that the material designated in literature variously as *O. sativa* var. fatua, *O. perennis* (Asian, American, African, *balunga* etc.) should all be designated as *O. rufipogon*. He observed that Asiatic forms of *O. rufipogon* were/are generally listed in the literature as *fatua* or *rufipogon* and the American ones as *perennis*, and that both Chevalier (1932) and Chatterjee (1948) had applied the name *O. perennis* to both the African and American taxa. Since these taxa are separated on the basis of spikelet and awn length, presence of rhizomes, and life history traits, Tateoka studied these characters in more than 250 herbarium sheets from important herbaria in the USA and Europe. He found that both the spikelet length and awn characters showed continuous variation in Asian, African, and American samples. For example, the spikelet length varied all the way from 6.75 mm to 11.25 mm and awn length from 2 cm to 16 cm. The mean spikelet length of the Asian samples was smaller than that of American samples and the awns of African samples were shorter than those of American samples. Rhizomes were usually lacking in Asian and American samples, while *O. longistaminata* of Africa had strong, creeping rhizomes. Tateoka (1964) observed that the life history habit (annual/perennial) of the cultivated and annual rices should not be used as a discriminating taxonomic character. In the tropics, most Asian rice varieties are able to live longer than one year, if water is not a delimiting factor, he observed. Thus, even *O. sativa* could not be considered strictly as an annual plant. He stated that "the so-called *spontanea* type and the perennial types found in Asia may hardly be separated at the species level". He presented the opinions of Backer (1946), Bor (1960), and others to support his contentions. He then gave detailed reasoning for why *O. perennis* could not be used as a valid name for the perennial floating wild rices of Asia and Oceania. His argued for rejecting the name *O. perennis* and using the epithet, *O. rufipogon* Grif. instead. This proposal has since been accepted by all the taxonomists.

The next opportunity for taking stock of the understanding of the origins of Asian rice came in 1985 in the session devoted to Systematics and Evolution in the first Rice Genetics Symposium held by the IRRI (IRRI 1986). Six papers on the topic were presented here. The IRRI subsequently organized five rice genetics symposia every five years, but interest in this topic has been steadily waning, and not even one paper on the topic was presented in Rice Genetics V (2007). In the last few years, there has, however, been a spurt of renewed interest in the subject, thanks to the advances being made in molecular and computational biology applications and archaeological studies.

Nayar (1973) has summarized the early literature on the origin and cytogenetics of rice. His review covers most studies carried out up to that period in botany, taxonomy, cytogenetics and genetics, linguistics, and archaeology. Molecular biology studies were then only just beginning to take off. More recently, Oka (1988) published a monograph, *The Origin of Cultivated Rice*, but this was primarily an account of the work done by him and his co-workers. They mainly used biometrical methods in their studies, but they had the big advantage of conducting extensive collection missions in most of the wild rice growing regions of the world. Experimental papers in the areas of molecular genetics had begun to appear by then, but they were not covered by Oka (1988). Archaeology had been just touched upon.

Nayar (1973) had observed that 'the progression' from wild to cultivated rice has involved changes from prostrate, procumbent, or floating habit to an erect and compact one, from shattering character of the grain to a non-shattering habit, from asynchronous tillering/branching and flowering to synchronous habit, an increase in the number and length of panicles, a change from lax to compact panicles, an increase in the weight and/or number of spikelets per panicle, a reduction in awn length, a change in spikelet color from black to brown or golden brown, a change in pericarp color from red to white, a change from partial cross-fertilization to nearly complete self-fertilization, from a distinctly perennial habit to near-obligate annual habit, and from partial to near-complete seed setting". Nayar(1973) stated also that rice had "a diffuse origin in both space and time". He stated also that rice had "a different origin in both space and time".

The changes indicated above are an idealized account. There are areas in the world where deep water and floating rices are grown, which may grow up to 6m or even more in length, and there are traditional varieties that possess long and sometimes stiff awns. Varieties possessing these characters are more often grown in areas which are exposed to various extreme conditions such as flooding, salinity, acidity, bird damage, etc. The bulu and some japonica varieties are also traditionally awned. In addition, there are regions in the world, such as the Malabar-Konkan coast of peninsular India, south-east India, and parts of Bangladesh, where there is a distinct preference for parboiled, red pericarped rice. The total population in this region may exceed 100 million people. It is interesting that all 63 improved cultivars released for cultivation in Kerala state (southwest India) by the state agencies possess a red pericarp, with the exception of just one; Triveni (Nair et al. 2012). In the late 1990s, when the Kerala Agricultural University released this white pericarped, Triveni cultivar, the farmer resistance was so high that the University was compelled to release a red pericarped Triveni type within a couple of years. Hence, it is felt inappropriate to assume awnness and red pericarp as components of the so-called domestication syndrome in rice.

The same observation applies to seed dormancy. In the widely heterogeneous conditions under which rice is grown in the high rainfall monsoon regions of tropical south and southeast Asia, there are low lying areas that are inundated with water during the summer monsoon season. Here, the farmers do not have any alternative to planting medium duration varieties (more than 135–150 days from sowing to harvesting), which will invariably show seed dormancy. The *aman* varieties of southeast India and Bangladesh, and *virippu* of the Malabar coast are examples of such varieties. There are also situations where the crop takes 6–11 months for ripening and harvest, for example, the *rayada* varieties of Bangladesh and adjacent West Bengal (India). Hence, it is also inappropriate to consider loss of dormancy as an advanced character, and dormancy as a primitive character. There are a few other characters that should be treated in a similar manner.

In interspecific crosses involving cultivated rice, *O. sativa* and *O. rufipogon*, the mode of inheritance was of the nature of complementary, modifying, and duplicate genes involving 2–3 genes (Nayar 1958). This is suggested to be a consequence of a phenomenon then described as "block transference" of characters (Harland 1936, Anderson 1949). The recovery of recessive characters in cultivated rices signifies that wild rices possess enough variability stored within them, and hence possess the potential to produce cultivated rices by recombination and segregation. A complementary type of inheritance in interspecific hybrids has been attributed to the pseudo-allelic action of small repeats, and they have been suggested as at a transition stage between duplicate and independent genes (Stephens 1950, 1951).

Various well known authors in the post Second World War period have pointed out the importance of major genes in crop plants evolution; e.g., Hutchinson (1965) in wheat, Mangelsdorf (1965) in maize, Takahashi (1955) in barley. In rice, "the single most desired and effective change would be from the easily shattering nature of spikelets to less shattering and more persistent spikelets. This character is controlled by a pair of complementary genes. A non-shattering nature could be obtained either by recombination, or by mutation at one of the two loci involved. With such a primary gain, selection for the other desirable characters could be taken up in course of time." (Nayar 1973).

We shall revisit this aspect later.

3. LATE PLEISTOCENE–HOLOCENE CLIMATE

The Holocene epoch is the most recent part of the Quaternary Period (from 1.8 Ma to present) of the Cenozoic Era (from 65 Ma to present). It is useful to get a general view of this period (from 0.01 Ma to present (cf., page xvii, Geological Timescale) – about the last 12,000 years) in the region while discussing crop plants origins. This is because all the plant domestications took place in this period. In tropical monsoon Asia, however, the Holocene climate appears to have had only a minor, rather than any crucial, effect.

Compared to west and north Africa, information about the Holocene climate of Asia is scanty. This may be because most of the region that we discuss – east, southeast, and south Asia – enjoys the advantages of a tropical steady monsoon climate, which does not change as unpredictably and drastically as the Saharan climate. The changes that take place in the monsoon are more often only quantitative in nature rather than qualitative.

Wang and Gong (1999) identified four periods within the Holocene epoch in China: the Megathermal, Medieval Warm Period, Little Ice Age, and Modern Warm Period, based on pollen deposits and pollen composition. The Megathermal period covered 8.5 to 3.0 ka BP, with two highs during 7.0 ka BP and 6.0–5.5 ka BP. The average temperatures were then about 2°C higher than the previous 100 years' average. The Medieval Warm Period lasted from the late 9th to 13th centuries CE. The temperatures were then warmer in east China, but 12th century CE was comparatively cold. The Little Ice Age lasted from 1550 to 1850 CE, and the Modern Warm Period covered the 20th century CE. During this period, the average temperatures were 0.5°C higher than the previous 1,200 years' average temperature. It was the warmest century in the last 12 centuries. The authors had observed also that overall, there were five cold spells in the last millennium – in the 12th, 14th, 15th, 17th, and 19th centuries.

Zhou et al. (2005) carbon-dated a peat core in southern China (Dingnan region, across the Taiwan straits) from about 18 ka. Their findings can be summarized as: 15,630–10,450 cal a BP: cool and fairly wet; The Younger Dryas period (12,900–11,600 cal a BP): variable but overall wet; 10,450–6,040 cal a BP: warmest climate and highest precipitation; Holocene Maximum, (6,040 – 3,800 cal a BP): moderately dry and cool; <3800 cal a BP: obscured by agricultural disturbances.

Naidu (1999) has summarized the Holocene climate of south Asia (Indian subcontinent) as follows: 13–9 ka: weak monsoon phase after the last Glacial Maximum, summer monsoon intensifies; 9–6 ka: intense monsoon phase; and 5 ka to present: changes from humid to dry phase; driest at 3.5 ka. He elaborated this further as: 12,000–10,000 BP: major glacial event; intensification of southwest monsoon; Younger Dryas: sea level 90 m lower than current level; 10,000–9,500 BP: rainfall 250 mm/year more than now; 9,000–5,000 BP: relatively high rainfall; 5,000–3,500 BP: moist conditions with good rainfall; 3,500–1,200 BP: dry, less rainfall, no pollen in lake beds; and 1200–present day: present condition.

Saxena et al. (2013) reconstructed the Holocene palaeo-climate of the Lahurdewa lake (26°46 N lat., 82°57 E long., in the eastern central Ganga plain, north-central India) using Poaceae phytoliths as the environmental indicator. In this methodology, dominance of Panicoideae morphotypes indicates warm and humid climate, of Chlorideae indicates a warm, dry climate, and of Festucoideae, a cool, moist climate. The 2.8 m thick sediment covered from 10,425 cal BP period. On the basis of the phytolith index, the authors found five wet periods alternating with four dry phases which occurred during 10,300–9,200 cal BP, 5,300–4,100 cal BP (most severe), 1,650–1,200 cal BP (severe), and 950–700 cal BP. The present annual precipitation at the lake (21 km^2 area) site is ca. 1,200 mm. The wet phases were during 10,600–10,300 cal BP (a strong monsoon after the cool and dry Younger Dryas), 9,200–5,300 cal BP with a peak at 7,000 cal BP), 4,100–1,650 cal BP (with several century-scale humid-arid oscillations) 1,200–950 cal BP (moderate wet phase), and 700 cal BP– present (increased humidity).

As regards southeast Asia, the region enjoys a unique climate. It receives rains almost throughout the year. This region did not appear to have had any climatic upheavals during the Holocene period, except for the sea level changes that took place during the early Holocene (ca. 9,650–5,000 BCE) (Glover and Bellwood 2004, Smith et al. 2011). We shall touch upon this later.

4. ARCHAEOLOGICAL STUDIES

Archaeological remains of rice and rice cultivation have been obtained from three regions: east Asia (southeast and southeast China, Korea, Japan), Indo-China and southeast Asia, and south Asia (India, Sri Lanka). Of which, the ones from China are the oldest and the most extensive. China may have had another advantage from a prehistoric perspective, in that writing was developed early here during the Late Shang dynasty (1,250–1,056 BC) (Liu and Chen 2012). This may explain the availability of several accounts from prehistoric times of many of the activities connected with rice culture and traditions. In south Asia, early archaeological studies from the middle Ganga valley suffered adversely for a time, because of using faulty sampling and dating methodologies. Compared to several other regions of the world, archaeological studies in the rice regions of south and southeast Asia have been fewer and more scattered.

i. East Asia

Rice and millets are the most ancient crop plants of China. China has had a long history and tradition of rice culture. Tang and Xuan (2010) have summarized the prehistory and archaeology of rice (Table 6.1) (also, Liu and Chen 2012).

Rice is mentioned in the document *Guan Zi* of the Shen-Nong Era (Xia dynasty, 41st century BCE), which indicates its importance in the traditions of China. The text mentions the five most important food crops: *shu* (millet, *Panicum miliaceum*), *ji* (millet, *Setaria italica*), *mai* (barley/wheat), *dao* (rice), and *shuu* (soybean). Japonica rice was known as *keng* or *jing*, indica rice as *hsien* and *xia*, and glutinous rice as *nuodao*. It can be inferred also from ancient writings and archaeological findings that wild rice, *O. rufipogon*, was recorded as growing in south and central China (middle and lower Yangtze river basins) since ancient times.

In China, the oldest archaeological evidence is of pottery remains from two sites in north and south China (Yuchanyan and Hutouliang), which appeared during the Pleistocene–Holocene transition period at 18,300–14,000 cal BP. The people were then hunter-foragers (Liu and Chen 2012). The earliest plant remains recovered from the various excavations consisted of acorns, nuts, and seeds of wild rices and grasses.

Fuller (2010) proposed that the first farmers initially cultivated wild rice, approximately during 10,000–8,000 ya, and that fixation of domestication characters was complete by about 6,000 ya (Fig. 6.1).

Fuller (2010, 2011) and his students have recently carried out detailed studies of rice archaeology in both east and south Asia. Based on their studies and also those of several others (cf., Zhao 2010, Lui and Chen 2012), four main centers of rice agriculture have been identified in China – middle Yangtze river basin, Huai river basin, southwest Henoi and Houli culture of Shandong and lower Yangtze river basin. Comparatively, the most detailed studies have been carried out in the lower Yangtze river basin (Fig. 6.1, Table 6.1).

The oldest archaeological evidence of rice in China has been obtained from three sites, all in the lower Yangtze river region: Shangshan, Yiaohuangshan, and Kuahugiao (11,000–7,000 cal BP) (Liu and Chen 2012). The original publications are in Chinese. Many scholars are now agreed that rice was first domesticated in the middle and lower Yangtze river basin and the upper Huai river area (Crawford and Shen 1998, Higham and Lu 1998, Crawford 2006, Fuller et al. 2007).

Tang and Xuan (2010) have listed 16 sites where rice archaeological remains have been found. They cover the periods from 12,300 ± 1,200 BP to 6,300–7,000 BP (Table 6.1, Fig. 6.1). Majority of the sites yielded large

Table 6.1 Major Archaeological Sites of Rice in China*

	Location – Village, Country, Province, Region	Date	Remarks
1.	Yuchanyan, Dao, Hunan, middle Yangtze river	12,300 ± 12,000 BP	Few carbonized grains and husk with mixed indica, japonica, rufipogon characters
2.	Shangshan, Pujiang, Zhejiang, lower Yangtze river	11,400–8,600 BP	Large number of carbonized husk in pottery
3.	Bashidang, Li, Hunan, middle Yangtze river	9,000–8,000 BP	Hundreds of carbonized grains
4.	Jiahu, Wuyang, Henan, upper Huai river	8,258 ± 7,450 BP	Hundreds of carbonized grains of indica and japonica
5.	Kuahujiao, Xiaoshan, Zhejiang, lower Yangtze river	8,220–7,660 BP	Thousands of carbonized grains and husk
6.	Pengtoushan, Li, Hunan, middle Yangtze river	8,200–7,450 BP	Bits of carbonized grains and husk on roughest pottery
7.	Chengbeixi, Zhicheng, Hubei, middle Yangtze river Zhichengbei Zhicheng, Hubei, middle Yangtze river	8,000–7,000 BP	Few carbonized husk on pottery
8.	Erjian, Lianyungang, Jeangsu, middle Huai river	7,885 ± 480 BP	Few carbonized grains
9.	Lijia, Xixiang, Shaanxi, middle Yangtze river	ca. 7,600 BP	Husk pieces on burned soil
10.	Luojiajiao, Tong uang, Zhejiang, lower Yangtze river	7,040 ± 150 BP	Hundreds of carbonized grains, mostly primitive
11.	Tongjiaao, Cixi, Zhejiang, lower Yangtze river	ca. 7,000 BP	Few carbonized rice forms husk
12.	Dulinau, Chalin, Hunan, middle Yangtze river	ca. 7,000 BP	Few carbonized grains
13.	Zhichengbei Zhicheng, Hubei, middle Yangtze	ca. 7,000 BP	Few carbonized rice hulls on pottery
14.	Longqiu, Gaoyou, Jiangsu, lower Yangtze river	7,000–6,300 BP	Hundreds of carbonized grains
15.	Tianluoshan, Yuyao, Zhejiang, lower Yangtze river	7,000–5,600 BP	Hundreds of carbonized grains
16.	Hemudu, Yuyao, Zhejiang, lower Yangtze river	6,950 ± 130 BP	Thousands of carbonized grains, few *O. rufipogon* grains

*Modified from Tang and Xuan (2010)

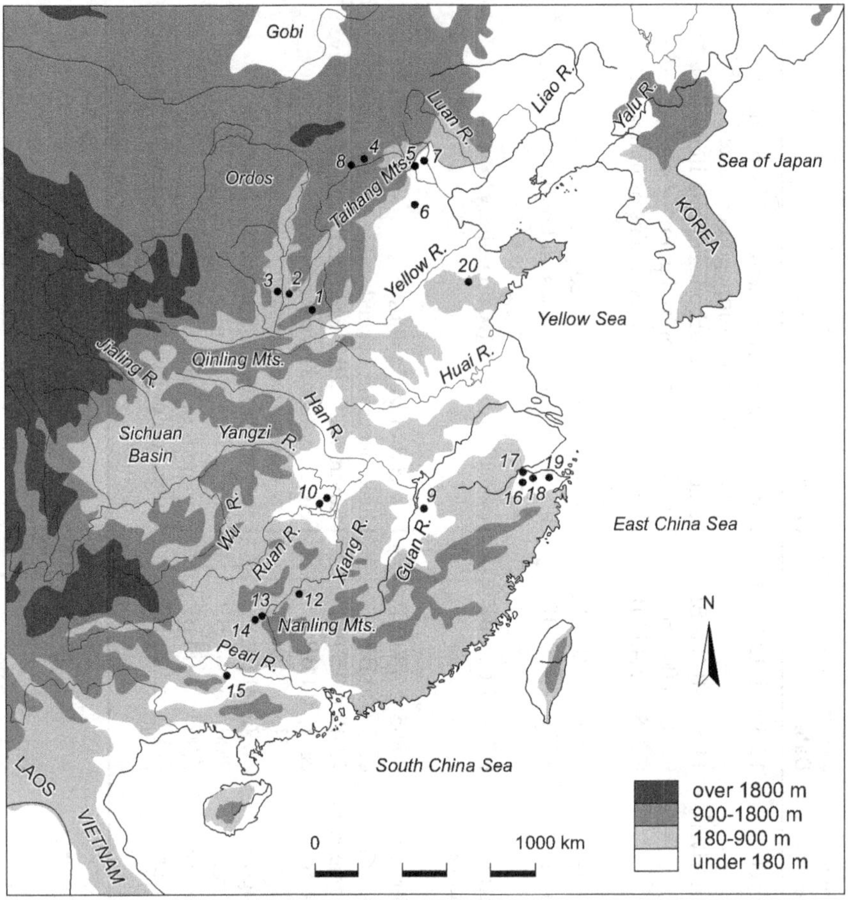

Figure 6.1. ***Major Archaeological Sites of China.*** 1. Xiachuan; 2. Zhizitan; 3. Longwangchan; 5. Donghulin; 6. Nangzhuangtou; 7. Zhuannian; 8. Yugiagou; 9. Xianrendong; 10. Diaotonghuan; 11. Pengtoushan; 12. Yuchanyan; 13. Zengpiyan, Miaoyan; 14. Dayan; 15. Dingshishan; 16. Shangohan; 17. Kuahugiao; 18. Xiaohuangshan; 19. Hemudu; 20. Bianbiandong.
Adapted from Fig. 3.1 of Liu and Chen(2012) with permission.

quantities of grain. All except two are located along the middle and lower Yangtze river basin in the provinces of Zhejiang, Jiangsu, Shaanxi, Hunan, and Hubu. These 16 locales are among the oldest of over 192 archaeological sites where rice remains (grains, husk, stem, leaf, etc.) have been obtained. These sites are located in southwest, south, and central China.

Hemudu (29° 58'N, 121° 22'E) (Zeijiang province, lower Yangtze river) was excavated in the 1970s and dates to 6,950 ± 130 BP. Thousands of carbonized rice grains have been obtained from this site, along with many tools made of bone or stone which were apparently used for land preparation and harvesting. The carbonized rice grains showed much variation in their dimensions, and appeared to be a mixture of primitive japonica, indica, and *O. rufipogon* rices (see Fig. 6.3).

Practically all the early publications on Chinese rice archaeology are written in Chinese, so it has not been possible to get a proper perspective on the role of rice in the early food economy of China. However, An (1989) has given a very brief account of prehistoric agriculture in China. Since the 1960s, the Chinese Academy of Social Sciences has discovered over 7,000 Neolithic sites, and the excavations of hundreds of them have helped to develop a chronology of agricultural origins for most regions of China. Most of the cultures are centered along the two major rivers of China, the Huang He (Yellow river) and the Yangtze. Along the Huang He river, the earliest Neolithic cultures are represented by the Peiligang, Cishan, and Dadswan cultures (all ca. 6,000–5,000 BCE). These were then followed by the Yangshao (5,000–3,000 BCE) and Longshan (3,000–2,000 BCE) cultures, which was in turn followed by the historical Shang period. In the Yangtze river basin, the sequences consisted of Hemudu (5,000–4,000 BCE), Majiabang (4,000–3,500 BCE), Songze (ca. 3,500–3,000 BCE), and Liangzhu (ca. 3,000–2,000 BCE). All these cultures were based on an agricultural economy; hence, agriculture has had an extremely long history in China (An 1989) (Fig. 6.1).

Post-glacial climatic conditions (warm and wet) prevailed from ca. 15,000 BP (16,500 BC), and this enabled the wild rice *O. rufipogon* to reach the northernmost Shandong region (Fuller et al. 2010). Then, during the early Holocene, the sea level rose by 60 m due to glacial melting (NAS 1990, Bellwood 2010, Smith et al. 2011, Clark et al. 2012). This continued up to about 6,000 Ya.

The farming systems in the two major river basins developed in response to the environments that prevailed in the two regions at the time. Most of the cereal remains obtained from the archaeological sites consisted of foxtail millet (*Setaria italica*), broomcorn millet (*Sorghum bicolor*), and rice. Very broadly, the millet and rice zones are separated by the Quinling mountains and Huai He river, with rice being grown in the southern region. Incidentally, wheat and sorghum have only been observed only from the beginning of the historical period (Fuller 2010, Liu and Chen 2012).

There is still some difference of opinion about the period when rice was domesticated in China. These differences are centered around the role of rice as a primary food, especially in relation to millets (Lu 2006, Balter 2009,

Guedes 2011), the availability of wild rice along with rice, and in the manner of distinguishing cultivated rice – the use of length/breadth ratio of grains, the phytolith morphology of wild and cultivated rices, the use of wild rice for food purposes, and the differences in the brittleness of the rachis of wild and cultivated rices.

Liu and Chen (2012) have opined that the rice remains from late Pleistocene Yuchanyan and Diaotonghuan possibly represent wild rice. In north China, the material from Jiahu (Huai river region) was a mixture of wild and cultivated rices. By 8,000 cal BP, rice had reached north of the lower Yellow river (Huang He) region at Yuezhuang in Shandong. Further, rice domestication had developed by 9,000–8,000 cal BP in both north and south China, as exemplified by Jiahu and Kuahugrao. The development of non-brittle rachis during the domestication process is thought to be an important domestication event for rice (Fuller et al. 2009, Liu and Chen 2012). While this character (rachis brittleness) is relevant in temperate cereals, in rice, there is an abscission layer immediately below the lemma and palea, and it is at this point that the mature rice spikelet abscisses and falls off at maturity. Hence, the extent of importance of the development of brittleness in rice may be a moot question.

Fuller et al. (2007) re-examined the data obtained from four excavated sites in the Yangtze river region to ascertain the claims made about very early agricultural developments in the region. The sites were Hemudu (Majrabang period, 5,000–4,000 BCE), Kuahuqiao (Liangzu period; 6,000–2,200 BCE), Bashidung (Pengtoushan culture, on Lishmi river, Hunan province, 6,200–5,450 BCE), and Chengbeixi (middle Yangtze river, Huber), and Chengloushan (Daxi culture, Huper province, 4,000–2,000 BCE). The economy appeared to have relied heavily on the seasonal collection and storage of nuts, especially acorns. Rice constituted only a small fraction of the plant food (also, Bellwood 2011). Based on the sequence of domestication, and comparing it with the initiation and development of agriculture in the Middle East, Fuller et al. (2008) argued that plant cultivation should have preceded animal domestication, and this did not appear to have happened in China, if the present proposition on very early agricultural origins and rice domestication were accepted. Further, the authors pointed out that the proponents of early agricultural origins in the Yangtze region have not identified the series of progressive subsistence systems changes, as had been done in the Near East. The authors therefore proposed that in sites such as Hemudu, wild rice might have been under incipient cultivation for very extended periods, and in sites such as Kuahuqiao, rice domestication might have taken place even later. In

Bashidung (Pengtoushan culture), there was some evidence of settled village communities during 6,500–6,000 BCE, and hence this site would have had some of the earliest dated rice grain assemblages (Fig. 6.1).

Nakamura (2010) has given some new insights into early rice culture in China. Rice cultivation began in lower Yangtze in the first half of the 6th millennium BCE (Kuahuquia culture). Then, during the 5th millennium BCE, the settlements expanded and became adapted to wetland agriculture (Hemudu culture). It then took another millennium to develop a civilization (Liangzhu culture). Although old rice remains have been obtained from Hemudu, the author suggested that rice cultivation might have begun during the Kuahuquio and Hemudu cultures. In the lower Yangtze region, rice remains began to appear in increasing quantities after 6,000 BCE. The importance of rice as a food item also began to increase steadily from 8.2% (4,900 BC) to 24% (4,600 BC), with the percentage of domesticated rice increasing from 27% to 39% during the same period.

Zhao (2010) has given his analysis of the earliest rice archaeological findings from China. The earliest evidence of rice agriculture "regardless of domestication" was observed at the Shangshan site (ca. 10,000 cal BP), and that the economy appeared to be based on fishing and hunting. Gathering of rice and animal husbandry appeared to have met only subsidiary needs. At Tianloushan (6,000–7,000 cal BP), rice was important, but was still only a part of a wider subsistence pattern. Rice domestication seemed to have taken place after 6,500 BP only.

Thus, the timing of the beginning of rice cultivation in China continues to be unclear. Zhao (2010) has proposed two subcenters of origin of rice culture in China, in the middle and lower Yangtze river basins. But, here rice has been found along with foxtail millet (*Setaria italica*) and broomcorn millet (*Panicum miliacerum*), and they both appeared to be more important than rice. There was possibly a third center of agricultural origins, but here, roots and tubers including colocasia appeared to have been the major crops (also, Crawford 2005).

Zhao (2010) then identified Jiahu (upper Huai river, 9,800–7,800 cal BC) as the earliest site of rice agriculture (Fig. 6.1). More than 400 charred rice grains have been recovered from the 125 samples collected at the site. However, the subsistence economy of the region appeared to be based on fishing and hunting, with rice and animal husbandry meeting only supplementary needs. Hemudu, on the southern side of Hangzhou bay and 20 km from the present sea coast (7,000–5,800 cal BP) was another ancient site, but here, the role of crops in the economy was not studied.

However, Tianloushan (6,000–5,000 BP) another ancient site, which is only 7 km from Hemudu appeared to be similar in the environmental context to the latter site. From here, 222 soil samples had been collected, and the rice material obtained contained a high proportion of wild rice material. Rice appeared to be only a subsistence food item here also. In the Liangzhu culture (6,000–5,000 BP), a late Neolithic site in the lower Yangtze basin, rice cultivation had become established and by 5,200–4,300 cal BP, rice had become the dominant food source. In the excavations undertaken at Xiaojin (Ziycean country, Guanxi province), > 10,000 charred rice grains had been recovered (6,000–4,000 cal BP). Also, in the same region, large numbers of rice phytoliths were recovered in Dingshishan (Yongning county, Guanxi province) from Phase IV of the excavations (ca. 6,000 BP). Zhao (2010) indicated that rice domestication was initiated about 10,000 BP, and the transition from harvesting and gathering to rice agriculture took place during 9,000–6,000 BP, with rice culture becoming established in China during 6,000–5,000 BP.

Based on the studies conducted by Fuller and associates, he (Fuller 2011) has further proposed that the earliest domesticate of Asian rice that possessed the various domestication traits was a japonica rice and that this took place ca. 6,500–6,000 ya. He proposed a "diffuse and multicentric model" for rice domestication, starting *de novo* in more than one region of China. He found evidence for intensive wet rice cultivation in two sites in lower Yangtze basin, Chuodun and Caoxieshan, dated to 4,800–3,800 BC, and one site in middle Yangtze basin at Chentoushan dated ca. 4,500 BC. Rice then spread both southwards and northwards and reached north of the Qinling mountains ca. 6,000 ya (Fig. 6.1). He supported the proposition of Bellwood (2010 and earlier) regarding the spread of rice to the south and southeast of China (Taiwan, Philippines, Indo-China, Indonesia, etc) based on the language/farming hypothesis.

Cattle domestication had preceded crop domestication in the Sahel region of west Africa also. Here, the first farmers might not have felt the urgency or need to domesticate rice, because of the adequate availability of other vegetal food sources in the form of grass seeds, fruits, and tubers. Further, taking up crop agriculture in Sahel is riskier than herding, because of their life on the edge. Cattle rearing allows greater and faster mobility if the prevailing climate turns suddenly hostile as a consequence of the overwhelming influence of Sahara. However, we do not have such data for early agriculture in China. Notwithstanding these comments, it is indisputable from present evidence that the Yangtze basin is the earliest center of Asian rice domestication. Well known archaeologists such as Smith (1998) and Bellwood (2004) have stated that China could have been an original center of plant domestication.

A. Korea and Japan

It is generally agreed that rice was taken to Korea and Japan from China in a domesticated state, alongside certain other cultivated crops, notably millets. But, the place from where they were introduced, and the time of their introductions are still unclear. Cultivated rice was possibly introduced from the Liaodong and Shendong peninsula, northeast China (Miyamoto 2009, Ahn 2010), or, more probably taken directly across the Chinese Yellow Sea from eastern-central China (which covers the Yangtze basin).

Ahn (2010) has reviewed the rice archaeology of Korea. The first charred rice grains were obtained from Hunanmi (1,000 BCE, Bronze Age) (Im 1978). However, water-soaked rice husks and millets had been recovered from an earlier date from peat from Gahycon-ri, Gimpo (4,010 ± 25 BP) and from two other sites in Ilasan district, Seongjco-ri (4,070 ± 80 BP) and Gahyeon-ri (4,330 ± 80 BP), located on the lower Han river in midwestern Korea (Ahn 2010). Earlier claims of finding domesticated rice remains in Sorori have been found to be incorrect. The available evidence from Chulman (Neolithic) are still insufficient. The rice obtained from Oun (3,610 ± 280 BP) has been ascribed to the late Chulmon period (Crawford and Shen 1998, Crawford and Lee 2003), but Ahn (2010) has indicated that the context appeared to be controversial.

In Japan, archaeological evidence indicates that rice was cultivated there 2,400–3,000 ya (Tanaka et al. 2010). Evidence for paddy cultivation has been obtained from the Maekawa archaeological sites in Yoyoi (2,100 BP) and Heian (1,100 BP) in the form of phytoliths, but not for the intervening period (2,100–1,100 BP). Paddy cultivation became widespread in Japan by 400 BCE (Tanaka et al. 2010).

ii. Southeast Asia

This vast region extends southwards from south China to the Malayan archipelago and eastwards up to the Philippines and New Guinea. It consists of a continental region, seas, and hundreds of islands. It is divided into Mainland southeast Asia and Island southeast Asia for convenience. The former region includes south China, Myanmar, Thailand, Laos, Cambodia, Vietnam, and western Malaysia, and Singapore, and the latter includes Taiwan, The Philippines, Indonesia, and New Guinea. The region's present population is estimated at 500 million. All the native peoples of southeast

Asia have common ancestries in terms of culture, language, and genotype (Bellwood and Glover 2004). The region has been greatly influenced by its extensive early contacts with China and India, and later, with Islam and Christianity. There is no evidence of a truly independent development of agriculture here (Bellwood and Glover 2004), but, their region contributed the earliest sea crossings and the greatest ethno-linguistic human dispersal in human history.

The earliest humans are thought to have arrived in Mainland Southeast Asia 1.5 Ma, and their spread throughout the region was completed by 35,000 BP (Bellwood and Glover 2004). This appears to have been facilitated by the last two Glacial Maxima that occurred 26–20 kya and 2.3 Mya, when the average land temperatures decreased by 5–6°C and the sea level fell by 60–80 m below present levels (Clark et al. 2006). In southeast Asia, this led to the formation of Sundaland and the Sahul Shelf. The former consisted of Taiwan, Hainan, Sumatra, Java, and Borneo, their outlying islands and also those of The Philippines, which became part of the Asian mainland. The latter comprised New Guinea, The Solomon Islands, and Tasmania joined with Australia. This process is thought to have facilitated the large-scale movement of people, plants, and animals across the region (Bellwood and Glover 2004).

The earliest people would have lived a subsistence life as hunters and gatherers. They mainly occupied the coastal regions and the numerous riverbanks and estuaries. Although elsewhere in the world, such people were constantly on the move in search of foraging areas, this did not appear to have been the case in southeast Asia. Based on archaeological evidence, Higham (2004) has observed that this region was one of the world's richest habitats in terms of natural bounty. The inhabitants did not appear to have to hunt land mammals, as the region was rich in plant and aquatic resources.

The sea level rise of 60–80 m consequent on the last glacial melt appears to have occurred during the late Pleistocene–early Holocene period. This led to the submersion of both the Sunda and Sahul Shelfs. Higham and Maloney (1989) have estimated that all the archaeological evidence all along the coast for the period 4,000–8,000 BCE has been lost permanently as a consequence of this (see also, Higham 2004).

The existing hunter-gatherer economy might have been disturbed about 5,000 years ago with the arrival of the first agricultural populations from central and southern China (Fig. 2.1). They were predominantly Austronesians. By 1,300–2,500 BC, the agricultural populations had steadily moved southwards and then to island southeast Asia up to Celebes and The

Philippines (Bellwood 2004). The archaeological assemblages that we associate with them have been widely discovered in the early Neolithic sites of southeast Asia spanning the period 3,000 to 1,500 BC. Based mainly on linguistic evidences, it is thought that the early farmers moved south and east, both across the sea and overland from the Yangtze basin, by two routes through Taiwan and The Philippines and through Guangdong province, into northern Vietnam (Bellwood 2004, 2011). They might have brought with them to Mainland southeast Asia at least the cultivation practices of rice and millets, domesticated pigs and dogs, pottery, etc. The early (from around 5,000 years ago) southeastern Asian assemblages in northern Vietnam, Thailand, and coastal regions of China cannot however be associated directly with evidence of agriculture.

One of the earliest mainstays of the Austronesian people was rice (Bellwood 2004, 2011, Higham 2004). The wide occurrence of rice materials in the excavated sites of Mainland southeast Asia were advanced to support this proposal. However, it is widely known that wild rices were very extensive in this region, at least until recent times (Oka 1988, Morishima 1994, Vaughan et al. 2008, etc.). As with elsewhere in the tropical rice regions of Asia, the local population must have collected wild rice seeds from their large stands for direct consumption, and it is not easy to distinguish between wild and cultivated rices in the archaeological material. Further, the local population appears to have been primarily dependent on the plentiful aquatic resources to meet their major food needs, as we have seen before. Both Bellwood and Higham have mentioned about this possibility in their extensive publications. The Austronesians, who had moved south from the lower Yangtze basin, had brought with them knowledge of rice cultivation. At that time, all the coastal regions and most of the river flood plains of mainland southeast Asia had already been occupied by the local foraging population. Hence, the incomers appeared to have settled in the interior regions of Indo-China, along the rivers and valley areas that were suitable for farming. They are thought to have come down to Indo-China through the Lingnan region in southeast China (Higham and Lu 1997).

The claims of very early rice finds made by Solheim (1969) and Gorman (1969, 1972) in Non Nok Tha, Ban Chiang and Spirit Caves have since been found to be exaggerated. The radiocarbon dating of rice chaff used to temper clay employed for making pottery at that period has a date of 1,500–1,000 BCE (Higham 2004).

Higham and Lu (1997) have reported evidence of rice cultivation between 4,300 and 3,500 BP from several sites spread over Vietnam (Red

River delta) and northeast and east Thailand. They have given also sites of slightly older chronology (near 4,500 BP), but none of them had any evidence of rice.

Incidentally, Bellwood (2004, 2011) has made certain observations about the origin and spread of agriculture, including rice culture, in New Guinea and beyond to the western Pacific Islands. They are based primarily on the spread of Austronesians and archaeological evidence. He observed that the Austronesians colonized Taiwan during 3,000–2,500 BCE by replacing or incorporating the earlier local populations. Then, by 2,500–1,500 BCE, they moved both southwards and eastwards to The Philippines and Indonesia. He has mentioned finding carbonized rice (dated "before 2,000 BC") in the Gua Sirch cave in Sarawak, Borneo. By 1,500 BCE, people had reached the Mariana Islands of western Micronesia (across an "absolutely astonishing" 2,500 km of ocean from Luzon, Bellwood), and soon after, they reached the islands of Bismarck, Solomon, New Caledonia, Vanuatu, and Fiji. The Austronesians reached western Polynesia (Tonga, Samoa, Wallis and Futuna Islands) during 1,000–800 BCE. The actual timing of the introduction of agriculture into Sumatra, Java, and Borneo is uncertain, but this might have happened between 2,000 and 1,000 BCE. There is no evidence for the introduction of agriculture into New Guinea and the Solomon Islands via the spread of Austronesians (Bellwood 2004, 2011). He felt that indica rices might have spread to southeast Asia only since 2,000 ya.

Bellwood (2005, 2010) has summarized the "archaeological chronology of development and outflow of Neolithic life styles in the eastern Asian region" as follows (see also Table 2.1):

(i) 8,000–6,000 BC: development of predomestication cereal agriculture in central Asia with japonica rice in the Yangtze, Han, Huai and lower Yellow river basins, and foxtail (*Setaria italica*) and broomcorn (*Sorghum bicolor*) millets to the north and west;

(ii) 6,000–3,000 BC: gradual spread of Neolithic lifestyles and fully domesticated rice through southern China along with increasing use, especially after 4,000 BC;

(iii) 3,500 BC: Neolithic settlement of Taiwan (Dabenkeng culture) possibly following developments in Fujian and/or Guangdong;

(iv) 3,000–2,000 BC: Neolithic settlement of Mainland southeast Asia from Guangdong and Guanxi into northern Vietnam and then into southern Vietnam and Thailand;

(v) 2,000–1,500 BC: Neolithic settlement of The Philippines and central Indonesia via Taiwan and Mariana from the northern Philippines.

Bellwood has related these Neolithic spreads to the southward movements of various language families. The prehistory of rice cultivation in Island southeast Asia was obscure until carbonized rice and millets dated to ca. 2,800 BC were recovered in large quantities in waterlogged conditions from Nanguanli (Tainan Science-based Industrial Park) by Tsang in 2004. Intriguingly, rice did not spread into and beyond New Guinea until after 1950.

Hoabinhian culture is a generic term given to all the artificial stone assemblages containing flaked and cobbled artifacts that have been found in southeast Asia. This term was coined originally in the 1920s by French archaeologists to refer to the rock shelters that they had discovered in Indo-China. These are dated between 18,000 and 3,000 BP (Bellwood 2007). Gorman (1969), who had excavated the Spirit Cave (north Thailand, 20°N, 98°E), found remains of several fruits such as *Prunus* (almond), *Terminalia, Areca* (betelnut), *Vicia* (bean), *Phaseolus* (bean), *Pisum* (pea), *Lagenaria* (bottle gourd), *Trapa* (water chestnut), etc. in the lowest two layers. No cereals were recovered. The oldest ^{14}C date of the biological material was 11,690 ± 560 to 7,622 ± 300 BP. Gorman then observed that the level of economic development of the people appeared to be limited to food gathering.

Pyraman (1989) discussed the excavations carried out by Surin Pookajoon in the Hoabinhian Project (Late Stone Age period) in Ban Kao, central Thailand (Kanchenabury Province, 14°0′N, 99°19′E). The excavations went down to 12 m in depth and covered the period 11,000–2,000 yr BP. Evidence of human occupation was discovered, but with only a very few plant remains. These included *Eugenia, Croton, Phaseolus lathyroides*, and *Citrullus lanatus*. Just one charred rice grain was obtained: from the level 2,800 ± 300 BP.

It is possible that the population subsisted primarily by hunting and fishing, and the rice specimens obtained of this period both here and elsewhere could have been of either wild or cultivated rice. Whyte (1989) had indicated that wild rices were widely present in the region, often in extensive stands.

Higham and Maloney (1989) studied the coastal archaeology of the Bay of Bangkok, and proposed that rice might have been cultivated in the area by the 4th millennium BC, but this could not be verified. A major intensification in agricultural methods occurred only after 500 BCE. The archaeological evidence obtained from this region (Nong Nor, Khok Phanom Di; Fig. 6.2) amply suggested that the inhabitants met their food needs mainly from marine sources. They might have possibly obtained rice by barter from the interior (Higham 2004).

Castillo (2011) has recently reviewed the rice archaeology of Thailand. A total of 29 archaeological sites from the Hoabinhian to Late Prehistoric

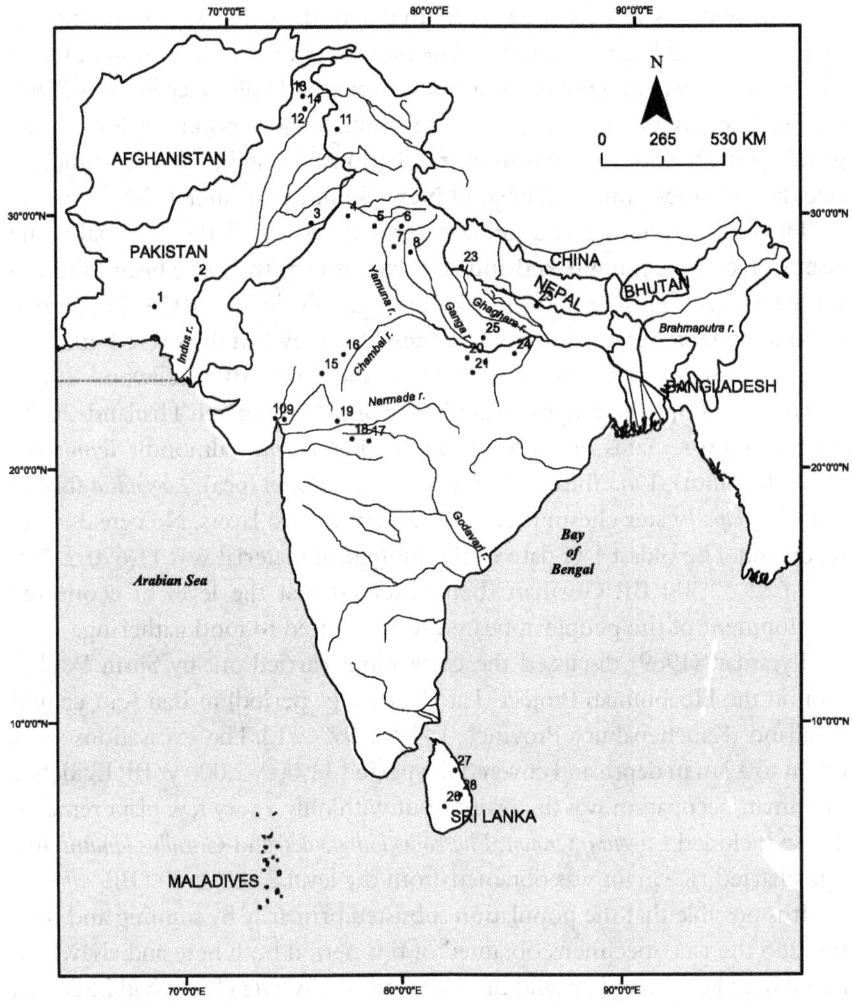

Figure 6.2 ***South Asia Showing Major Archaeological Sites.*** 1. Mehrgarh; 2. Mohanjodaro; 3. Harappa; 4. Kalibangan; 5. Mitathal; 6. Hulas; 7. Lal Quila; 8. Atranjikhera; 9. Lothal; 10. Rangpur; 11. Semthan; 12. Ghaleghay; 13. Bir-khot; 14. Loebandhr; 15. Ahar; 16. Dangwada; 17. Imamgaon; 18. Daimabad; 19. Navdatoli; 20. Chapani-Mando; 21. Koldihwa; 22. Mahagara; Lahurdewa; 24. Tokwa; 25. Jhuse; 26. Kilalghata; 27. Anuradhapura; 28. Horton Plains.

periods have reported finds of rice. Phytolith studies indicated that rice might have been cultivated from mid-Holocene period. The first evidence of domesticated rice was obtained in the form of macroremains from the coastal site of Khok Phanom Di, 2,000–1,500 BCE. It appeared to be japonica rice, but opinions have differed about the language group that introduced it: either Miao Yao, Austro-Asiatic, Austric, or Austronesians. Khao Sam Kaeo is a site in peninsular Thailand (ca. 400–100 BCE) which appeared to have been a point of contact with India.

The highlands of PNG (Papua New Guinea) have been known since long as a center of ancient farming (Golson 1977). The physiography and climate of the region, with its year-round rainfall, suggest that the life of the people would have been based on fishing, hunting, and collecting tubers of taro and sweet potato and fruits, especially banana. Recently, banana phytoliths have been discovered in the Kuk swamps in the Waligi valley in the Holocene stratography (beginning 10,200–9,910 BP), that have been dated to 6,950–6,440 BP (Denham et al. 2004, 2007a,b). Rice is only a recent introduction to New Guinea, but it is now the second largest component of the food economy of the island (Golson 1997, Bourke et al. 2000).

iii. South Asia

Politically, South Asia covers Afghanistan, Pakistan, India, Sri Lanka, The Maldives, Bangladesh, Nepal, and Bhutan.

A. Indus Valley Civilization

South Asian archaeology is usually considered synonymously with the Indus Valley (Harappan) Civilization, which is reckoned as one of the major civilizations of humankind. This covered a vast area of the Indian subcontinent, including practically the entire river basins of the Indus and Saraswathi rivers. The latter has since dried up, but it used to run parallel to the Indus river to its east. Only the remnants of the river bed are now noticeable in places.

The Indus valley civilization was first reported in 1826 CE, but was first opened only in 1856–1872 CE, with systematic excavation beginning much later, in 1920 CE (Possehl 2002). The excavations have yielded a lot of seals containing inscriptions, but it has not yet been possible to decipher the script. Consequently, the dating of the Civilization has been done indirectly, including via radiocarbon dating. The Harappan civilization is now estimated to have lasted from ca. 3,200 to 1,400 BCE, and is divided into Early (ca. 3,200–2,600 BCE), Mature (ca. 2,600–1,900 BCE), and post-Harappan (ca. 1,900–1,400 BCE) phases (Possehl 1997). Mainly because of our inability to decipher the Indus script, much information about the command, control, and management of the system remains in the realm of the unknown. It has been suggested that this civilization might not have "disappeared" due to some catastrophic event, as has been suggested by some authors, but possibly faded away because of de-urbanization and the eastward shift of the population (Possehl 1997).

The Harappan civilization was apparently an urbanized phenomenon involving much manufacturing and trading. It met its food, clothing, and other requirements from the rich and fertile valley systems of the two rivers, the Indus and the Saraswati (Higham 2004). Higham has opined that the Indus civilization might have developed on the foundations of a long preceding cultural sequence during which agriculture, craft skills, and trade had grown in complexity and range. According to him, the settlement of Mehrgarh (Baluchistan, about 250 km north of Mohenjo-daro) is an example of this phenomenon (Fig. 6.2). Mehrgarh existed from 7,000 BCE to 2,600 BCE. Even the earliest settlement showed remains of barley and wheat, faunal remains of various wild animals, and domesticated cattle (Higham 2004).

Barring a couple of major excavations including Mohenjo-daro and Harappa, that had been carried out before the beginning of the Second World War, much of the excavations (more than 1,100 in Pakistan and India) were done after the separation of Pakistan from India in 1947. Kajale (1991) and Fuller (2002) have summarized the archaeobotanical work done in the last five decades (Fig. 6.2).

The first archaeological remains of rice were obtained from Khokrakot (Rohtak, Haryana) by Birbal Sahni in 1936 from the early historical period. This was followed by a large number of excavations, some of which yielded rice. Most of them have been dated to between 2,500 BCE and 1,500 BCE. A couple of them are older, dated to ca. 3,000 BCE (e.g., from Ghaleghay (Swat, Pakistan) by Constantini in 1987). Incidentally, most of the area covered by the Harappan civilization falls outside the present traditional rice growing area (Table 6.2).

B. Ganga Valley Excavations: Koldihwa, Lahuradewa, Tokwa, Jhusa

In the last few decades, several excavations have been done in the middle Ganga region, south of present-day Allahabad. The initial sites were Chopani-Mando, Koldihwa, and Mahagara on Belan river, and Mahadaha in the Ganga valley (Sharma et al. 1980a,b, Sharma 1985). They were dated to upper Paleolithic (ca. 17,765 BCE), pre-Neolithic (ca. 8,080 BCE), and early Neolithic (ca. 6,570 BCE, 5,400 BCE) periods. Impressions of rice straw were also observed on pottery sherds. However, subsequent workers have found it difficult to accept these findings, alleging that questionable sampling and dating procedures were used in the study (Glover and Higham 1996, Fuller 2002, and so on).

Lahuradewa is one of the major archaeological sites of the region. It was a lakeside settlement in Sant Kabir district, Uttar Pradesh (26°46′ N lat.,

Table 6.2 Major Archaeological Sites for Rice in South Asia

S No.	Site, Location	Period in BCE	Material Observed
i.	Harappan sites		
1.	Rangpur II A, Gujarat	2,600–2,200	Impressions
2.	Harappa, Punjab, Pakistan	2,250–2,000	Seed remains
3.	Daulatpur, Punjab, Pakistan	2,000–1,700	Seed remains
4.	Lothal, Gujarat	2,000–1,700	Impressions
5.	Rangpur III. Gujarat	2,000–1,700	Impressions
6.	Mitathal, Punjab, India	2,000–1,400	Seed remains
7.	Pirak 1, Baluchistan	1,950–1,550	Seed remains
8.	Atranjikhera, Upper Ganga, UP	1,800–1,200	Seed remains
9.	Hulas, Upper Ganga, UP	1,800–1,200	Seed remains
10.	Lal Quila, Upper Ganga, UP	1,800–1,200	Seed remains
ii.	Harappan outlying sites		
1.	Ghaleghay I, Swat valley, Pakistan	3,000–2,500	Impressions
2.	Ahar, Banas, Rajasthan	2,600–1,500	Seed remains
3.	Ghaleghay II, Swat valley, Pakistan	2,500–2,000	Impressions
4.	Dangwada, Banas, Rajasthan	2,000–1,500	Seed remains
5.	Ghaleghay III, Swat valley, Pakistan	1,900–1,700	Impressions
6.	Inamgaon, Malwan, Maharashtra	1,700–1,500	Seed remains
7.	Bir-Khot-Ghawandai, Swat, Pakistan	1,700–1,400	Seed remains
8.	Loebandhr 3, Swat Pakistan	1,700–1,400	Seed remains
9.	Daimabad IV, Ahmednagar, Maharashtra	1,500–1,100	Seed remains
10.	Navdatoli, MP	1,500–1,100	Seed remains
11.	Semthan I, Ananthanag, Kashmir	1,500–600	Seed remains
iii.	Ganga Valley		
1.	Jhuse, near Allahabad UP	7,110–5,932 BP	Grains
2.	Mahadaha, Ganga valley, UP	6,320–4,010 BP	Impressions
3.	Chopani – Mando, Belan river, UP	5,440–4,530 BC	Wild rice
4.	Lahurdewa, SK Nagar dist., UP	3090–6442 cal BC	Lake sediments
5.	Tokwa, Mirzapur dist., UP	3,810–3,414 BP	Charcoal
6.	Koldihwa, Belan river, UP	3rd century BC	Wild rice
7.	Mahagara, Belan river, UP	3rd century BC	Wild rice

82°57′ E long.). It is bounded by the Sarayu River on the west and south, the Gandak River in the east, and the Nepalese *terai* region to the north. The site was excavated from 2001–2006 (Tewari et al. 2006). The lake sediments of micro-charcoal indicated continuous human activity over more than 10,000 years.

The site revealed the presence of wild (*O. rufipogon*, *O. officinalis*) and cultivated rices, foxtail millet (*Setaria italica*), Job's tear (*Coix lachryma-jobi*), black and red potteryware with coden patterns, etc. A lump containing husks of domesticated rice was AMS-dated to 6,409 BC (8,359 cal. BP). Material (including rice) in the lowest (1A) layer gave dates of 3,090 cal BC to 6,442 cal BC. Tewari et al. (2006) indicated a radiocarbon age of 9,230 ± 100 yr BP to 9,510 ± 100 yr BP and 9,590 ± 110 to 9,880 ± 110 yr BP for the site (Fig. 6.3).

Saxena et al. (2006) obtained phytolith-based evidence of rice and rice cultivation from about 10 kyr BP from Lake Lahuradewa, which is close to the archaeological site of Lahuradewa. The area receives about 1,400 mm annual precipitation. Palaeo-vegetation studies have indicated that during the Holocene period, the Ganga plain was a grassland with occasional thickets (Singh 2005). They determined the age of wild rice phytoliths as 10,300 cal. yr BP and that of cultivated rice phytoliths as 8,350 cal. yr BP. They stated that this corresponded to the dates for cultivated rice obtained from Lahuradewa. Upwards in the profile, the frequency of cultivated rice phytoliths progressively increased in frequency and wild rice phytoliths correspondingly reduced. The authors proposed that this indicated that rice cultivation started in the Ganga plain ca. 8,650 cal yr BP (Fig. 6.3).

Phytoliths are microscopic opal particles that occur in many plants, when monosilicic acid that is absorbed into the plant with the ground water gets precipitated within and between the cells. The phytoliths of every species or genus have their characteristic shape. *Oryza* is a genus that possesses a characteristic phytolith morphology (Pearsell et al. 1995, Zhao et al. 1998). Phytoliths in wild and cultivated rices are distinguished on the basis of the number of scales present on the edges of the fan in bulliform phytoliths (Zhao et al. 1998).

Pokharia (2008) studied the palaeobotanical records of cultivated crops and associated weeds at a Neolithic site, Tokwa (Mirzapur district, Uttar Pradesh, India, 24°54′N, 83°21′E). It is situated at the confluence of two small rivers, Belan and Adwa. Both ultimately join the River Ganges. The region was excavated during 2000–2003 and showed evidence of three cultures, Neolithic, Chalcolithic, and Iron Age. The limited radiocarbon

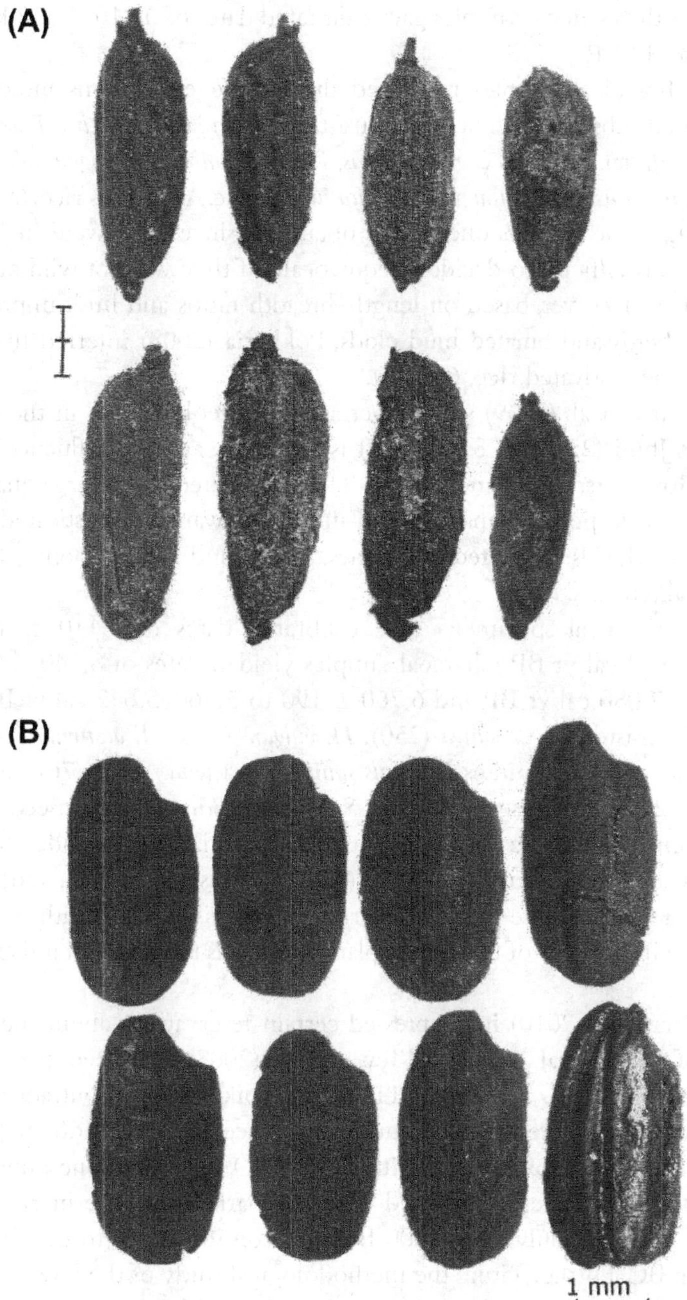

Figure 6.3 ***Rice Archaeological Specimens Recovered from Hemudu (China) and Lahurdewa (India).*** (A) Ricegrains excavated from Hemudu relic sites in lower Yangtze(6950 +/− 130 BP) (B) Grains of domesticated rice from the Neolithic phase of Lahurdewa, India (6th – 5th millennium BCE).

dating of the charcoal samples gave calibrated dates of 3,810–3,572 BP and 3,635–3,414 BP.

The botanical samples recovered during the excavations indicated a crop-based subsistence economy consisting of cereals (*O. sativa, T. aestivum,* and *H. vulgare*), legumes (*Lens culinaris, Pisum arvense,* and *Vigna radiata*), oil crops (*Linum usitaltissimum* and *Brassica juncea*), etc. As regards rice, 69 grains including some broken ones, were obtained. Since they were in husked form, it was difficult to decide unequivocally if they were of wild or cultivated rice. However, based on length–breadth ratios and husk impressions on pot sherds and burned mud clods, Pokharia (2008) inferred that they were of the cultivated rice, *O. sativa.*

Pokharia et al. (2009) studied yet another Neolithic site in the Ganga plains at Jhusi (25°26′N, 81°54′E). It is located near the confluence of the two major rivers, Ganga and Yamuna. The authors recovered large quantities of hand-made pottery (mainly cord-impressed ware and rusticated ware) bone arrowheads, semiprecious stones, beads, and animal bones, besides botanical remains.

The botanical specimens gave calibrated dates of 7,110 ± 170 to 5,990–5,932 cal yr BP. Charcoal samples yielded dates of 8,140 ± 220 to 7,105 to 7,080 cal yr BP and 6,760 ± 190 to 5,560–5,642 cal yr BP. The cereals consisted of *O. sativa* (250), *H. vulgare* (146), *T. aestivum* (14) and *T. sphaerococcum* (12), pulses *Lathyrus sativus* (6), *Vigna radiata* (76), and *Lens culinaris* (210), and oil seeds (sesame, *Sesamum indicum,* 1 and linseed, *Linum usitatisssimum,* 24). The dates and botanical material corresponded broadly to those obtained from Lahurdewa (Ganga plains), Koldihwa, and Tokwa (Vindhya region). The authors inferred from this that Neolithic culture prevailed in this part of the Ganga plain as early as the 9th–8th millennium BCE.

Fuller (2006, 2010) has expressed certain reservations about the inferences of Saxena et al. (2006) and Tewari et al. (2008). He stated that spikelet measurements (L-B/T ratios) and illustrated spikelet bases contradicted the claims of domesticated rice at Lahuradewa, since the grain ratios and absolute measurements overlapped with those of *O. nivara* and one population of *O. officinalis.* Fuller has opined that the marked increase in rice grain breadth occurred only after 2,000 BC, and continuing up to the first millennium BC. Further, citing the methodological study of the lower Yangtze studies (done by Lu et al. 2002) using number of fan-edged chips of rice bulliforms), he questioned the correctness of using this criterion for identifying domesticated rice by Saxena et al. (2006). Fuller (2010) then concluded, based on "unambiguous domestication traits", that domesticated rice was present from ca. 2,000 cal BC in the Ganga plain.

C. Sri Lanka

Sri Lanka is an island nation that lies to the southeast of peninsular India (ca. 6–10°N). It is 65,000 km^2 in area and enjoys a tropical climate with more than 2,500 mm annual precipitation. Broadly, the central region is a plateau and its northern and southern regions are flatlands. Only very fragmentary archaeological work has been done in this island nation.

Kajale (1989) studied plant remains recovered from the cave site of Beli-Lena (Kilalgata, 38 km east of Colombo). The charred plant remains gave dates of 11,570–10,200 bp based on Libby's half-life of 5,570 ± 30 yr. They did not contain any cultivated plants. There were instead charred remains of wild banana (*Musa acuminata, M. balbisiana*, possibly), wild breadfruit (*Artocarpus nobilis*?), and *Canarium zeylanicum* (a tree whose kernels are consumed even today). The author inferred that the human population consisted of foragers.

In a review of the archaeological work done in Sri Lanka to date, Prematileke (2006) mentioned about another two sites where rice might have been found. In an excavation done in the Horton Plains (central Sri Lanka, 2,000 m amsl), he observed that land use changes indicated evidence of human settlement and possible evidence of rice cultivation. The radiocarbon-dated pollen analyses gave the date as 5,500 BC. The second evidence was obtained from Anuradhapura (300 m amsl, central northern plains), dated to 900 BC. The site showed evidence of iron production, horse breeding, and paddy cultivation. The author did not give any further details of this work.

From the foregoing review, it follows that the oldest rice remains obtained so far have been from the Yangtze river basin. The oldest sites have been dated between the 8th and 12th centuries BP. Although large quantities of rice have been recovered from some of the sites, it is interesting that overall rice was only one of numerous plant and animal remains recovered from various sites. The other major plant remains consisted of millets (*Setaria*), acorns, aquatic foxnut (*Euryale ferox*, Nymphaeaceae), etc. Bellwood (2011) has estimated that rice constituted only 30% of the plant food consumed in the region. Thus, the prevailing economy did not appear to have been a rice-based one. The situation in the middle Ganga plain appears to be also comparable to this, in that rice was only one of several plant foods available in the region.

This calls for further excavations in south and southeast Asia, in southeast and northeast India, Bangladesh, and the Konkan-Malabar coast. Wild rices used to abound in this area until about five decades ago. Rice constitutes the predominant staple food of the people, it has great cultural and religious significance for the local people, the variety and species diversity is comparatively high, the local populace has a unique, distinct preference for red pericarped, parboiled rice, and local landraces contain higher numbers of forms with so-called wild/primitive characters such as awns, red apiculus, and so on.

5. ORIGIN AND EVOLUTION

We shall now consider the following aspects of the origin of Asian rice: the species ancestral to Asian rice; its place and time of origin; and its mode of evolution. Since these aspects were comprehensively reviewed 40 years ago (Nayar 1973), substantial work has been done on all these aspects, especially in the areas of molecular biology, archaeology, and phylogeny. However, a few short reviews and perspectives have been published during the intervening period (Morishima et al. 1962, Chang 1976a,b, 2003, Oka 1988, Morishima 2001, Nanda and Sharma 2003, Nayar 2004, Sharma 2005, 2010, Kovach et al. 2007, Sweeny and McCouch 2007, Vaughan et al. 2008).

i. The Ancestral Species

There is now general agreement that Asian wild rice is the progenitor of Asian cultivated rice. Possibly, no other putative ancestral taxon of a cultivated crop shares such an extensive interface with its derived cultivated species as the wild rices do with Asian rice. This situation appears to have endowed the Asian rice with a very wide range of adaptation and variability, possibly not seen in any other cultivated species. This is possibly because the genetic isolation required by the derived incipient "proto-species" could not be realized in many regions because of the extensive presence until recently of the ancestral species, *O. rufipogon*. Unfortunately, the drastic changes in land use patterns which have decimated most of the wild rices in Asia, and the varietal homogeneity that has arisen since the 1950s consequent to the wide use of improved cultivars have resulted in the loss and disappearance of most of the wild rices throughout south, southeast, and east Asia. The overall loss may be more than three-fourths of the pre-1950 population.

The wild rices show a tremendous range of both heterogeneity and heterozygosity. Their life history habits range all the way from floating perennial forms to obligate annual ones. Generally, they occur within and in the vicinity of paddy fields, often in wetlands, both seasonal and perennial. They grow also in rice fields, on bunds, and bodies of water. Some populations have developed into self-sustaining stands away from paddy fields in both upland and wetland areas. Morphologically, they range from *O. rufipogon* to *O. sativa*. Most of the intermediate forms show heterozygosity, as numerous authors have pointed out (see for example Watt 1892, Roy 1921, Bhalerao 1928, Chatterjee 1948, Sampath and Rao 1951, Ramiah and Ghose 1951, Sampath and Govindaswamy 1958, Nayar 1958, 1973, Bor 1960, Chang 1976, Morishima et al. 1961, Oka 1988, and so on).

There are two opinions regarding the progenitor taxon. Some authors have proposed that cultivated rice originated directly from the perennial, *O. rufipogon,* and the others, from the annual form variously designated as *O. nivara,* spontanea, fatua, etc. The earlier differing perceptions about the progenitor species were caused by the overwhelmingly wide diversity present in the wild rices and the diverse viewpoints and different nomenclature used by earlier workers (Watt 1892, Roy 1921, Roschevicz 1931, Burkill 1953, etc.). Watt (1892), for instance, had identified four varieties of *O. sativa*; var. *rufipogon,* var. *coarctata*, var. *benghalensis,* and var. *abuensis*, with a caution that even these four forms should be viewed as purely provisional; and Roschevicz (1931) identified two forms; *f. spontanea* and *f. aquatica*. Roy (1921) identified as many as 32 forms in the wild rices of Madhya Pradesh (which is just one of about 24 states in India).

A. Annual Wild Rices

It was possibly Chatterjee (1951), who initially proposed that Asian rice originated from annual wild rice. Ramiah and Ghose (1951), Porteres (1956), Kihara (1958), and Shastry (1964b) also held similar views. More recently, Chang (1976a,b,c) and Sharma (2003, 2010) supported this proposal. Earlier, Sharma and Shastry (1965b) had designated the annual wild rices occurring in the upland regions of south and southeast Asia as *O. nivara,* and restricted the application of *O. rufipogon* to the perennial wild rices. Sharma (2010) suggested this on the basis of this entity (his *O. nivara*) having already acquired an annual, gregarious habit, self-pollination, synchronous flowering and seed setting, higher productivity, and bolder seeds – all characteristic of the cultivated rices. Sang and Ge (2007) have observed that "it is more parsimonious to consider the origin of indica rice from the annual ancestor", but gave no reasons for this statement.

Some authors have continued to state that an annual wild rice taxon exists that is directly descended from the perennial wild rice, *O. rufipogon* (Chang

1965a,b,c, Sharma 2010). Annual wild rices and red rice are ubiquitous in all the rice growing areas of the world. Those occurring in south, east, and southeast Asia demonstrate tremendous polymorphism. Their morphological features and genetic characteristics are intermediate between those of the wild perennial *O. rufipogon* and the cultivated rice, *O. sativa* (Sampath and Govindawsmy 1958, Nayar 1958, 1973, Chang 1965a,b,c, Morishima et al. 1981, Oka 1988, Vaughan 2003, and so on). Possibly, during the progression from the perennial wild species, *O. rufipogon,* to the cultivated rice, *O. sativa* some forms might have formed a transient intermediate state. But, the question to be considered is whether such intermediate forms could exist in a natural self-reproducing condition as an autonomous entity, or only transiently during the domestication process. We do see annual rices in profusion in south and southeast Asia as weeds in cultivated rice fields, along bunds and waterways, in abandoned fields, and seasonal wetlands. From the nature of plant architecture, plant vigor, panicle characters, high spikelet fertility, and boldness of the spikelets, they appear to be hybrids. Genetic studies of such annual wild rices have shown that they are both heterogeneous and heterozygous (Govindaswamy and Krishnamurthy 1958, Nayar 1958, 1973, Sampath 1966, 1973, Oka and Chang 1961, Oka and Morishima 1971, and so on).

Until very recently, it was usual to see indigent people collecting seeds of wild rices for consumption in several regions of south and east Asia. This observation has been recorded by workers such as Watt (1892), Oka (1973, 1988), and so on. The most common method used for collecting the grains was to sway open baskets over the panicles in the fields so that the ripened or ripening grains would fall into them. Sometimes, the panicles were tied together into small bunches to ensure that the ripened spikelets did not fall to the ground through the effect of wind, movement of cattle or people in the fields, or similar disturbances. The collection used to be done at approximately weekly intervals.

Sharma and Shastry (1965b) (also Sharma 2010) have used the following attributes to separate *O. nivara* from *O. rufipogon*.

Characters	*O. nivara*	*O. rufipogon*
Spikelets		
Length	6.0–8.0 mm	7.0–8.0 mm
Width	2.5–3.0 mm	2.0–2.4 mm
Color	Often black	Not given
Awns	6.0–11.0 cm	> 4½ times longer than lemma: 2.3–3.2 cm

—Cont'd

Characters	*O. nivara*	*O. rufipogon*
Habitat	Plateau regions in seasonal pools	Not given
Photoperiod	Insensitive	Sensitive
Flowering period	June–November	November–December

We can see that there is only a marginal morphological difference between *O. nivara* and *O. rufipogon*. The most constant difference between *O. nivara* and *O. sativa* may only be in easier shedding of spikelets, and presence of robust and longer awns, and consistently, red pericarp color in *O. nivara*.

Sharma (2000) has given the distribution of *O. rufipogon* as: "The coastal lowlands of south India, in the lower valleys of the Ganga, Brahmaputra, Irawady and Mekong rivers. It is also available in southern China, southern islands of The Philippines, Indonesia, Papua New Guinea and northern Australia".

He has given that of *O. nivara* as seasonal pools that occur in south and southeast Asia, including the Deccan peninsula, middle Ganga valley, plateau regions of Myanmar, southeast Asia, and southern and southwestern China. It has been described as photo-insensitive, with June to November as its flowering season. Sharma termed *O. rufipogon* as photosensitive, with flowering occurring in November–December.

In the literature, some authors seem to assume that *all* the annual wild rices are *O. nivara* (Cai et al. 2007, for instance), but Sharma (1965b, 2003, 2005, 2010) has distinguished *O. nivara* and the annual wild rices (i.e., *O. nivara* and, what he terms as fatua/spontanea rices) as two entities, though he has not given any features to separate or distinguish them. He has further observed that: "in nature, plants of *rufipogon* or *nivara* with some features of *sativa* are not uncommon and ecologically they are very similar to the recurring parent (sic). They occur in east, south, and southeast Asia."

Importantly, all the characters/dimensions of both *O. nivara* and *O. rufipogon* overlap, except for awn length and spikelet boldness. Other than Vaughan (1994, 2003, 2010), no other taxonomist, who has revised the genus *Oryza* (Tateoka 1963, Clayton et al. 2011) has recognized *O. nivara* as a valid taxon/species (cf. Tables 4.1 and 4.2). Oka (1988), Morishima et al. (1992), and Cai et al. (2008) have termed the entire annual rices as an ecotype of *O. rufipogon*. Other taxonomists like Backer (1946), Bor (1960), and Duistermaat (1987) also recognized only one species for wild rices. Sharma (2010) has gone further and has stated that *O. nivara* occurs in only upland situations and the upland types (*aus* of eastern India) of

O. sativa originated from *O. nivara*. Later, when it spread to lowland areas, it crossed with *O. rufipogon* and the lowland types (*aman* of eastern India) were developed. Sharma has not given a rationale for this proposition.

If it is assumed that *O. sativa* evolved from the annual *O. nivara,* then certain aspects will need to be explained further. First, by all accounts, *O. sativa* is not an annual species, at least, not the tropical landraces. Sharma (2003a,b) in his various writings has observed that *O. nivara* is a photo-insensitive annual species. It is widely accepted that perenniality is a primitive character vis-à-vis annualness (Stebbins 1950). Secondly, it is widely accepted that in the tropics, agriculture originated and the (staple) crop plants were domesticated alongside bodies of water, and not in the upland regions (Sauer 1952, Denham et al. 2007, for instance). In the context of *O. nivara,* Sharma (2010) has stated that rice was domesticated in the upland regions, as according to the author, the early hunter-gathering humans would have found the lowland regions (where *O. rufipogon* generally occurs) inhospitable and insecure to settle in. Incidentally, this statement is totally at variance with the observations of archaeologists, especially with respect to rice in Asia. Thirdly, several authors (Morishima, Oka, Nayar, for instance) have observed that most, if not all, annual wild rice forms are found in typical weedy habitats – along bunds of paddy fields, water/drainage canals, seasonal water bodies, etc. Further, the bigger and bolder spikelets, robust growth habit, and larger panicle size of *O. nivara* could be an expression of hybrid vigor. Numerous authors, beginning with Sampath and Govindaswamy (1958) and Nayar (1958, 1973), have suggested that almost all the wild rices are the products of natural crossings between the perennial *O. rufipogon* and the cultivated Asian rice, *O. sativa.* All these points make it improbable that *O. sativa* originated from *O. nivara* in an upland situation.

B. The Perennial Wild Rice, O. rufipogon

As stated before, there has been much confusion and diffidence in classifying and naming the Asian wild rices because of the tremendous variability in their life history traits, breeding behavior, and morphology. Griffith, working in south Asia in 1851, designated them as *O. rufipogon.* Watt (1892) classified them into four forms of *O. sativa,* viz. *f. spontanea*, *f. benghalensis*, *f. coarctata,* and *f. abuensis.* Roschevicz (1931) classified them as *O. sativa*, *f. spontanea* and *f. aquatica,* Chevalier (1932) as *O. fatua,* Backer (1946) as *O. fatua,* and Chatterjee (1951) as *O. sativa* var. *fatua,* which he subsequently named as *O. rufipogon.* Tateoka (1963) and Clayton et al. (2011) designated them as *O. rufipogon,* but Clayton et al. (2011) restricted the use of *O. rufipogon* to only the perennial forms and did not mention anything about the annual forms. Sharma and Shastry (1965), on the other hand, limited *O. rufipogon* to only the perennial forms, and named some of the annual forms as *O. nivara*. While Vaughan (1994, 2001) initially recognized *O. nivara* as a valid species, he appears to have become somewhat ambivalent about its species status in his subsequent publications (Vaughan et al. 2005, 2008a,b).

Sampath and Rao (1951) were possibly the first authors to state unequivocally that the cultivated rices of Asia evolved by human selection from *O. rufipogon* (then, *O. perennis*). They gave the following points in support of this position:

(a) It has the widest distribution of all the *Oryza* species, occurring in Asia, Oceania, and America;

(b) Its chromosome number is the same as that of *O. sativa* (2n = 24);

(c) Its progeny shows great variation, and significantly, some of the variants closely resemble 'spontanea' forms;

(d) Since its grains are edible, it would have been worth harvesting by the early hunter-gathering people, as is done even now; and

(e) It is gregarious and often occurs in large stands.

They had observed further that fatua or spontanea rices were widely distributed in Asia, and that they appeared to accompany rice cultivation. They cross-fertilized with cultivated types and their progenies occurred in both fields and swamps.

Several workers have studied the nature of Asian wild rices. For a list of references and a summary of early studies, please see Nayar (1973). Both the perennial and annual forms are characterized by heterogeneity, heterozygosity, and semi-sterility. In experimental studies, as many as 17 out of 20 single plant progenies segregated for various plant characters. Different characters showed complementary, modifying, and duplicate modes of inheritance (Nayar 1958, 1973). This observation was further strengthened by Sampath and Govindaswamy (1958) based on their field observations in the Jeypore tract of Orissa (southeast India; 18°–20°N, 82°–84°E).

Scientists of the National Institute of Genetics, Misima, Japan had undertaken extensive field surveys since 1958, almost throughout the tropical rice world, for collecting and characterizing wild *Oryza* species. The surveys and studies were led by the late Dr. H I Oka. Oka, Morishima, and their co-workers had carried out extensive experimental studies on various aspects of rice genetics, primarily using biometric analyses. For their most notable publications, please refer to Iyama and Morishima (1980) and Oka (1988).

In one of the earliest studies to determine the nature and extent of the sterility barriers between and among the cultivated and perennial wild rices, Hinata and Oka (1962) found that the perennial wild strains were mostly interfertile and they gave largely fertile hybrids with *O. sativa*, but at the same time, some of the inter-*sativa* hybrids were partially sterile. The Asian wild rices showed the potential to differentiate into indica and japonica rices, as the wild rices looked more similar to the cultivated rice. The sterility of the perennial wild rices was genetic in nature showing diplontic and/or zygotic effects. The authors proposed that wild rices carried double dominant combinations of duplicate genes.

Morishima et al. (1961) had proposed that the intermediate forms of wild and cultivated plants were the products of introgression, which occurred whenever wild and cultivated rices occurred sympatrically, which was then the situation across much of the rice growing regions of Asia. They found (Morishima et al. 1962a) that *O. sativa* varieties also showed a tendency to differentiate themselves into japonica and indica forms, which could be distinguished by their different patterns of character associations and gene combinations. Sano and co-workers (Sano et al. 1980, Oka and Sano 1981, and so on) had observed that the Thai wild rices could be divided into three populations; perennial, intermediate, and annual (Table 6.5). Oka (1974a,b) proposed that the perennial wild rices are the progenitors of Asian rice. He based his inferences on:

(a) The presence of intermediate forms of wild rice in some regions of Asia, such as Thailand (Ayuthaya) and southeast India (Jeypore tract, Orissa), where they had conducted some field experiments;

(b) The perennial rices possessed more genetic variation, and consequently, had higher evolutionary potential than the annual forms;

(c) *O. sativa* was essentially perennial in nature.

Based on the foregoing discussions, we can see that the Asian rice, *O. sativa* has evolved from the Asian perennial wild rice, *O. rufipogon*.

Incidentally, some early authors (Watt 1892, Roschevicz 1931, for instance) had proposed that certain other species like *O. minuta, O. officinalis,* and *O. coarctata* had also contributed to the development of certain forms of *O. sativa*. This is because of the great diversity displayed by the Asian rice in various morphological and physiological characters – e.g., spikelet form and size, life history variations, very wide adaptability to variations in temperature, water regime and salinity tolerance, etc. Rice is also grown in Kuttanad, Kerala (southwest India) at 1–3 meters below sea level, where a single crop of 90–110 days is traditionally grown after building dykes all around large blocks of lowlands and pumping out water. However, it has now been shown that none of these above species are likely to have contributed to the evolution of Asian rice.

ii. Place and Time of Origin

Inferences about the spatial and temporal aspects of the origin of a cultivated crop are drawn from historical, archaeological, linguistic, and biological evidences.

All south, southeast, and east Asia have been witnessing intense human activity during the late Quaternary Period through to the Holocene epoch. Most of this region receives monsoon rains and possesses generally good, fertile soil. The region is watered by several major and numerous minor river systems, such as the Huang Ho, Yangtze, Mekong, Irrawady, Ganges, Brahmaputra, and Indus rivers. However, unlike the Middle East and north Africa, this region has seen comparatively fewer archaeological excavations except perhaps in selected areas of the Indo-Gangetic plains and east Asia. From the archaeological viewpoint, the submergence of Sunda and Sahul lands ca. 15–45 ka, when the ice melted after the last Glacial Maximum period (ca. <20 ka), was another disadvantage. It has been estimated that the sea level then rose by 60–80 m (Hanebuth et al. 2000, Glover and Bellwood 2004, Clark et al. 2012). This has submerged all evidence of past human activity in the low lying or coastal areas of southeast and east Asia and Island southeast Asia.

East Asia, particularly China, has seen some detailed archaeological excavations over about the last six decades (Liu and Chen 2012; Table 6.1). The oldest evidence of rice anywhere has been obtained from Bashidung (Pengtoushan culture), ca. 6,500–6,000 BP, and Hemudu (Majrabang culture), ca 6,950 ± 130 BP, both in the lower Yangtze river region. As has been the case with most of the excavated sites in east Asia, these sites have yielded significant quantities of rice.

In southeast Asia, the very limited excavations carried out so far have provided evidence of Hoabinhian culture in widely different sites (Bellwood and Glover 2004). In Taiwan, carbonized rice was obtained from Nan-kuan-li, near Taisian, ca. 3,000 BC (Bellwood 2004). Similar carbonized rice has been obtained from the Gua Sireh cave (Sarawak, Borneo) dated to before 2,000 BC (Bellwood 2004).

In Mainland southeast Asia, farmers of rice, cattle, and pigs from the north infiltrated and settled in the interior river valleys between 1,500–1,000 BC (Higham 2004). Khok Phanom Di in south Thailand has shown evidences of rice cultivation, but rice did not appear to be a major component of the local diet then. Since marine and aquatic food appeared to be available in adequate quantities, the local Neolithic societies did not appear to have had much stimulus to change. The other sites where rice remains have been recovered are Phung Nguyen and Co Loa (on the Song da river, north Vietnam), 3rd to mid-2nd millennium BCE. In Thailand, the earliest settlements, where rice has been found, are at Non Nok Tha (northeast Thailand, 2,500–2,000 BCE), and Ban Non Wat (2,000–1,300 BC). The controversy over the claimed very early date of Non Nok Tha and Ban Chiang has now been resolved, since the fragments of rice chaff used to temper clay used for making mortuary vessels have now been AMS-dated to 1,500–1,000 BC (Higham 2004).

In south Asia until about five decades ago, practically all archaeological excavations were related to the Indus Civilization. The oldest rice remains were obtained from Ghalegay (Swat, north Pakistan) and dated to about 6,000 BCE. The Harappan civilization sites are all located outside the traditional rice growing regions, and, as stated already, this civilization was an urbanized, industry-based economy (Poessehl 2002) (Table 6.2, Fig. 6.2).

About 40 years ago, Sharma and co-workers of the University of Allahabad (UP, India) carried out excavations in the Ganga valley at Chopani Mando, Koldihwa, and Mahagara (Sharma et al. 1980, Sharma 1985). They obtained some very early material belonging to the pre- and early Neolithic eras. Unfortunately, the accuracy of their data has been questioned because of the faulty sampling and dating procedures used. More recently, some newer sites have been excavated in the Ganga valley region, as stated before (Saxena et al. 2006, Pokharia 2008). These authors have obtained rice phytoliths dated to 8,350 cal yr BP from Lahuradewa (Saxena et al. 2006), and rice grains along with the seeds of several other crop plants from Takwa (Pokharia 2008), which have been dated to 3,810–3,572 cal BP. Incidentally, Fuller (2010) has expressed some reservations about certain methodologies used in Lahuradewa.

Thus, we have evidence from east and south Asia of the possible use of wild rice around 8–10,000 years ago, and rice culture from about 2,000 years later in east Asia, and later than this in southeast Asia. This is understandable, because we can infer that 10,000 years ago, wild rices were present very much more widely than was the case in the 1950s, when scientists in Asia began to look at the wild rices scientifically. We have also noted that the climate in early Holocene in east Asia favored wild rices extending their habitats further north than is now possible. In addition, the human population at that time would have been only a fraction of what it is now. Based on this, some scientists from China, beginning with Ting (1949), have claimed that rice first originated in China. However, in archaeology, the presence of evidence for a certain thing in one location does not preclude its presence elsewhere.

China or India have been proposed as the center of origin of Asian rice by early workers such as De Candolle, Watt, and Vavilov. While Watt (1892) proposed that rice originated in peninsular India, Gustin (1938) suggested that it arose on the slopes of the Himalayas on both the Chinese and Indian sides. Hamada (1949) and Burkill (1953) stated that rice could have originated in India or China because of the prevalence of maximum species diversity in the region. Similar points of view have been expressed by Ting (1949), Ramiah and Ghose (1951), and Chatterjee (1951). Please see Nayar (1973) for an overview of earlier work.

More recently, Chang (1976b) suggested that Asian rice evolved in a vast stretch of land extending across the foothills of the Himalayas in south Asia and its associated mountain ranges in Mainland southeast Asia and southwest China. Since this reference is one of the most quoted ones, we shall go into this aspect in some detail.

Whyte (1972) stated that the alternating periods of drought and temperature variations during the Neothermal age (10,000–15,000 years ago) that prevailed in east India, northern southeast Asia, and southwest China favored the development of annuals (from perennials). The annual forms survived better and produced more seeds. However, he did not explain how this could have happened. This enabled the newly developed forms to move further northwards than their perennial progenitor. The increased aridity that prevailed at that time also forced people to move into more humid areas. According to Chang (1976b), "Asian rice (*O. sativa*) evolved from an annual progenitor over a broad belt that extended from the Ganges plains along the foothills of the Himalayas, across upper Burma, northern Thailand, and Laos to north Vietnam and south China. Domestication could have occurred independently and concurrently, and consequently, at many sites inside or bordering the belt. From this aggregate of centers and noncenters, rice was most likely introduced from the Nepal–Assam–Burma–Yunnan area into the Yellow river (Huang He) valley, and from Vietnam via a coastal route into the lower Yangtze river basin where a cold-tolerant race (keng or japonica) became established... From China, rice was introduced into Korea and Japan in the third century BC, more appropriately designated as sinica than japonica... The tropical race (indica or hsien) spread southward into the Malay archipelago and northward into central China...".

To support this contention, Chang (1976b) cited Ramiah (1937), Ramiah and Ghose (1951), and Chatterjee (1951). The following is what these authors stated in their publications. Ramiah (1937) had stated that: "Researches point unmistakably to the fact that southeast Asia, India, China or Indo-China must have been the original home of rice... Whether rice was introduced into India from China or it existed in India at the same time as in China cannot be accurately determined..."

Ramiah and Ghose (1951) had observed that: "The cultivated species, *O. sativa,* is considered to be a native of India and Indo-China. From south India, where probably rice first originated,... rice spread east and west... In China also, rice probably spread from India in very early times..."

And Chatterjee (1951) stated that: "...it seems clear that the eastern part of India, Indo-China, and part of China could reasonably be the areas where the cultivated rice (*O.sativa*) originated."

Chang (1976a,c) cited Roschevicz (1931) to support his claim. On this topic, Roschevicz's (1931) observation was that: "there are solid reasons for supposing that it (rice) was not initiated in Africa, but in Asia, namely, India, China, and Indo-China".

Roschevicz had stated further that: "it is still China, which must be recognized as the center from where cultivation of rice probably spread to other countries..."

From the above, it is clear that much of what Chang stated in his papers were his own deductions. The detailed studies carried out in China since Chang's above observations have shown that the putative ancestral species of rice, *O. rufipogon*, was present extensively in south China (Tang and Xuan 2010) until recently. Although most of it have since disappeared, some populations are being maintained under protection. Presently, *O. rufipogon* grows in eight provinces in southwest and south China, with the northernmost site located at Dong-Xiang (Jiangxi province; 28.1°N), south of the Yangtze river. Archaeological and paleoclimatic evidence indicate that 10,000–7,000 years ago, rice had reached the southern Huang He river (37°N). Archaeological remains of wild rices (12,300–7,000 yr cal BP) have been obtained from 16 sites along the Yangtze river (ca. 30°N) (cf., Tang and Xuan 2010) (Table 6.1). More evidence of the ancient presence of both wild and cultivated rices in southeast and south central China have been obtained in the form of substantial archaeological remains from numerous sites dated to 6,000–4,000 yr cal BP, as we have noted already. Hence, there may not be any need to invoke a south to north movement of rice as proposed by Chang (1976a,b,c), since rice could have originated there also. Further, this viewpoint is totally at variance with the present widely held view (Bellwood 2004, 2010) that rice had been taken south from the lower Yangtze valley.

iii. Domestication Process

A. Introduction: Historical

Various authors have used different terms to refer to the domestication process of rice, such as dynamics of domestication, domestication syndrome, domestication process, and so on. The term "domestication syndrome", attributed by some authors to Harlan, appears to have gained greater currency in recent literature, but it does not seem to appear in the second revised edition of Harlan's *Crops and Man*, which appeared in 1992. Early evolutionary biology authors such as Stebbins (1950), and Harlan (1992), do not appear to have used this term either. Incidentally, Hammer (1984) may have been the first author to use this terminology. According to the Oxford English Dictionary (2006) a syndrome is a "group of symptoms which *consistently* (emphasis added) occur together". We can see that other than the non-shedding character (or reduced level of shattering) and synchronous ripening of seeds, no other character is universally present in all the cultivated rices, even including their annual habit. All the other characters such as awnness, white pericarp, waxy rice, floating habit, smaller spikelets (Jeerakasala), low tiller numbers (e.g., javanica) are only regional preferences of various linguistic, ethnic, religious, or cultural groups, at least until recently (say, up to about the mid 1950s).

It is now universally accepted that domestication is a process and not an event. Further, in cultivated plants, the direction and pace of evolutionary processes are largely determined by humans. The mode of evolution of rice has been sympatric in nature, as noted already. In nature, however, sympatric speciation occurs only rarely (Dobzhansky 1970). It is more so in plants, because the initial protection required by the mutant form against pollination from the mother species would be difficult to achieve under natural conditions because of their stationary nature.

The mode of evolution of cultivated rice may have been the simplest of all the cereals, because the morphological and physiological characters that distinguish the cultivated rice, *O. sativa,* from its putative ancestral species, *O. rufipogon,* are mostly only quantitative in nature. There are of course some strong disagreements on certain details – the number of times the domestication process might have taken place in the evolution of indica rices, and the places where this was done.

There is general agreement that Asian rice, *O. sativa,* evolved from the wild rices that occur almost sympatrically throughout the regions of traditional rice cultivation in Asia. Opinions differ over the immediate progenitor – whether the annual one, part of which is named *O. nivara*, or the perennial species, *O. rufipogon*, or an intermediate annual–perennial form. In this context, we need to keep in mind that *O. sativa* is not strictly an annual species, but an 'annual–perennial' form, one that is capable of perennating, if suitable growing conditions – primarily water – are made available.

As early as 120 years ago, Watt (1892) had stated that the progression from the wild to the cultivated forms seemed to have been in the reduction of the size and alteration of receptacle shape, in the loss of the awn, and in the shortening, but widening, of the combined outline of the inner glumes. These changes have been elaborated further (Nayar 1973) as a change in habit from prostrate or procumbent habit to an erect and compact one, from shattering nature of the grains to non-shattering habit, from asynchronous flowering and panicle maturation to synchronous flowering and panicle maturation, and to an increase in the number and weight of spikelets per panicle, reduction in awns, reduction in dormancy, from black lemma-palea to brown or golden color, and from red to white pericarp. This is the generalized situation in most regions of the world. Of these changes, the four most significant ones in the domestication process appear to be:

(i) The non-shattering habit of spikelets,

(ii) Synchronous flowering and grain ripening in the plants,

(iii) Development of a more annual or determinate habit from the perennial state, and

(iv) Changes from a partly cross-fertilizing nature to predominant self-fertilization.

Even among these, the single most decisive and effective change has been the change from the highly shattering nature of the spikelets that is present in the wild rices to less shattering and more persistent spikelets in domesticated rice. This is followed by synchronized flowering and ripening of grains.

Seed shattering in rice is controlled by a pair of complementary genes (Kinoshita 1995). Now, quantitative trait loci (QTL) have been identified for this character. A non-shattering nature could have been obtained either by recombination or by mutation at one of the loci involved in the character. With such a primary gain, selection for other desirable characters could have been taken up by early humans in due course (Nayar 1958, 1966, 1973, Doebley 2006). We can see that even without acquiring this gain, poor, but not primitive people, have been harvesting grains from wild rices in south China and India, wherever their large stands were present.

Morishima, Oka, and co-workers have done some pioneering studies on the domestication process in rice. They found that positive responsiveness of wild and cultivated rices to improved cultural practices were directly correlated with their degree of domestication (Oka and Chang 1959, 1964b, Hinata and Oka 1962b, Oka and Morishima 1971). They found further through experimentation that, because of their partially allogamous nature, asexual reproduction, and pronounced seed dormancy, perennial wild rice was able to accumulate variation brought about by mutation and hybridization. When plants approached cultivated forms, the extent of selfing increased and seed dormancy was weakened. When intermediate forms of wild rices were grown in cultivated conditions, they rapidly tended to become homozygous and show cultivated characters (Morishima et al. 1961). Rice does not show any adverse effects of inbreeding.

All the evidence favors a multiple *in situ* origin of Asian rice, since the cultivated species has had an extensive interface with wild rices throughout most of the regions of its cultivation in east, southeast, and south Asia. Thus, the mode of speciation in rice has obviously been sympatric in nature.

However, sympatric speciation (i.e., formation of new species within a freely interbreeding population) has always been considered as a difficult, controversial, and rare event (Coyne and Orr 2004). Mayr (1970) had stated earlier that: "the essential component of speciation, that of genetic repatterning of populations can take place only if these populations are protected from the disturbing inflow of alien genes".

In rice, this protection would have been given by the first farmers of that period, 8,000–12,000 years ago. They would have selected apparently superior forms which possessed some resistance to shedding, and then grown them protectively elsewhere. This would also have been aided by the additional protection provided by the changes in the breeding system from partial cross-fertilization (30–51%; average ca, 35%) in wild rices to self-fertilization in cultivated rices (0–6.8%, usually $<3.0\%$; Oka 1988).

Asian rice could have evolved by disruptive selection (Nayar 1973). This "process involves the selection of divergent phenotype extremes in a population, until, after several generations of selection, two discontinuous strains are obtained" (King et al. 2006). In rice, the process of "nobilization" brought about by the first farmers may be analogous to an experiment conducted by Thoday and Gibson (1962) on bristle number in *Drosophila,* in which low and high numbered forms were obtained after just 12 generations. In rice, however, the conditions would have been less favorable for random mating, and increasingly adapted to assortive mating, consequent on human intervention. This would have hastened the process of speciation. When the "first farmers" harvested grains from wild rice fields by swaying baskets over them, what they would have been retrieved would have been what was left over after the most easily shed grains had fallen down by various factors such as wind, human and cattle movements within the fields, and so on (Doebley 2006). They would have then saved a part of this harvest for further planting. This progeny population would have been less susceptible to shedding than that of the "parent" population. Competition between the two populations was excluded and gene flow too would have been restricted, consequent on spatial isolation provided by human intervention.

Based on the above observations, Nayar (1973) had proposed "that rice has had a diffuse origin both in space and time", implying that rice originated more than once and at more than one place. This may explain the different centers/regions suggested for the origin of Asian rice – southern India (Watt 1892, Ramiah and Ghose 1951, Nair et al. 1964), Jeypore (Orissa in southeast India) (Ramiah and Ghose 1951), the Ganga valley (Fuller 2010), The Philippines (Ramiah and Ghose 1951), China (de Candolle 1886, Roschevicz 1931, Ting 1949), Indo-China (Hamada 1949), the western Pacific region (Denham et al. 2007), and so on. Subsequent to Ting (1949), numerous authors have claimed that China is the center of origin of Asian rice. In the light of present evidence, this appears to be proven as of now because of the recovery of substantial quantities of rice older than 4,000 BCE from several archaeological sites in China (Liu and Chen 2012).

a. K and r Selection

Oka and co-workers have studied reproductive allocation in wild rices as an adaptive strategy and as a contributory component of domestication (Table 6.3). Reproductive allocation was higher in annual rices (38–50%) than in perennial types (3–12%), with intermediate forms showing intermediate values (10–44%) (Sano and Morishima 1982, Morishima et al. 1984). Further, reproductive allocation was negatively correlated with vegetative regenerating ability of excised stem cuttings ($r = -0.75$). These authors found further that populations living in an environment that

Table 6.3 Life History Traits of *O. rufipogon* vis-à-vis: Reproductive Allocation and Pollinating Allocation*

Trait	Mean ± sd[1]	Correlation with RA	Correlation with PA
Seed no./plant	481 ± 243	.94★★	− .69★★
Seed weight/plant (g)	11 ± 6.4	.93★★	−.70★★
Panicle no./stem no. (%)	69 ± 29	.87★★	−.67★★
Seed setting (%)	73 ± 25	.81★★	−.61★★
Spikelet no./plant	618 ± 222	.78★★	−.72★★
Selfing rate estimate (%)	71 ± 21	.71★★	−.79★★
Single seed weight (mg)	21.6 ± 4.5	.61★★	−.64★★
Panicle no./plant	14 ± 7.3	.59★★	−.65★★
Awn development index	1.15 ± 1.64	.51★★	−.47★★
Submergence tolerance (%)	29 ± 29	.47★★	−.36★
Awn weight/seed weight (%)	9.7 ± 4.0	.45★★	−.38★
Seed dormancy index	2.3 ± 0.3	.43★★	−.11
Drought tolerance index	2.1 ± 1.2	.38★	−.36★
Spikelet no./panicle	50 ± 22	.04	.01
Regenerating ability	1.4 ± 0.7	−.48★★	.76★★
Tillering angle	56 ± 21	−.49★★	.52★★

[1]sd: standard deviation
*Adapted from Oka and Sano 1981

imposed high density-dependent mortality would be selected for allocating a higher proportion of resources to reproductive activities at the cost of their capacity to propagate under crowded conditions, and conversely, populations living in another environment that imposed higher density-dependent mortality would be selected for allocating a greater proportion of resources to nonreproductive activities at the cost of their capacity to utilize unstable habitats (Oka 1988).

Oka (1988) went on to suggest that this might be an instance of K and r selection, in which annual and perennial wild rice forms followed K and r strategies respectively Oka (1976) (Table 6.3). K and r strategy is a theory in population ecology that attempts to establish whether environmental conditions favor maximization of r (the intrinsic rate of natural increase) or K (the carrying capacity of the environment) (King et al. 2006). When the populations can expand without food reserves limiting their growth, then r selection is in control. When food reserves limit the population size, K selection takes over, and increases in one genotype must be at the expense of the other, whereas r selection operates in ecological situations where food reserves fluctuate rapidly and species are favored which reproduce slowly and generate a few offspring that are well adapted to a relatively stable environment (King et al. 2006).

Oka (1976) had earlier considered that this theory was too deterministic in the context of the evolution of rice. For instance, experimental plantings of germinating seeds of several wild rice strains in seminatural weedy habitats showed that annual forms showed higher juvenile mortality than perennial forms. However, the author was unsure if the observed mortality was density-dependent or independent.

Since rice domestication is heavily influenced by anthropogenic factors, it is not sure if this theoretical model would be applicable to the domestication process. In the 30–50 years after Oka and associates undertook these field studies in Thailand (central) and India (central and southeast), practically the entire wild rice population in these regions has disappeared, and what now exists are for all purposes feral populations carrying various levels of *O. rufipogon* and *O. sativa* genetic content.

b. Phenotypic Plasticity

Phenotypic plasticity is a phenomenon in which a given genotype may develop different states of a character or group of characters in different environments (King et al. 2006). This can include all types of environment-induced changes. Even though phenotype plasticity is expressed in all organisms, it is more widely expressed in plants. This is ascribed to the sedentism of plants (animals can move away from an adverse environment) and their presence in an open ontogeny in which functionally equivalent organs are produced sequentially through the growing season (e.g., variations observed in the size, shape and frequency of appearance of leaves depending on the environment, especially the availability of nutrients and moisture, population density, etc.) (Bradshaw 1965, Niklas 1997).

Character variation within an individual species is typically more determined by environmental factors than by genotype differences. The ability of a species to adaptively alter morphological and physiological functional traits to environmental variations is well expressed (Niklas 1997, Agrawal 2001). This attribute can then easily explain the tremendous variations in morphological and physiological characters that we notice, both in wild and cultivated rices, including in their life history traits. However, this phenomenon does not appear to have been studied in rice.

Eiguchi et al. (1993) observed that the phenomenon of rapid elongation of internodes that takes place in deep water rice varieties, when the flood waters may rise by a meter or more in 24 hr, is an instance of this phenomenon.

They used a perennial *O. rufipogon* strain possessing deep water tolerance together with a non-deep water cultivated rice variety. The authors determined that the tolerance for deep water was controlled by a recessive gene with major effects, *dw3*.

Shimizu et al. (2010) investigated the extent of genetic control of phenotypic plasticity in a cultivated rice (a japonica landrace) and a wild rice (an annual *O. rufipogon*) by studying their responses to five environmental conditions consisting of nutrient and plant population changes. They detected QTL clusters producing large effects on the studied characters on chromosome 7. They felt that these QTLs might have contributed to bringing about changes to the plant's architecture during domestication. They proposed that allele-specific environmental specificity could be controlling plasticity; further that both cultivated and wild rices showed distinct genetic differences for phenotypic plasticity, and these must have contributed to adaptation under contrasting environmental heterogeneity during domestication.

Zhang et al. (2012) indentified a gene *rice plasticity 1 (RPL1)*, that is responsible for causing environment-dependent plant height variations. A BLAST analysis showed that the gene was located in a putative gene (*LOC–Os 06g 13640*) on chromosome 6 (BLAST: Basic Local Alignment Search Tool, finds regions of local similarity between sequences).

The few studies that have been carried out on phenotypic plasticity in rices have been one-off studies using only limited experimental conditions. Other than confirming the prevalence of phenotypic plasticity, these studies do not appear to have contributed to estimating the extent, direction, and manner of its effect on the domestication process in rice.

B. General Studies on Domestication-Related Characters

We shall now consider the experimental studies which have been carried out on various characters of rice that are relevant to its origin.

Xiong et al. (1999)'s study was one of the earliest in this respect. They studied 19 characters – seven qualitative and 12 quantitative – in an F_2 population of *O. sativa* (indica) x *O. rufipogon*. The loci controlling inheritance were identified from a molecular linkage map having 348 molecular marked loci (313 RFLP + 12 SSR + 23 AFLP). All the qualitative characters were controlled by simple Mendelian genes. The analyses of 12 quantitative traits showed 44 QTLs, with the amount of variation/QTL ranging from 6.9% to 59.8% per trait. Most domestication-related characters were located in a few chromosomal blocks. Regeneration ability was controlled by four QTLs, and together, they accounted for only 36.7% variation. Several of the other characters studied by the authors then did not seem to have had any direct bearing on rice domestication.

Cai and Morishima (2002) carried out a similar study using *O. sativa* (indica, Taiwan) and *O. rufipogon* (China) as parents. By raising single plant progeny from 189 F_2 plants, they selected 125 RILs (recombinant inbred lines). They examined 147 polymorphic markers (mostly RFLPs, few isozymes, RAPD, etc.) for 37 traits distributed on 12 chromosomes. They detected QTLs for 24 characters. The QTLs showed a general tendency to cluster, but the QTLs for annual–perennial differentiation did not cluster.

Li et al. (2006) studied the molecular biology of domestication-related characters using 304 F_2 plants of an *O. sativa* (indica, China) x *O. nivara* cross, using single sequence repeats (SSRs) covering all the 12 rice chromosomes. They detected 115 polymorphic markers. Synchronous seed maturation (*sm 9*) was correlated with change of hull color. They proposed that, compared to *O. rufipogon*, domestication from *O. nivara* would have involved "QTLs with a greater degree of chromosomal co-localization" and would have required fewer genetic changes associated with life history and mating system transitions; they also observed that "development of rice domestication syndrome involved co-localized QTL and QTLs of large effect".

Grillo et al. (2009) observed that more than 80% QTL alleles of *O. nivara* acted in the same direction in phenotypic evolution, and "they were fixed under directional selection". They felt that "loss of photoperiod sensitivity might have been essential for the survival of the ancestral populations of *O. nivara*". According to Sharma (2010), *O. nivara* is a short-day adapted, photo-insensitive annual, photoperiod sensitivity was controlled by a QTL of relatively large effect, and mating system evolution from cross- to self-fertilization and modifications in panicle and floral morphology might have been controlled by QTLs of small effect. "The lack of segregation of recessive habit in the F_2 mapping population suggested that the evolution of the annual from perennial forms had a complex genetic basis" (Grillo et al. 2009).

Rice is often classified as a short-day adapted plant (Gao et al. 2013), but this does not appear to be correct, as we have noted before. A thumbrule is that short duration varieties are photo-insensitive and long duration ones photosensitive. There may be only a few exceptions to this. Earlier studies have shown that the *CO-FT* module in the long-day plant *Arabidopsis* is evolutionarily conserved in the short-day rice plant as *Hd1-Hd3a*. Gao et al. (2013) observed also that there is another flowering pathway in rice that is regulated by the gene *Edh-4,* it functions independently of *CO-FT.* Plants possessing the *edh-4* gene do not flower under long-day conditions. The authors identified a rice mutant possessing the *edh-4* gene that never flowers under long-day conditions. This gene, *edh-4,* is a novel transcriptional regulator that promotes flowering through activation of two florigen genes, *Hd3a* and *RFT1* through *Edh1,* which signals flowering initiation. *Edh-4* was found to have two major haplotypes: *Hap-2* is the major haplotype in indica accessions, mostly distributed in lower latitudes and lower elevation zones, and *Hap-3* is the major haplotype in japonica accessions, mostly distributed in higher latitudes and higher elevations. *Hap-3* was functionally more important in promoting flowering under long-day conditions. This suggested to the authors that *Ehd-4* might have contributed to the initial expansion in ancient times of rice northward and also to higher altitudes.

C. Seed Shattering

We have stated before that imparting resistance to seed shedding could be considered as the most important single step in the domestication process of rice. Seed shedding in cereals is known to be under relatively simple genetic control (Harlan 1975). In rice, Cai and Morishima (2002) found four QTLs responsible for shattering. They were *sh1* (chromosome 11), *sh2* (chromosome 2), *sh3* (chromosome 4), and *sh4* (chromosome 3). Earlier, Eiguchi and Sano (1990) had discovered a gene complex responsible for seed shattering and open panicle in common wild rices.

Li et al. (2006) studied this character using an F_2 population of indica x *O. nivara.* Of the three alleles studied by them, *sh4* accounted for 69% phenotypic variance, *sh3* caused 6% variance, and *sh8,* 3.1% variance. The analysis located the QTL *sh4* between the SSR markers, *RC 4-123* and *RM 280.* They were 1360 kb apart in the *sativa* genome. F_2 plants heterozygous at *sh4* and having *ns* and *nn* genotypes shed all the grains when tapped, while homozygous recessive plants with *ss* genotypes did not either shed grains upon tapping, or if they did, did only partially. By examining SSR and single nucleotide polymorphism (SNP) markers between *RC4-123* and *RM 280,* the authors were able to map the mutation responsible for nonshedding to a 1.7 kb region of a gene with a previously unknown function. This was found to be a transcription factor. The authors proposed that human selection of an amino acid in the predicted DNA binding domain encoded by the gene of previously unknown function was primarily responsible for the reduction of grain shedding.

The authors then sequenced this region in 45 accessions covering all the AA genome species (including 14 *O. sativa*) to assess its polymorphism and evolutionary history. All the cultivars were polymorphic for the mutation, i.e., some of the cultivars had the same sequences as all the accessions of *O. nivara*. At the remaining six mutation sites, all the cultivars shared the same sequences, and they were different from those of the *O. nivara* parent. Further studies showed that the expression of *sh4* was seen at the pedical and flower junction. The *sh4* gene showed increased expression in the later stages of seed maturation (Figs 6.4 and 6.5).

Lin et al. (2007) used a backcrossed population of a cultivar, Tequing, and an introgression line, IL 105, with seed shattering effect derived from a perennial wild rice, YJCWR. They found that seed shattering was controlled by a single dominant gene, *Shattering1* (*SHA1*), encoding a member of the trihelix family of plant-specific transcription factors. *SHA1* was mapped to a 5.5 kb genomic fragment which contained a single reading frame. The predicted amino acid sequence was distinguished by a single amino acid substitution caused by a single nucleotide change, *g237t*. The authors found that this mutation was present in all the 92 indica and 108 japonica cultivars studied by them, but in none of the 24 wild rice accessions. Based on these results, they concluded that all the domesticated rice cultivars harbored the mutant *sha1* gene, and consequently, did not shed their seeds on maturity. The authors proposed further that the non-shattering mutation occurred prior to japonica–indica differentiation.

Konishi et al. (2006) also carried out a detailed study on the molecular genetics of seed shattering. They used initially Nipponbare, a non-shattering japonica landrace, and Kasalath, a shattering indica landrace, and their F_2 population. QTL analysis showed the presence of five QTLs, three of them on chromosomes 1, 2, and 5 of Nipponbare and two on chromosomes 11 and 12 of Kasalath. The QTL with the largest effect, *qSH1*, with 68.6% of the total phenotypic variation, was located on chromosome 1 in Nipponbare. The authors then prepared a near-isogenic line (NIL) with the Nipponbare genetic background, but containing a short chromosomal segment of Kasalath that included the *qSH1* region. To fine-map the *qSH1* gene, the authors conducted a linkage analysis using a segregating population of 10,388 plants.

These authors found that the *qSH1* gene, encoding a *BEL1* type homebox gene and a SNP in the 5′ regulatory region of the *qSH1* gene, caused loss of seed shattering, owing to the absence of abscission layer formation. Further, haplotype and association analyses in various rice collections

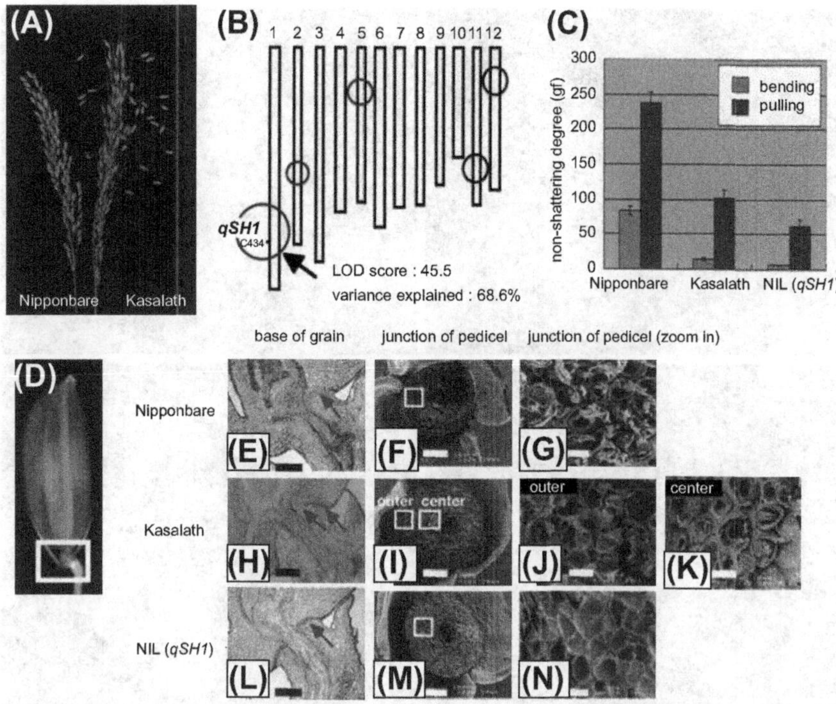

Figure 6.4 ***Formation of Abscission Layer in Rice.*** (A) Seed-shattering habit of rice panicle. Nipponbare: nonshattering cultivar; Kasalath: shattering cultivar; (B) Chromosomal locations of QTLs for seed-shattering. Positions of circles indicate QTL positions. Circle size indicates relative contribution of each circle; (C) Non-seed –shattering habit of Nipponbare, Kasalath, and NIL(*qSH1*). Breaking tensile strength upon detachment of seeds from the pedicels by bending and pulling was measured (910). Higher value indicates loss of shattering. (D) A rice grain. Box indicates position of abscissin layer. (E) to (G) Nipponbare. (H) to (K) Kasalath. (I) to (N) NIL (*qSH1*). (E) (H), and (L) indicate longitudinal sections of positions corresponding to white box in (D). Arrows point to the partial abscission layer of Kasalath in (H), the complete abscission layer of NIL(*qSH1*) in (L), and the corresponding region of Nipponbare in (D). (F), (I), and (M) Scanning electron microscope (ESM) photos of pedicel junctions after detachment of seeds. (G), (J), (K), and (N) Close-up SEM photos corresponding to white boxes in (F), (I), and (M). (G). Broken and rough surface of Nipponbare when forcedly detached. (N). Peeled-off and smooth surface of NIL(*qSH1*) upon spontaneous detachment in Kasalath, rough centre surface (K) and smooth outer surface (J) are observed. Scale bars: 500 micro m in (E), (H) and (L); 100 micro m in (F), (I), (M); 1o micro m in (G), (J), (K), and (N).
Reproduced from Konishi et al. (2006) with permission.

revealed that the SNP was highly associated with loss of seed shattering in japonica rices. The NIL (near isogenic line) of Nipponbare having the *qSH1* gene of Kasalath exhibited a complete abscission layer between the spikelet

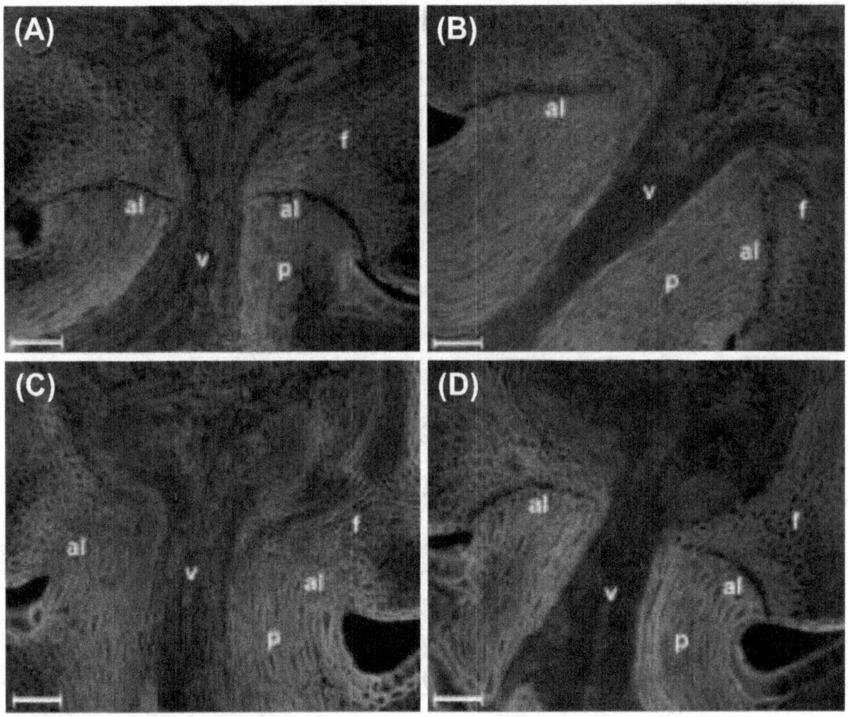

Figure 6.5 ***Fluorescent Images of Longitudinal Section of Flower and Pedicel Junction Showing Variations in Abscission Formation in Different Wild Rices.***
Reproduced from Li et al. (2006) with permission.

and the pedicel. Nipponbare did not show any evidence of an abscission layer. This also confirmed that a mutation in the *qSH1* gene alone resulted in the loss of abscission layer in Nipponbare, and re-introducing the gene from Kasalath could restore it. Kasalath could form a partial abscission layer only at the peripheries, possibly because of the presence in it of other minor QTLs for seed shattering (Fig. 6.4). The large-scale linkage analysis of 10,388 plants segregating at the *qSH1* region enabled the mapping of the functional natural variation in a 612 bp portion between the flanking markers *qSH-F* and *qSH-H,* with only one SNP within this region. However, located away from the SNP, the authors found one ORF (open reading frame) for a rice ortholog of the *Arabidopsis REPLUMLESS (RPL)* gene. This gene encodes a *BEL1* type homebox that is involved in the formation of a dehiscence zone in *Arabidopsis.* The authors were subsequently able to confirm that this ORF ortholog was the *qSH1* gene, and that this SNP only affected the spatial mRNA expression of *qSH1* at the abscission layer.

Zhang LB et al. (2009) studied nucleotide variation at the shattering loci, *sh4* and *qSH1* in indica (17 accessions), japonica (14 accessions), *O. nivara* (16 accessions), and *O. rufipogon* (28 accessions). Earlier, *sh4* had been found to account for 69% of the phenotype variation between indica and *O. nivara* (Li et al. 2006), and *qSH1*, 69% shattering difference between indica and japonica (Konishi et al. 2006). The causative mutation was a nucleotide substitution approximately 12 kb upstream of a homeobox gene, with the mutation substantially reducing gene expression and disrupting the development of the abscission zone. In *sh4* gene, a G to T substitution which resulted in amino acid change from lysine to asparagine was responsible for grain shattering reduction. The *sh4* phylogeny together with neutrality tests and coalescent simulations suggested that *sh4* had a single origin. It was further estimated that with strong selection pressure, the fixation of *sh4* gene could have been completed in 118 years in japonica and 97 years in indica rices.

The results obtained by Lin et al. (2006) are at variance with the other three studies. Zhang LB et al. (2009) concluded that only about 100 generations of intensive selection would be enough to impart seed shattering resistance during the domestication process, which is very much at variance with the proposition of Fuller et al. (2009) that the process of rice domestication in the lower Yangtze river basin took several millenia.

Ji et al. (2006) studied the nature and genetics of seed shattering in a mutagen-induced (N-nitroso-N-methylurea) mutant named Hsh obtained from a japonica variety, Huracheong. Earlier, Ji and coworkers had studied the biology of seed shattering in rice and found that the morphology of the abscission layer differed between varieties. Some varieties did not possess an abscission layer and they were resistant to shattering, some others had well-developed abscission layer, and their seeds were shed easily, and yet others had a partially-developed abscission layer and they showed moderate levels of seed shattering. They studied the breaking tensile strength (BTH) of the pedical at five day intervals during 5–40 days after flowering and measured the gravitational unit of force (gf). In the japonica variety, it was 180–240 gf, while in the mutant Hsh, it was 50 gf. Its early shattering nature was controlled by a single recessive gene *sh-h* located on chromosome 7.

In practice, if we need a benchmark to designate a wild form as being at least partly domesticated, this could be when the wild form has acquired a certain extent of resistance to seed shattering. This is so for two reasons. Firstly, even without the wild rices possessing this characteristic, poor people, not primitive people, in many regions of the world had considered it worthwhile to harvest the easily shedding wild rice grains for food. This has been documented by many authors (e.g., Watt 1892). It is also mentioned in some ancient texts of China and India (cf., Kumar 1988, Oka 1988, for details). Secondly, with this primary gene, the early farmers could have used the seeds of their selected material for further multiplication for use partly as food and also for carrying out further multiplication (and ennoblement) of this material.

D. Synchronous Flowering and Panicle Ripening

Possibly, the second most desirable gain that the first farmers would have liked to have in their domesticated rice would be synchronous flowering and ripening of all the panicles in the plant.

In a normal rice crop, flowering and seed ripening in a plant are completed within 7–10 days each. However, in the progenitor species, *O. rufipogon,* the production of flowering panicles, their opening and seed ripening is a process that continues for several weeks. In south and southeast Asia most of the flowering and seed setting takes place in *O. rufipogon* during August–November, but some stray particles continue to emerge until the following March–April, if moisture availability is not a limiting factor in the development of the plant. Usually, if *O. rufipogon* grows in a pond or other wetland that has water/moisture throughout the year, the stubbles of the mother plant remain viable indefinitely. In seasonal wetlands, new shoots begin to appear in March–April with the onset of the first rains. In such conditions, the survival of the stubbles during summer will depend on the availability of at least some residual moisture in the soil. While the main shoot flowers and produces panicles, new shoots carrying panicles will continue to emerge periodically from inside the leaf sheaths of successively younger leaves of the main shoot. Two to four such shoots may appear from a main shoot. Seed setting will be low (say, <50%) in the main panicle and generally, this declines in the secondary and tertiary panicles.

If the annual spontanea/fatua/*O. nivara* rices grow on field bunds, they flower at almost the same time as the cultivated crops, or if they grow as a weed in paddy crop, this occurs usually a week or so before the crop comes to flower. If the annual rice grows in pure stands, it follows its own timetable. This mode of flowering is attributed to the adaptive reproductive strategy of a successful weed plant (Baker and Stebbins 1965, Harper 1977).

In cultivated rice, practically all the tillers come to flower and are harvested at the same time. But in annual wild rices, often some tillers do not flower at all, while the others will flower at different times (see also Morishima et al. 1962b, Oka 1988).

Oka and co-workers (mainly Morishima and Sano) have worked on the synchronized appearance of tillering and flowering and seed maturity in relation to rice domestication. They had the advantage of visiting all the wild rice growing regions of the world before carrying out some *in situ* studies in Taiwan, Thailand, and India, lasting up to five seasons in some sites. Based on their extensive field observations and experimental studies, the authors recognized a continuum of wild rices ranging from perennial wild rices (*O. rufipogon*) to annual rices (spontanea, fatua, *O. nivara*, etc.) including intermediate annual-perennial forms (Oka and Chang 1959, Morishima et al. 1961, Oka and Morishima 1981). According to them, this situation was prevailing throughout south and southeast Asia, except possibly in Indonesia and the Philippines, with heterogeneous wild rices occurring in the vicinity of rice fields. The life habits of wild rices are conditioned by the availability of water and extent of disturbance in the fields. The annual wild rices occur also as weeds in rice fields, on bunds and channels, and in seasonal wetlands. The annual–perennial forms are found in intermediate habitats. Several field experiments and numerous field observations have shown that the intermediate annual–perennial forms are the intermediate forms of domestication of *O. sativa*. Presence of perennial, intermediate, and annual forms closely related to the depth of water table early in the season, the extent of human disturbance, and percentage of plant cover (Table 6.4).

E. Development of Annual Habit/Determinancy

This character is relevant to synchronized flowering and ripening of grains discussed above. However, we should recall that rice is not an obligately annual crop. Most landraces/varieties are capable of regenerating and producing new shoots and panicles, if minimum conditions for growth, especially water, are made available. Incidentally, in the 1950s, there was some interest in raising a ratoon crop of rice (Grist 1975), since such a crop would reduce cultivation costs substantially. A ratoon crop can give nearly two-thirds the yield of the "mother" crop.

Most of the life history traits of wild rice appear to have a bearing on its breeding system and resource allocation. The annual types are generally early flowering, short in stature, short lived, produce more seeds and are better adapted to disturbed habitats. Many of these traits show continuous variation and interrelationships (Sano and Morishima 1982, Oka 1983) (Figs. 6.3–6.5).

Oka (1988) opined that water stress on perennial wild rice possibly played a significant role in the differentiation of annual forms. This may be an exaggerated view, because perennial wild rice is unlikely to have

Table 6.4 Habitat Conditions of Wild Rices in Thailand and India*

Condition	Plant Type			No. Habitats Studied	G Values[a]
i. Water depth in early dry season					
0	–	3	9	12	49.7**(df:4)
<50 cm	11	16	16	43	
>50 cm	24	4	–	28	
ii. Grade of disturbance					
0	6	–	1	7	
1	11	1	8	20	12.7*df:3
2	16	15	14	45	
3	6	5	5	16	
iii. Per cent plant cover					
40	–	–	2	2	
60	4	3	2	9	11.5*(df:6)
80	7	3	1	11	
100	6	3	–	9	
iv. Per cent drymass of wild rice					
20	2	1	1	4	
40	2	2	–	4	7.2(df:8)
60	3	3	1	7	
80	7	1	2	10	
100	–	3	2	5	

[a]G-test for independence from randomness (Sokal and Rohlf 1981); * $p<0.05$; ** $p<0.01$.
*Adapted from Morishima et al. (1984)

suffered severe water stress in Asia 7–10 millennia ago throughout its growth range. The global weather conditions prevailing at that time were generally more moderate than now, and hence more conducive to wild rices. Further, we know that plant populations are regulated by mortality and phenotypic plasticity (Harper 1977, Schlichting 1986). Phenotype plasticity implies the ability of an individual to alter its physiology and morphology in response to changes in environmental conditions, as we have noted already. This phenomenon may have played a major role in the evolution of annual forms from perennial rice, even if in a transient manner and as a precursor to the development of cultivated rice.

Ebana et al. (2011) studied the molecular basis of the variations observed under the long-day conditions prevailing in summer in Tsukuba, Japan

Table 6.5 Variation in 'Germination Activity' of Husked and Unhusked Seeds in Wild and Cultivated Rices*

Seeds, Condition	Days After Pollination								No. of Tested Strains
Unhusked (20% germination									
Wild perennial	2	2	3	1	3	3	–	–	14
Wild intermediate	6	2	–	1	–	2	1	–	12
Wild annual	3	1	3	1	3	1	1	4	17
Cultivated	86	4	–	2	–	–	–	–	92
Husked (40% germination									
Wild perennial	1	–	1	2	1	2	1	6	14
Wild intermediate	1	1	–	2	1	2	1	6	12
Wild annual	–	1	–	–	–	1	–	15	17
Cultivated	67	22	1	2	–	–	–	–	92

[1]Germination activity: formula from % germination and mean days to germination (cf., Morishima et al. 1961)
*Modified from Morishima et al. (1961)

(36°N). For this, they studied F_2 populations using the japonica variety, Koshihikari (flowering duration: 109 days) as the common parent and 12 rice varieties (japonica, indica, aus; flowering duration 91–191 days). Ten F_2 populations showed transgressive segregation. The authors observed 1–4 QTLs in each cross. Some of them were common to all the crosses, and some others had been known earlier – *Hd1, Hd3a, Hd6, RFT1,* and *GHD7*. They also identified several functional polymorphisms, which indicated that allelic variations of these loci probably contributed to variations in heading time, which had been generated by a combination of different alleles. *Hd1* is a major QTL for photoperiod sensitivity. It encodes a protein with zinc finger (a DNA-binding protein motif characterized by two closely spaced cysteine and two histidine residues that serve as ligands for a single Zn^{2+} ion). It is an ortholog of *CONSTANS*, which is involved in flowering promotion in *Arabidopsis*. Incidentally, *Arabidopsis* is a long-day adapted plant, while photoperiod-sensitive rices are short-day adapted (Table 6.5).

Dreni et al. (2011) studied the genetic interaction among four *AGAMOUS (AG)* subfamily members, *MADS 3, MADS 13, MADS 21,* and *MADS 58*. All of them generally play some role in determining various reproductive functions or meristem determinacy. But in this study, *MADS 21* was not seen to play any role. The *MADS* domain transcription factors belonging to the *AGAMOUS (AG)* subfamily regulate the identity of stamens, carpels, and ovules and are further important in floral meristem determinacy (*MADS* box is a highly conserved DNA sequence motif found in a large family of transcription factors). Most prominently, the *MADS* box genes known in flowering plants are intimately involved in the control of flower morphogenesis (Glossary, FAO 2001). There are 75 *MADS* box genes in rice, four of which belong to AG subfamily; *MADS 3* and *MADS 58* are in the AG lineage, and *MADS 13* and *MADS 21* in the AG II lineage (Arora et al. 2007). The *agamous* subfamily is one of three classes of floral organ identity mutations, the other two being *APETALA 2* and *APETALA 3*. These kinds of mutations allow the genes that are normally inactivated to express themselves (King et al. 2006).

These two studies on floral determinacy do not give us a proper perspective on their relevance in rice domestication. However, a series of elaborate experimentation carried out by Tian et al. (2010) on determinate growth in soybean, *Glycine max* (Family Leguminosae) may be instructive in rice also. Under cultivated conditions, soybean varieties may have determinate (*dt1/dt1*) or indeterminate growth. In cultivars with indeterminate growth, growth gradually slows down and ceases as the pods begin to develop, and the plant increasingly diverts nutrients and water to the developing pods. The putative ancestor of cultivated soybean is *Glycine soja*, which has an indeterminate growth habit (*Dt1/Dt1*). This character is monogenic in inheritance.

Dt1 is a homolog (designated as *GmTfl1*) of *Arabidopsis* terminal *flr 1 (Tfl 1)*, which is a regulatory gene encoding a signaling protein of the shoot meristem. The transition from indeterminate to determinate phenotypes appears to be associated with four single nucleotide substitutions in the *GmTfl*1 gene, each leading to an amino acid change. This may have happened several millennia ago as a consequence of selection by the first farmers. The authors then confirmed by suitable experimentation that the *GmTfl1* gene complements the functions of *TFL1* in *Arabidopsis*. However, the *GmTfl1* homeolog (in soybean) "appeared to be sub- or neofunctionalized". This was revealed by the differential expression of the two genes at the multiple plant development stages and by allelic analysis of both the loci.

In the soybean, stem growth is controlled by the *Dt1* locus and indeterminate growth is expressed by *Dt1/Dt1*. This character may be expressed as fully or partially dominant, due to the operation of other genes that are also responsible for this character, *Dt2*, and possibly also, *dt1–1*. Incidentally, the authors have fine-mapped the *Dt1* locus as a major QTL. It is located between the two SSR markers, *Sa-999* and *Sat-229* on linkage group L (now called chromosome 19). They have also indicated that *GmTfl1* may be a candidate gene for *Dt1*.

There do not appear to have been any critical studies in rice on the transition from the perennial life history of the putative ancestral species to the current annual habit. However, as we have seen, barring a few landraces, cultivated rice is not a true annual, and if adequate moisture and nutrients are available, most cultivated rices will put out new tillers and even produce a fair crop. They have an annual–perennial intermediate life history trait that appears to be somewhat akin to the situation studied above in soybeans by Tian et al. (2010).

It is often observed that the straggling, floating habit of the perennial *O. rufipogon* tillers that are observed when the plants grow in standing water is modified they are grown in upland situations. They then develop into semi-open bushy plants and give the appearance of the annual spontaneas. This implies that during the domestication process, the transition from floating to erect habit could have happened naturally due to phenotype plasticity, in a water-deficient environment. A couple of authors have nevertheless studied the molecular genetics of this transition and came to identical conclusions.

Tan et al. (2008) observed that the transition of the rice plant from a prostrate to an erect habit brought about an increase in grain yield. They found that the prostrate growth habit in rice was controlled by a semi-dominant gene, *PROG1* (Prostrate growth), on chromosome 7 that encoded a single *Cys-His*$_2$ zinc-finger protein. Variants of *prog1* disrupted its function, leading to erect growth, increased spikelet number, and higher grain yields in cultivated rice. The sequence comparison of 182 varieties (87 indica, 95 japonica) from 17 countries showed that all of them carried identical mutations in the *prg1* coding region. The number of primary and secondary branches and grains in *pPROG-1–1* transgenic line per panicle were 7.3, 3.7, and 44.3, respectively while in the control, they were 11.3, 28.0, and 149.3, respectively. When checked with 15 SNPs and six indels, all the domesticated varieties were found to possess the same haplotype. To identify the location of the *PROG1* gene, the authors constructed a set of introgression lines using a *O. rufipogon* line from Yuanjiang county (Yunnan, China) and an indica cultivar, Tequing, as receipient. From these, they selected an introgression line with prostrate habit, YIL 18. This had two YJCWR chromosomal segments from the long arm of chromosome 3 and the short arm of chromosome 7. The grain number of YIL 18 was only 57.6% of the recipient, Tequing, due to smaller number of primary and secondary branches on the main panicle. A linkage analysis of 246 F_2s showed that prostrate habit was controlled by a semi-dominant gene, *PROG1*, located between the SSR markers, *RM298* and *RM481*, on the short arm of chromosome 7. The location was then further delimited to an 8.6 kb region between *pr5* and *pr7* markers. Here, there was only one hypothetical gene, *LOC-Os 07.05900,* in the Nipponbare genome.

Jin et al. (2008) also studied the genetic control of rice plant architecture and arrived at similar conclusions. They fine-mapped the location of the *PROG1* gene in an *O. rufipogon* line from Hainan to a 14 kb region located between the markers *S3204* and *P71*.

F. Alteration in the Breeding System

The domestication process involved a change in the breeding system from partially cross-fertilizing to a predominantly self-fertilizing system. The putative ancestral species (*O. rufipogon*) of the Asian rice is partially cross-fertilizing in habit (30.7–50.6%) (Table 4.6).

Usually in areas where the annual wild rices occur in the vicinity of *O. rufipogon,* the outcrossing rates are higher, possibly somewhat higher in indica than in japonica rices. In evolutionary biology, self-fertilization is assumed to have derived from cross-fertilization (Stebbins 1950). Adoption of self-fertilization enables the species to allocate all its resources to flowering and seed production, thus contributing toward higher productivity and consequent higher yields (Harper 1977).

It is difficult to visualize how intensely the first farmers would have attempted to bring about self-fertilization in the early stages of rice

Table 6.6 Variation in Floral Characters of Wild and Cultivated Rices*

		O. rufipogon		*O. sativa*	
Organ	**Trait**	**Perennial (n = 32)**	**Annual (n = 21)**	**Indica (n = 72)**	**Japonica (n = 56)**
Pistil	Angle of exserted stigma	5.0 ± 0.0	5.0 ± 0.0	2.7 ± 1.4	2.3 ± 1.3
	Stigma length, mm	1.67 ± 0.25	1.41 ± 0.20	1.16 ± 0.18	1.06 ± 0.19
Stamen	Anther length, mm	4.47 ± 0.53	2.42 ± 0.45	2.18 ± 0.21	2.15 ± 0.27

Note: The other flower characters studied – length, breadth, and thickness of lemma and palea – did not show much differences.
*Modified from Ebana 2003a

domestication. They were unlikely to have had the awareness and/or ability to select plants which would have shown only incremental improvements towards self-fertilization. Hence, it may be more rational to assume that cultivated rice acquired this character gradually and in association with some other recognizable character.

Uga et al. (2003a) studied variations in several flower traits of 128 *O. sativa* (71 indica + 56 japonica) and 53 *O. rufipogon* (32 perennial + 21 annual) accessions. They found that the annual wild rices have shorter stigma and anthers than the perennial wild rices (Table 6.6). Cultivated rice showed lower stigma exsertion, shorter stigmas, shorter anthers, and thicker and wider lemma and palea than the wild rice accessions. They opined that these characteristics would have played a decisive role in promoting selfing and increasing seed setting. The japonica and indica rices did not show any differences between them for these characters. However, it is difficult to associate the effect of lemma-palea dimensions on selfing.

In another study, Uga et al. (2003b) studied the QTLs influencing floral morphology. They used a recombinant inbred line population derived from an *O. sativa–O. rufipogon* hybrid. QTLs were detected using a linkage map having 147 markers consisting mostly of RFLPs. The authors detected seven QTLs for pistil characters, four for stamen length, 14 for glume size and six for glume shape (total: 31 QTLs). While 10/31 QTLs were located in four adjacent regions on chromosomes 2, 4, 5, and 10, the others were dispersed, with most QTLs showing only minor effects. The authors observed that the quantitative variations observed for pistil, stamen, and glume traits were controlled by several genes, with each producing small effects.

The limited studies conducted on breeding systems in cultivated and wild rice confirm that the changeover from a partially cross-fertilizing system to a predominantly self-fertilizing system was brought about gradually during the domestication process, and it is possible that no direct selections were carried out for this.

G. Seed Dormancy

In recent times, reduction in seed dormancy has been suggested as a component of the rice domestication process. This had been suggested by some early authors such as Roschevicz, Ramiah, Kihara, etc. also.

Seed dormancy in plants prevents the germination of seeds when they are exposed to optimal environmental conditions for germination, even when the embryos are viable. In wild species, and more particularly in weeds, seed germination is characterized by heterogeneity in the time of germination within a season, and some seeds may continue to germinate in this manner for two or more seasons or years. This appears to be nature's way of ensuring the survival of the species in uncertain times. It does not appear to be any more than an adaptive strategy for the survival of the wild and weedy species.

This phenomenon of staggered germination is brought out well in rice by the studies of Morishima et al. (1961) (Table 6.5). While wild rices show only staggered germination, the cultivated rices completed their germination almost synchronously.

Seed dormancy is more intense in annual wild rices than in perennial wild rices. Further, husking the seeds removes dormancy in many species. Most of the dormancy is induced by the seed coat – in rice, by lemma and palea – or the embryo. In most species, good germination can be obtained if the seeds are kept for germination after removing lemma and palea/seed coat.

An important favorable attribute of seed dormancy in rice is the protection provided from immediate germination to the maturing panicle and the harvested crop and grain if the rains come unexpectedly, while the crop is being harvested, threshed, readied for storage, or sale. This is particularly relevant in monsoon Asia. Pre-harvest sprouting of panicles, if the rains fall after the crop reaches maturity can cause much distress and loss to farmers, because this results in lower yields and reduces the quality of the harvested crop, consequently reducing both the grain yield and the market value of the produce.

In monsoon Asia, three main crops are raised in rice, depending on land elevation and the availability of water for raising the crop. Of these, the first crop comes to harvest during the monsoon season, and hence a minimum two or three weeks' protection from rain is necessary to prevent the germination of the ripening crop, while standing in the field. The ripened crop is then harvested manually and transported to a threshing yard, threshed by feet or sticks, winnowed and sun-dried to attain 12–13% seed moisture. In this case, a three to four week seed dormancy in the growth variety appears to be a decided advantage for the cultivars used for growing in this season. Normally, this crop is of short duration–up to 120 days. A thumb rule is that short duration– varieties have no or only low dormancy, but there are exceptions to this. Hence, the variety chosen for sowing in the first season has to also consider this aspect, but this choice is secondary to the suitability of the variety to the elevation of the field and the water level.

A medium period of dormancy will be a desirable attribute for a deep water paddy crop also, as the flood waters could rise by a meter or more in

24 hr, and if the crop is maturing at that time, it will be necessary for the variety to possess a medium level of seed dormancy.

In subtropics or temperate regions, where only one medium duration crop is grown as in northeast Asia, southern Europe, north America, etc. and also for the main season irrigated crop in rest of Asia, a short period of dormancy will be an advantage, in case the crop lodges following gale or storm, since varieties with some level of dormancy will not germinate on the panicle or on the threshing floor. In any case, seed dormancy in this crop is unlikely to pose a disadvantage, since the harvested crop is unlikely to be used as seed for raising another crop immediately after harvest, and seed dormancy does not adversely affect its quality. For all these reasons, it is felt unnecessary to assume that removal of dormancy in rice is a domestication-related attribute. It is to be taken as only an ecological adaptation.

In classical genetic studies, the inheritance of seed dormancy has been found to be monogenically and digenically controlled with complementary gene action (cf. Ramiah and Rao 1953). Seshu and Sorrels (1986) also obtained similar results in experiments carried out in glasshouses under short day conditions. They found that preharvest sprouting was negatively correlated ($r = -0.88^{**}$) with dormancy period.

There have been a small number of studies to determine the molecular basis of seed dormancy. Lin et al. (1998) studied the possible relationship between seed dormancy and heading date. The authors used 98 *BC1* F_5 backcross inbred lines (BIL) derived from Nipponbare (japonica, dormant) – Kasalath (indica, nondorment) hybrids employing 245 RFLP markers for a framework linkage map of Kasalath x Nipponbare/Nipponbare. Interestingly, the segregation ratios of the two genotype classes in most loci fitted the expected Mendelian ratio of 3 (Nipponbare): 1 (Kasalath). (The percentage of Nipponbare genome in each line had varied from 49.3% to 91.3% average: 74.2%.) They identified five putative QTLs affecting seed dormancy, located on chromosomes 3, 5, 7 (two regions), and 8. They showed 6.7–22.5% phenotypic variation. The five QTLs together explained 48% of the total phenotypic variation for this character in the population. The map location of the QTL for seed dormancy on chromosome 3 coincided with that of a QTL for heading date, but not of the other QTLs.

Cai and Morishima (2000) carried out a similar study. They used 125 recombinant inbred lines involving a cultivated and a wild rice hybrid, and employed 147 markers distributed on all 12 chromosomes. They found that all the chromosomes except chromosomes 4 and 10 harbored large numbers of genes responsible for seed dormancy.

Gu et al. (2004) used a weedy rice strain ('red rice') to determine the QTLs for seed dormancy. They observed six dormancy QTLs, *qSDs −4, −6, −7 −1, −7 −2, −8,* and *−12*. Further, *qSD−7−1* was tightly linked to the red pericarp color gene, *Rc*. The authors detected two, three, and four orders of epistases (the nonreciprocal interaction of nonallelic genes). In this, one gene masks the expression of another with four, six, and six QTLs, respectively (King et al. 2006). They felt that the high order epistasis strongly suggested the presence of genetically complex networks in the regulation of variation for seed dormancy in natural populations.

Li et al. (2004) compared the operation of seed dormancy in three cereals, wheat, barley, and rice, vis-à-vis pre-harvest sprouting. They detected a major QTL controlling both seed dormancy and pre-harvest sprouting in barley on the long arm of chromosome 5H that explained 70% of the phenotypic variation. This locus showed good synteny with the "terminal end" of the long arm of chromosome 3 in rice. The gene sequence of rice was annotated and they identified a gene encoding *gibberellic acid-20 oxidase* as a candidate gene. In wheat, chromosome 4AL (long arm) QTL region was syntenic to rice chromosomes 3 and 11. Incidentally, this is a good example of a syntenic relationship among the three cereals.

Sugimoto et al. (2010) cloned one of the five QTLs for seed dormancy that has been discovered earlier by the same group (Lin et al. 1998), viz., *Sd-4* (*Seed dormancy* 4). The authors used also the same material as used by Li et al. (2004) viz., the backcross inbred lines (BIL) derived from Nipponbare (japonica) x Kasalath (indica) cross derivatives. The expression of *Sdr4* was positively regulated by *OsVP1*, a "global regulator of seed maturation", which in turn regulated seed dormancy by "repressing the expression of post-germinative genes". By haplotype analysis, the authors found that *Sdr4* acts as an intermediate regulator of seed dormancy. *Sd4* encodes a novel protein with an amino acid sequence that has no similarity to any of the proteins with known functions. Japonica cultivars possess only the Nipponbare allele, *Sd4-n*, while both the Kasalath alleles, *Sd4-k* and *Sd4-n*, are widely present in indica cultivars.

Xie et al. (2010) studied seed dormancy and pre-harvest sprouting in rice using 122 backcrossed inbred lines of a cross between N22 (indica, India, strongly dormant) and Nanjing 35 (japonica, non-dormant). They found three QTLs on chromosomes 1, 2, and 3. They accounted for 18%, 8%, and 7% phenotypic variance. All the QTLs came from N22. The authors then developed an advanced backcross line ($BC6\ F_2$) to precisely map the locus *qSD-1* on chromosome 1. It was found that *qSD-1* consists of two distinctly linked loci of *qSD–1–1* and *qSD–1–2,* which act additively to promote seed dormancy. The QTL *qSD-1* on chromosome 1 was mapped in the vicinity of RM 493 and RM 1297 with the nearest SSR locus at RM 488. The second locus, *qSD-2,* was flanked by RM 525 and RM 240, and its nearest SSR locus was RM 318. The authors concluded that strong dormancy was under the control of just one or two major genes only, with regulatory involvement of a few minor genes (as had been proposed by Seshu and Sorrells (1986)).

From the foregoing account, it is apparent that a short period of seed dormancy of about three weeks will be a decided advantage to the rice crop everywhere, especially in monsoon Asia. A slightly longer period of seed dormancy may be beneficial in regions where long duration and floating rices are grown. It has been pointed out that in barley, seed dormancy is a disadvantage if the harvested grain is to be immediately used for malting, because dormant barley seeds do not malt satisfactorily.

Some authors have suggested that the preharvest sprouting of panicles is separate from seed dormancy, but logistically, it may be difficult to separate pre-harvest sprouting and seed dormancy. Further, it may be more rational to begin calculating dormancy from the time the spikelets in a panicle become mature. We could consider seed dormancy as only an adaptive trait, and not a component of the domestication process. Further, seed dormancy in rice is apparently controlled by a number of genes/QTLs each having minor effect.

H. Weedy Red Rices

Red rice is one of the most serious weed problems of rice cultivation worldwide (FAO 2007). They are commonly known as weedy red rice, or simply, red rice. They possess red pericarp and shed their grains easily. We have seen before that red pericarped rice is preferred as a food by certain language/culture groups in south Asia.

Weedy red rices are the most serious and most ubiquitous of all the weeds in rice. Their presence in the cultivated fields of south, southeast, and east Asia has been known for a long period (Watt 1892, FAO 2007). In west Africa too, several wild rice species such as *O. barthii, O. longistaminata,* and *O. punctata* and also red rices occur as as weeds in paddy fields (Oka and Chang 1965, FAO 2007). All of them have red pericarp. In Central America, *O. latifolia* occurs as a weed in rice fields (FAO 2007).

Weedy red rices occur wherever rice is grown in the world, even in regions/continents like Europe, North America, and northeast Asia (Korea, Japan), where wild rice species, such as *O. rufipogon,* do not occur. The most common assumption regarding the origin of weedy red rices is that they are the products of natural hybridization between cultivated rices and wild rices, either *O. rufipogon* or spontanea rices. This had been proven experimentally time and again by several early workers (Roy 1921, Nayar 1958, 1973, Oka and Chang 1961, Sampath and Govindaswamy 1958, Oka and Morishima 1971, Oka 1988, etc.). The most commonly accepted assumption regarding the occurrence of red rices as weeds in rice fields, areas of the world where wild rice species do not occur, is that they were accidental introductions brought in along with rice seeds. Some authors have also suggested also they are the products of reversion from cultivated varieties. We shall go into these aspects in more detail.

Bres-Patry et al. (2001) studied a double-haploid population derived from a cross between a japonica variety and a weedy rice plant collected from Camargue (France) and "set up" a linkage map using 68 SSRs and 31 AFLPs. Then, they studied five pigmentation and 15 morphological characters. In spite of showing good fertility in the population, they observed segregation distribution. The sterility genes corresponded with the QTLs for sterility observed in japonica–indica crosses. The main specific traits of weedy forms also revealed the same genes/QTLs as observed in progeny populations of *O. sativa* × *O. rufipogon*. This is one of the earliest studies using molecular biology techniques to understand the nature of red rices.

Weedy red rices occur in all the states of the USA where rice is cultivated commercially. The tropical japonicas are the most commonly grown rice in the southern states of the USA, and the temperate japonica rices are grown in California. Red rice is a menace in all US states (Langewin et al. 1990, Kane and Baack 2007, Londo and Schaal 2007, Olsen et al. 2007). Langevin et al. (1990) used 2,400 single plant progenies of red rice collected from six cultivars grown in Louisiana and employed morphological and electrophoretic characters to study their nature, determine the extent of natural hybridization with cultivated rices, and possible convergence of red rice with cultivated rice. The percentage of natural hybridization between cultivated rice and wild rice ranged from 1 to 52%. The F_1 hybrids tended to flower later and grew more robustly than either parent. The authors admitted that their conclusions were very tentative, but they felt that red rices showed a greater tendency to converge with cultivated rices if the latter is of longer maturity.

Londo and Schaal (2007) studied 29 accessions of weedy rice collected from across the USA. Sixteen of them were straw-hulled, awnless types, ten were black-hulled, awnless types, and three were straw-hulled, awned ones.

They were subjected to microsatellite and DNA sequence data analyses along with 50 *O. sativa* accessions, 54 *O. rufipogon,* five *O. nivara,* two spontanea, and one each of *O. glaberrima* and *O. meridionalis,* using population assignment of multilocus genotype signatures with PCA and STRUCTURE analyses. The authors inferred that the US weedy red rices originated from Asia. Their results suggested that in the USA, two major strains of weedy rices exist: straw-hulled and black-hulled. Both of these not only hybridizes with the US rices, but also between themselves, giving rise to new weedy forms. According to anecdotal evidence, the first rice varieties were imported into the USA (North and South Carolina) from India in 1698 and the first red rice was reported in 1846 (Allston 1846). Usually, red rices are characterized by a red/rouge pericarp and cultivated rices by a tan/beige pericarp.

Similar accounts are repeated in other regions of the world as well, where wild rice species are absent in the native vegetation. In Europe, the first accounts of red rice date from the early 19th century. It became a significant problem from the 1950s, and a major one since about the mid-1980s. The problem has now worsened the world over as more farmers have resorted to direct seeding in preference to transplanting to save labor costs. It has been estimated that 40–75% of the area under rice in Europe is infested by red rice (FAO 2007, Ferraro and Vidotto 2010, Fogliatto et al. 2011).

The situation is similar in Korea and Japan also. Akasaka et al. (2009) studied polymorphisms in 27 weedy accessions and 88 rice cultivars from the Okayama Prefecture (Japan) using 15 sequence-tagged site primers. They found that the band patterns of 13 japonica weedy accessions corresponded exactly with those of local japonica landraces for all morphological and physiological characteristics except the shattering habit. On the basis of these close similarities, they proposed that weedy rices originated from cultivated rice as off-types by gene mutation.

Red rices are a more severe problem in south and southeast Asia, especially in upland areas, because wild rices occur almost sympatrically with cultivated rices in most rice growing areas here. Until recently, the menace was kept under partial control by resorting to transplanting and weeding. As labor increases in cost and decreases in availability, farmers are steadily abandoning the practice of transplanting and substituting it with direct seeding. In the 1940s, in Maharashtra and Madhya Pradesh (now Chattisgarh), India, the State Agriculture Departments released some improved rice varieties possessing deep purple leaf and sheath pigmentation. The assumption was that the purple leaf would make it easier to mark out the weedy red rices at the seedling stage, and thereby facilitate their removal. But it was found that the associated weed rices also developed purple leaves within 3–4 years, and this made them indistinguishable from the cultivated rices from the seedling stage itself, so this program was given up.

The FAO has been aware of the menace of wild rices as weeds in paddy fields. It organized a workshop in Cuba during 1999 (FAO 1999) to take stock of the situation. Then, it produced a set of recommendations (FAO 2007) suggesting the adoption of one or more of the following five, long-term, well-tested, control strategies *plus* a new one for adoption depending on location, prevailing cropping system, and other factors. The recommended controlled measures were:

(i) Use clean healthy seed;
(ii) Practice hygiene and appropriate cultural practices;
(iii) Practice flooding and water management to kill red rice seeds;
(iv) Use the herbicide molinate in standing water fields; and
(v) Rotate with other crops.

The sixth recommendation is the use of herbicide-resistant rice cultivars.

Since most of the rice in tropical Asia is cultivated by marginal resource-poor farmers, these recommendations are hard to practice. As for using herbicide-resistant cultivars, this is fraught with the risk of herbicide resistance transferring itself to red rices, mainly because of their widespread presence and their breeding behavior. This is known to be happening in areas which are growing GM crops with enhanced herbicide tolerance.

I. Red-Pericarped Rice

There are some regions in south Asia where red-pericarped rice is preferred for consumption by the local communities. These include the Kerala-Konkan coast in peninsular India, Orissa and West Bengal in southeast India, and Bangladesh. Red pericarped rice is incidentally more nutritious than white rice (Table 1.7). These regions possibly contain up to 200 million people. We have already mentioned the release of a red pericarped mutant cultivar, Red Triveni, by Kerala Agricultural University (India), after the originally released cultivar Triveni, that had white pericarp, was rejected by farmers (Rosamma et al. 1991, Nayar et al. 2011). Hence, it is apparent that the evolution of a white pericarp in Asian rice need not be considered as a domestication advance, as has been reported in some publications (Fukuwara et al. 2006, Sweeney et al. 2006, 2007, for instance).

In classical genetics, red pericarp color has been found to be simple dominant over white color. Red coloration is governed by two complementary genes, *Rc* and *Rd* (Kinoshita 1995). While *Rc* is responsible for pigment accumulation, *Rd* helps in increasing the pigment content. Red pericarp has two forms, red and brown color. *RcRd* produces a red pericarp, *Rcrd* a brown pericarp, and *rcRd* and *rdrd* produce a white pericarp. The locus *Rc* is located on chromosome 7, and *Rd,* on chromosome 1 (Kinoshita 1995).

Furukawa et al. (2006) identified three anthocyanin pigmentation genes in a *RcRd* rice line and introduced them into a *rcrd* line. They obtained brown-colored seeds which demonstrated that the *Rc* gene encoded a basic helix-loop-helix (*bHLH*) protein. Comparison of the *Rc* locus among rice accessions showed that a 14-bp deletion in the *rc* locus resulted in the development of white pericarp. Sweeney et al. (2006, 2007) observed that the *Rc* gene, which imparts the red pericarp in rice, was located on chromosome 7, that this red color was present in all the wild ancestors of rice, and that it was closely associated with seed shattering and dormancy. They found that *Rc* encoded a basic helix-loop-helix (bHLP) protein, which they fine-mapped to a 18.5 kb region on chromosome 7. They found further that the dominant red allele differed from the recessive white allele by a 14-bp deletion within exon 6 that knocked out the bHLP domain of the protein. RT-PCR experiments confirmed that the *Rc* gene was expressed in both red- and white-pericarped rice, but that in the white-pericarped varieties, only a shortened transcript was present.

In the context of red-pericarped rice, Sweeney and McCouch (2007) stated that many QTLs responsible for various domestication traits have been identified and several of them clustered in the same genomic regions, suggesting that a single gene with pleiotropic effects, or one or more clusters of closely linked genes, underlie these QTLs. They observed such a cluster in the chromomere region of chromosome 7 (seed color, panicle structure, dormancy, and shattering). They found similar other clusters on chromosomes 3, 4, 6, 8, 9, 11, and 12. Thus, these QTLs are present in as many as eight out of 12 chromosomes of rice.

J. Glutinous and Starchy Rices

Glutinous rices are found most frequently in southeast and east Asia. Such rices become sticky upon cooking. They have a special place in the culture and lives of the people of the region, especially in the hilly regions of southeast Asia – Myanmar, Thailand, and Indo-China. In Laos, for instance, almost 85% of the rice consumed by the people was glutinous in nature (Golomb 1976). Elsewhere in east and southeast Asia, it is used for preparing special dishes and puddings, especially on festive occasions. Presently, its consumption is mostly limited to Laos and Myanmar. Elsewhere in east and southeast Asia, it is used only on special festive occasions, for making puddings, and sometimes, as cooked rice. It is used also for making beer.

Glutinous rice does not contain any dietary gluten. The carbohydrate in the glutinous rice grain does not contain any amylose, and if present, in only small quantities, but it is high in amylopectin (Juliano 1993). The uncooked grains of glutinous rice are softer and dextrinous, while those of normal rices are hard and starchy. The consumption of glutinous rice has decreased drastically in the last few decades. Waxy (glutinous) rices are differentiated from normal non-waxy rices on the basis of the relative levels of the polysaccharides, amylopectin, and amylose. Broadly, non-waxy rice starch contains 1–30% amylose and the rest is amylopectin, while the waxy (glutinous) rice starch contains only amylopectin.

The oldest record of glutinous (waxy) rice is in a Chinese classic text of the early Han period (200 BCE) (Sakamoto 1997). In 1967, Watabe observed that glutinous rice cultivation was centered around northern Thailand and Laos, and later its spread the surrounding areas of Myanmar, Vietnam, Cambodia, and Yunnan (China). Nakagahra et al. (1975) studied the nature of 1,042 landraces of south, southeast, and east Asia. He found that almost all the rices were non-glutinous in south Asia, but glutinous varieties were predominant in southeast Asia, and their frequencies were low in northeast Asia, viz., China, Korea, and Japan. MAFF (The Ministry of Agriculture, Fisheries, and Forestry), Japan conducted a similar study in the 1960s with 1,259 Japanese landraces and found that 42.8% of them were glutinous (Sakamoto 1997). Kinoshita (1995) has observed that glutinous endosperm and amylase content are controlled by three genes, *wx*, Wx^a, and Wx^b, all located on chromosome 6. Hirand et al. (1998) found that a single base change of the *Waxy* gene at the post-transcriptional level altered its regulation (also, Yamanaka et al. 2003, Cheng et al. 2010).

The presence of glutinous rice has been assumed by some authors as a domestication trait, but it does not have to be so considered. It is no more than a variant form of rice selected by certain cultural or language groups based on their social, cultural or religious preferences and beliefs. This is akin to the preference for parboiled, red-pericarped rice prevalent in certain regions of south Asia and west Africa, as we have seen already.

Olsen and Purugannan (2002) studied 105 cultivars belonging to various countries in southeast Asia to determine the origin of the waxy endosperm. From their results, they proposed that a mutation in an intron splice donor site was responsible for the change of endosperm to glutinous rice. They found that this mutation had only a single origin in southeast Asia, from where it diffused into the other cultivars of southeast and east Asia. According to the authors, the *Wx* haptotype tree could reveal the number of times the splice donor mutation had occurred. The glutinous rice accessions were restricted to a cluster of five closely related haplotypes with the splice donor type mutations directly leading to the haplotypes. The authors then argued that the mutations responsible for the disruption of amylose synthesis appeared to have had only a single origin in the glutinous rice landraces examined by them. The origin of glutinous rice was associated with reduced genetic variation at the *Waxy* locus. The authors then postulated that many non-glutinous varieties in southeast Asia also carried the split donor site mutation, and that a partial suppression of this mutation might have played a role in the development of non-glutinous rice is northeast Asia.

Olsen et al. (2006) studied 96 diverse accessions covering the entire distribution range of *O. sativa* and *O. rufipogon* along with one accession each of *O. nivara, O. barthii,* and *O. meridionalis* as outliers. Their objective was to look for evidences of a sweep in the *Waxy* genomic region. These accessions were then sequenced at the *Waxy* loci along with portions of 12 flanking loci that are located at ~40 kb intervals upstream and downstream of the *Waxy* locus. Then, they carried out a population-genetic analysis in both *O. sativa* and *O. rufipogon* in a 5.3 kb region covering the entire *Waxy* region. They observed 101 polymorphisms in the *O. sativa Wx* gene and the silent site regions, and 165 SNPs in *O. rufipogon*. The extent of diversity in the indica varieties was about five times more than in the japonicas, both tropical and temperate. The splice donor starchy mutation was observed with 29% frequency in *O. sativa,* and at a reduced level in the *Wx*-carrying *O. sativa*. This splice donor mutation was prevalent in temperate japonica, but was rare or absent in tropical japonica, aus, and aromatic rices. The sequence analysis of a 500 kb genomic region covering the *Waxy* locus revealed patterns consistent with a selective sweep in temperate japonicas. The size of the selective sweep (>250 kb) indicated very strong selection, with an inferred selection coefficient that was higher than similar estimates for maize domestication genes or wild species. This suggested to the authors that the selection pressure associated with crop domestication regimes could exceed by one or two orders of magnitude than those observed for genes under even strong selection pressure in nature.

Recent studies have shown that the *waxy* locus may contain five alleles, Wx^{a}, Wx^{in}, Wx^{b}, Wx^{op}, and *wx* (Mikami et al. 2008). They confirmed that a mutation at an intron/splice donor site of the *Waxy* gene was responsible for the change of endosperm to glutinous rice. In a molecular phylogenetic tree using *Wx* gene data, the authors found that japonica rices formed a single clade, while indica rices formed a different clade. They then suggested that allelic diversification appeared to have occurred after the differentiation of japonicas and indicas.

There are two main functional alleles at the waxy locus, Wx^a and Wx^b. They encode different levels of granule-bound starch synthesis, and thus control amylose content in rice starch. Generally, Wx^a is predominant in indica rices and Wx^b in japonica rices (Wang et al. 1995, Cai et al. 1998). The difference between the two alleles is due to alternate splicing caused by a single base substitution, AGGT to AGTT, at a donor site of the first intron. Phenotypic analysis cannot indicate, if in the waxy rice, the *wx* gene originated from Wx^a or Wx^b locus (Watanabe et al. 2004). The authors therefore identified a mismatch primer (*dCAPS*: derived Cleaved Amplified Polymorphic Sequence) that generated a restriction site in Wx^a (AGGT) allele, but not in Wx^b (AGTT) allele. Then, they studied 353 varieties from Asia using this mismatch primer. They found that while waxy strains had both Wx^a and Wx^b derived alleles, the latter allele was predominant and that its distribution was independent of japonica–indica differentiation. All the wild relatives of cultivated rice also possessed the AGGT allele. This suggested to them that polymorphism occurred in the japonicas during the domestication process. In addition, the nonwaxy japonica carrying AGTT might have been preferentially selected by the Asian people, they observed.

Incidentally, as already stated, there does not appear to be any rationale to implicate glutinous rice formation in the domestication process of Asian rice. This character need not be considered any more than the result of a regionally (ethnically or linguistically, or culturally) developed preference for a certain attribute, such as the preference for red-pericarped rice, scented rice, long, slender and flaky rices, and so on.

Chen et al. (2008) studied apparent amylose content (AAC) in 171 accessions of Asian rice belonging to 43 countries. The combination of SNP markers in the *Wx* allele identified three marker haplotypes in the gene. The first SNP was found at the leader intron splice site (in 1 SNP) and the second one in exon 6. Together, they accounted for 86.7% of the variation in AAC and discriminated the three marker classes of low, intermediate, and high AACs. They found further that higher air temperature during grain development was associated with decrease in AAC of low and intermediate types and increase of high AAC types. They classed the AAC as: glutinous/waxy, 0–5%; low AAC, 6–18%; intermediate AAC, 19–23%; and high AAC, > 23%.

The proposition of a single origin of waxy character in a japonica rice in southeast Asia, and of this character then diffusing gradually throughout southeast and east Asia is to be considered by factoring the following points.

(i) A mutation at a single nucleotide from AGGT to AGTT is enough to transform normal rice into glutinous rice;

(ii) Glutinous rice was widely prevalent in the past in southeast and east Asia across ethnic and language barriers, cultural and religious differences, nature of the cultivated rices, and ecological differences; and

(iii) There is much merit in assuming that Asian rices have multiple origins.

The proposed single origin of glutinous rice in southeast Asia and of its subsequent spread to the north to east Asia also needs to be considered in the light of our present understanding of the spread of rice and the movement of early humans in southeast and east Asia. A diffusion of *waxy* character to the north is theoretically possible, but during prehistoric times, the movement of people, along with rice and possessing knowledge of rice cultivation techniques, is assumed to have taken place southwards from the Yangtze basin (Bellwood 2004, 2010, 2011).

K. Aromatic, Fragrant, and Medicinal Rices

The very long history and traditions of rice, and its uses by people living across an enormous geographical and cultural range, means that a number of uses would have developed for rice beyond that for food. Consequently, landraces of rice now exist that are aromatic and fragrant, or possess medicinal properties. The website of the Royal Botanic Garden Kew states that rice has been used to treat skin conditions, such as boils, sores, swellings, and skin blemishes, that glutinous rices are employed to treat stomach upsets, heart burn, and indigestion, and that brown rice is used to treat breast cancer and warts. In Kerala (southwest India), there is a medicinal rice, *njavara,* that is very widely used in ayurveda, the Indian system of medicine, for a range of neuromuscular problems including arthritis and rheumatism (*Aashtanga Hridayam*, 8th century text, Murthy 2001).

There are about 300 aromatic rices used worldwide Singh et al. 1995, Weber et al. 2000). Khush (2000) proposed that aromatic rices originated in the central Himalayan foothills and valleys and then dispersed throughout Asia. He identified 45 aromatic rices belonging to three of the groups of

Glaszmann (1986, 1987): I (8 nos.), V (30 nos.) and VI (7 nos.). These originated in various countries: Group I: China (1 no.), Thailand (4 nos.), Vietnam (2 nos.), Cambodia (1 no.): Group V: Thailand (1 no.), India (11 nos.), Myanmar (6 nos.), Bangladesh (2 nos.), Iran (2 nos.), Afghanistan (2 nos.), Iraq (1 no.), China (2 nos.) Indonesia (2 nos.); and Group VI: Indonesia (3 nos.), China (1 no.), The Philippines (3 no.). These three groups of Glaszmann are: group I: typical indica, hsien, aman, tjereh; group V: short duration (120 days) aus, some boro, many from Himalayan foothills, several aromatic rices such as basmati, adri, etc.; and group VI: typical japonica keng, bulu, gundil, most high altitude varieties of eastern Himalayas (Nepal, Sikkim, Bhutan).

Singh et al. (1995) has given a list of 141 scented varieties from India, coming from 14 states. They consisted of 86 short, 25 medium, and 30 long-grained ones. A good number of them are from the central Ganga valley, which adjoins the Nepal terai region.

Interestingly, two of the scented rices listed here show some unique features. They are *Jeerkasala* and *Gandhakasala*. Traditionally, both are cultivated in the hilly district of Wayanad, Kerala. Unlike the other landraces of Kerala, all of which have a red pericarp, these two landraces possess white pericarp. Both have a long duration (150–180 days), their spikelets are only 5.0–5.5 mm long, and single grain weight is only 12.4–13.3 mg (average, 16–18 mg). Studies carried out by Thomas and associates show that there are two main forms of *njavara* with yellow and black bull colors. Each of them can be further divided into two groups (Sreejayan et al. 2010, Jose et al. 2011).

Much work has been done in the past to understand the biochemical nature of fragrance (Weber et al. 2000). Although more than 100–114 volatiles have been isolated and studied, it has not yet been possible to identify the chemical specifically responsible for imparting the characteristic odor to cooked, scented rices. However, the chemical, 2-acetyl-1-pyroline has been found to impart odor to several aromatic rices. The present odor might be the result of a blend of volatile compounds (Weber et al. 2000). Studies on the genetic basis of fragrance have indicated that a gene *fgr* (chromosome 8) is responsible for the character (Ahn et al. 1992, Lorieux et al. 1996). Further studies have shown that the betaine aldehyde dehydrogenase gene (*BADH2r*) is primarily responsible for imparting fragrance to rice, including to all the basmati and jasmine rices.

Detailed studies carried out by Kovach et al. (2009) have identified eight putatively nonfunctional alleles of the *BADH2* gene (*aldehyde dehydrogenase*), which are primarily responsible for fragrance in rice. These alleles have distinct genetic and geographic origins. Among them, a single allele, *badh 2.1* was the predominant allele present in fragrant rice, including in basmati and jasmine rices. Haplotype analysis indicated a single origin for the *badh 2.1* allele within japonica, and its possible transfer to indica rices by introgression. Basmati-like accessions were nearly identical to the ancestral japonica haplotypes in a 5.3 Mb region flanking *BADH2*, regardless of their fragrance phenotype, demonstrating a close evolutionary relationship between basmati varieties and the japonica gene pool. They found that fragrant rice germplasm from diverse sources showed an association with *badh 2.1*, and the transformation of the fragrant variety with the dominant non-fragrant gene masked the fragrance.

The findings summarized above (Kovack et al. 2009) raise certain questions: the possible manner in which a single mutation that imparts fragrance occurred in a japonica variety; the manner by which such a specialized character managed to spread very widely geographically, but very thinly, throughout the entire southeast, south, east and central Asia and came to establish itself in about 300 varieties; the possible reason for the relatively wide occurrence of fragrant landraces in the central Ganga plains; and also for its presence in two landraces of the remote Wayanad district of Kerala (which incidentally possess very small spikelets [ca. 5.0–5.5 mm length]).

Definitive studies on the origin and nature of medicinal rices such as *njavara* are also called for.

L. Subspecies Variability

The phenomenon of hybrid inviability, weakness, and sterility is the most common expression of remoteness in intertaxa relationships in plants (Stebbins 1958). Both inter- and intra-specific sterility is widely prevalent in the genus *Oryza*.

Kato et al. (1928) were the first to observe the occurrence of hybrid sterility in rice. They crossed two variant groups of Asian rice, which they later termed indica and japonica. Terao and Mizushima (1939) conducted extensive studies and concluded that sterility relationships among Asian rices were much more complicated than merely grouping them into the two forms of indica and japonica. This was followed up by more detailed studies by several groups, mainly in Japan, but a clear and complete explanation has not yet been found – a situation that continues even now. The results have been summarized by Oka (1988) and Sano (1997).

The sterility may be chromosomal and/or genic, but we shall confine our discussion to the latter, since we are only concerned with the sterility observed in hybrids within the A-genome species and taxa. Genic F_1 sterility may be diplontic or haplontic. Oka (1988) had broadly classified them into duplicate gametic lethals, one locus sporo-gametophytic interaction, one locus allelic interaction, and complimentary sporophytic partial sterility.

Incidentally, the existence of two forms of rice was recognized in China from the Han dynasty (ca. 200 CE), long before Kato et al. (1928) recognized two ecotypes in *O. sativa*, indica and japonica. They were known as *hsien* (indica) and *keng* (japonica) rice. They correspond broadly to the above two ecotypes. After extensive studies conducted by numerous workers, it can be seen that although some landraces are identifiable as typical indica (hsien) and typical japonica (keng) rice, many are intermediate between them. One such group is known as javanica (Matsuo 1952, Nakagahra 1978). Further studies on its relationship have shown that javanica may be more aptly designated as tropical japonica (Oka 1958). Sato (1987) had characterized japonica as a panicle number type and javanica as a panicle length type. Compared to temperate japonicas, tropical japonicas are taller, they have thicker culms, fewer and longer panicles, longer flag leaves, and higher spikelet length/breadth ratio.

In present-day rice literature, it is common to see authors designating japonica and indica, and sometimes javanica as subspecies. Almost all the authors italicize them, which is untenable. No taxonomic revision of the genus has assigned a subspecific status to these entities. Considering the nature of differences shown by them, their most appropriate categorization may be as ecotypes. This will become apparent from the following account.

In spite of the most extensive morphological, biochemical, and biometrical studies carried out to discriminate these ecotypes, it has not been possible to classify them with complete confidence. Visually, it is possible to distinguish typical indica and japonicas by the following characters: **japonica**: spikelets broad and short, apiculus hairy; **indica**: spikelets long and slender, apiculus hair few or absent. After extensive studies, Oka and co-workers (Oka 1958, 1988, Morishima and Oka 1981) proposed that a combination of the following four characters would be able to differentiate them with reasonable robustness: phenol reaction, potassium chlorate resistance, response to cold and drought, as **japonica:** grains do not stain with phenol, show higher potassium chlorate resistance, show lower drought resistance, and higher cold resistance; **indica:** grains stain with phenol, show lower potassium chlorate resistance, show higher drought resistance, and lower cold resistance. Very broadly, tropical and temperate japonicas can be differentiated by the following morphological attributes: **tropical japonica:** tall plants, fewer tillers, long and heavier panicles, slender grains; **temperate japonica**: shorter plants, more tillers, shorter, normal panicles, thicker grains.

In an extensive study carried out on the prevalence of variability in Asian rice, Glaszmann (1986, 1987) studied 1,688 varieties using 10 enzymes at 15 presumed loci. Based on the initial results, the author selected 120 varieties representing the entire range of variation and analyzed them using 14 enzymes at 21 polymorphic loci. The study grouped the varieties into six clusters, designated group I to VI. They were: **Group I**: typical indica group, includes aman (Bangladesh, southeast and east India), tjereh (Indonesia), hsien (China); **Group II**: many originating in the Himalayan foothills, short duration (120 days), grown under irrigated (in Pakistan) and rainfed (Bangladesh, southeast India) conditions: corresponding to aus and some boro varieties; **Group III**: includes the two deep water Bangladesh forms, bhadoia and aswina; "peculiarly" early maturing and nonphotosensitive; **Group IV**: long duration up to 12 months, floating varieties, rayada and some boro; **Group V**: "a very diverse group", but includes also the high quality rices, adri (Iran), basmati (Pakistan, India, Nepal), some special rices from Myanmar; **Group VI**: typical japonica, keng, and bulu, most high altitude varieties grown in the Himalayas (mainly in Nepal, Sikkim, India, Bhutan).

In a dendrogram given in Glaszmann (1986, 1987) groups 1, 2, and 3 formed one clade and groups 4, 5, and 6 formed a second clade. In these two clades, groups 1 and 2, and 5 and 6 were sister to each other. The nature of phytogenetic relationships of these biochemical groups revealed by the dendrogram is difficult to understand from the viewpoint of varietal differentiation and their natural geographical distribution.

In a study, Garris et al. (2005) studied the genetic structure and diversity of Asian rice using SSRs and chloroplast sequences. SSRs are able to detect the fine structure of diversity more efficiently than RFLP, AFLP or SNP loci, because of their high level of allelic diversity. The authors used 234 accessions representing all the continents and employed 169 nuclear SSRs and two chloroplast sequences. When genetic distance and population structure were studied, most accessions grouped themselves into five groups, corresponding to indica (79 nos.), aus (20 nos.), aromatic (19 nos.), temperate japonica (45 nos.), and tropical japonica (44 nos.). In all these groups, >80% of the inferred ancestry was derived from a single population. A further 24 accessions were categorized as having admixed ancestry. Most of them were admixtures of tropical and temperate japonicas. The AMOVA showed that 37.5% of the variation was due to differences among groups, and 62.5% was due to differences within groups. The authors found further that the amount and organization of genetic diversity differed among the five groups, with the indicas, tropical japonicas (99% polymorphic loci each), and aus (98% polymorphic loci) showing the highest polymorphism. These three groups also showed the highest heterozygosity values: 0.55 in indica, 0.54 in aus, and 0.47 in tropical japonica.

In addition to the above two studies, there have been several other investigations to assess the variability in Asian rice, but, as with the earlier studies, none of them appears to have been able to fully comprehend the nature and extent of variability and group the varieties in a manner that relates them to their geographical, ecological, and/or photoperiodic adaptations. As was stated with regard to the japonica–indica classifications, while the majority of the varieties are amenable to being included in one or the other grouping, there are always a significant number of outliers. Further, as much as two-thirds of the total variability is accounted for by within-group diversity (Garris et al. 2005). This also indicates that no group can be taken to be even reasonably homogeneous. The reason for this may be that the Asian rice is grown over a very wide range of latitudes and altitudes, and under most diverse water regimes, temperatures, soil types, ecological conditions, and so on. The period of maturity of the landraces ranges from 60–65 to 330 days! Possibly no other graminaceous crop species is grown under such diverse ecological and edaphic conditions as rice is. The traditionally grown landraces in a region have developed in response to the ecological situations prevailing there, or the locality where they are grown. The existence of large numbers of ecotypes and variety groups in Asian rice based on the ecological and climatic conditions of their cultivation (like aman, boro, tjerah, viruppu, kuruvai, kole, etc), the prevalence of low levels of cross-fertilization, and the wide presence of semi-sterility even in varietal group crosses, can be considered as providing the checks and balances to impart an appropriate level of variability to the group even while maintaining their respective varietal or genetic identity and integrity.

The process of nobilization is a continuing process. We can only set an arbitrary benchmark in a population that is being nobilized to mark when it has crossed a threshold to be considered as a domesticated form. Before the 1960s, when rice improvement programs started in the vast regions of Asia, where wild and cultivated rices used to occur sympatrically, rice breeding used to be done by pure line selection. The landraces and varieties released during this period used to possess what the present-day rice scientists may term as wild/primitive characters, e.g., "awnness", black spikelet color, red pericarp color, easy "sheddability", seed dormancy, and so on. The publications of that period bear this out (Watt 1892, Ramiah 1937, Ramiah and Rao 1953, Sahadevan 1967).

High total spikelet number (TSN) per panicle is one of the components of high yield, which can be brought about by an increase in panicle length and/or by increasing primary and/or secondary branches. However, it is difficult to describe this as a domestication-related character. Fujita et al. (2012) have however studied the biology of this character by examining QTLs. The authors used a set of 334 chromosomal segments introgression lines consisting of a BC 3-derived population developed from five new plant type cultivars as donor parents and IR 64 (indica) as the background parent. From this, they selected five plants with higher spikelet numbers than IR 64 all derived from the progeny of one of the five donor parents. When the QTL analyses were done in their F2s with IR 64 (indica), a QTL for high TSN was detected. It was located on the long arm of chromosome 4. The variations in TSN were attributed to higher numbers of primary, secondary, or tertiary branches. The authors also identified several other QTLs for high spikelet number in *O. rufipogon, O. nivara,* and *O. glumipatula.* They were located on all the 12 chromosomes.

Most present-day rice biologists may be unaware that until the introduction of modern agricultural practices in Asia from the 1950s, several japonica rice landraces were generally awned or at least tip-awned; further, most landraces grown in adverse environments such as saline and acid sulfate soils, etc were stiff-awned to provide protection from predatory birds and insect pests.

iv. Mode of Evolution of Asian Rice

All the early workers on the origin of the Asian rice proposed that cultivated rice originated from the wild rices of Asia, known variously as *O. rufipogon, O. nivara,* spontanea, or fatua rices. Kato et al. (1928) then proposed that Asian rice consisted of two subgroups, japonica and indica, and that they were separated by sterility barriers. Matsuo (1952) and Morinaga (1954) later designated a third ecotype, javanica. It included bulu and tjerah forms of rice grown in Indonesia. Oka (1958, 1983) proposed that javanicas might be identified as tropical japonica based on their morphological and physiological affinities with japonicas. This viewpoint has been now accepted. Javanica rices are now termed tropical japonicas.

Two forms of rice had been recognized in China from the time of the Han dynasty (2nd century CE), and they were known as keng and hsien.

Ting (1949) proposed that the Chinese keng rices originated from the wild rices occurring in China, which also showed incipient differentiation. Subsequent studies have shown that keng and hsien broadly correspond to japonica and indica rices, and that keng rices are grown in the cool northern China and the hilly regions of southwest China, and hsien rices in the warmer tropical and subtropical regions and lower valleys of the hills in south and southeast China.

Glaszmann (1986, 1987), studied 1,688 traditional rice varieties of Asia, and showed that they clustered into six groups, Group I to Group VI. Group I is found throughout tropical Asia and includes the aman rices of Bangladesh (and southeast India), tjerah rices of Indonesia, and hsien of China; Group II is found in south Asia, and includes mainly the rices of Bangladesh and southeast India; Groups III and IV consist of the deep water rices of Bangladesh, southeast India, and northeast India (mainly the lower Brahmaputra and Ganges river flood plains and deltas); with Group III containing short duration forms; and group IV, very long duration forms. Group V consists of the scented rices of south Asia, mainly the basmati rices of Pakistan, Nepal, and India; and Group VI contains the japonica rices of temperate regions, hill rices of subtropical and tropical regions on either side of the Himalayas, including the bulu of Indonesia and the keng of China.

Garris et al. (2005) studied the genetic structure and diversity in Asian rice using 234 accessions employing 167 SSRs and two chloroplast genes. They detected five groups which corresponded to indica, aus, aromatic, temperate japonica, and tropical japonicas. Nuclear and chloroplast data supported close evolutionary relationships between indica and aus rices, and among tropical japonica, temperate japonica, and aromatic rices. They proposed that the primary divergence between indica and japonica might have been due in part to the differentiation of ancestral *O. rufipogon* populations in different locations and at different times. The aus and indica rices showed high diversity values and several chloroplast haplotypes. At the same time, the temperate japonicas showed close genetic relationships and lower genetic diversity than indica rices. Further, the aromatic rices showed a loose relationship to the japonica rices. They revealed a high proportion of monomorphic loci suggestive of a severe bottleneck. In this study, about 10% accessions showed mixed ancestry.

Parsons et al. (1997) studied the variation and diversity of 115 aus, aman, and boro rice landraces of Bangladesh using 35 RAPD markers. While aus belonged to Group I of Glaszmann (1986) and aman to his Group II, the boro rices did not match with any group. Varieties belonging to all the six groups of Glaszmann were found within Bangladesh itself. The authors suggested that aus rices might have given rise to japonica rices. The gene diversity indices (H) of the three forms of rice were aus: 0.126, aman: 0.154, and boro: 0.221.

Incidentally, there is some misinformation in recent literature on the aus and aman rices of south Asia. These ecotypes are grown in southeast and east India and Bangladesh. They are not grown in the Himalayan foothills, aus is not also grown during the fall season (Watt 1891). Broadly, aus rices are rainfed, upland, non-irrigated rices. They are sown by broadcast in April–May with the onset of monsoon and harvested in July–August. They have a short duration. On the other hand, aman varieties are late rices. They are grown during May to October. They are often transplanted and are photosensitive (Watt 1891, Ramiah and Rao 1953). They are longer in duration, 120–160 days (Parsons et al. 1999).

Up to the 1960s, the prevailing assumption was that Asian rice had only a single origin from one of the two wild rices, *O. rufipogon* or *O. nivara* (Nayar 1958, Sampath and Govindaswamy 1958, Oka and Morishima 1970, Chang 1976b). Chang (1985) postulated that Asian rice originated from the annual taxon, *O. nivara,* which in its turn had originated from *O. rufipogon.* The others were of the opinion that *O. sativa* had arisen directly from *O. rufipogon.* Chang, Oka and Morishima concurred that the indica rices then gave rise to japonica rices. Nayar (1973) observed that "rice had a diffuse origin both in space and time". The far greater diversity present in indica compared to japonica and its (indica) closer relationship with *O. rufipogon* were pointed out as evidence supporting the proposition. A further observation supporting this is that while indica rices are found in the warmer lowland tropics, japonica rices are found in the cooler climes of the subtropics and temperate regions, and the hilly regions of the tropics (where the climate is generally cooler). This is very apparent in the rices of Bhutan, Nepal, Sikkim (India), and Yunnan (southwest China) (Morinaga and Kuriyama 1955, 1958, Kihara and Nakao 1960, Kihara 1964, Glaszmann et al. 1989, etc.).

Incidentally, the physical, climatic, edaphic, and also the ethnic, linguistic, and cultural variations prevailing in the rice regions of east, south, and southeast Asia are indeed very great. The tremendous amount of variability shown by the traditional rice landraces of the region are a reflection and response to this environmental, anthropological, and cultural diversity. In rice, there are landraces that mature in 60–330 days, grow in very saline and acid sulfate soils, grow as far north as 45°N (Hokkaido island, Japan,

42–46°N), even under snow, and in flooded conditions where the water may be as deep as 3–4 m (in Kuttanad region, southwest India, Bangladesh, etc.). The rice quality also shows tremendous diversity, from very flaky to glutinous, and from medicinal rices to different kinds of scented rices, as we have observed before. The milled grains also show much variability in length, from 3–10 mm. In a cultivated crop, both nature and nurture would contribute to creating such a tremendous diversity. No other cultivated crop species may be possessing diversity. To a large extent, this may have happened and continues to happen because of the existence of the very wide interface between the putative ancestral wild rices and cultivated rices throughout east, south, and southeast Asia. This region accounts for more than three-fourths of rice cultivation and consumption. The divisions prevalent in the Asian rice into japonica and indica rices, and further into five or six subgroups are obviously a consequence of this situation.

Studies of the patterns of genome-wide variation in wild and cultivated rice and the number and origins of alleles associated with domestication genes may help to understand the domestication process (Sweeny and McCouch 2007, Perugannan 2010).

Chen et al. (1993a,b) were possibly the first to determine that wild rices tended to show characters of both indica and japonica varieties. They found that japonica and indica rices possessed types of chloroplast DNA (cpDNA) that differed in the ORF region showed a deletion of 69 bp within the Pst-12 fragment in most indica rices, while no such deletion was found in japonica rices, as had been found earlier by Nakamura et al. (1991) and Kanno and Hirai (1992). (ORF) Open Reading Frame: a nucleotide sequence that starts with an initiation codon, partitions the subsequent nucleotides into amino acid encoding triplets, and ends with a termination codon.) The interval between the start and stop codons is called the ORF. King et al. (2006).

Chen et al. found further that wild rices also showed this characteristic. They tested this with PCR analysis using a representative collection of 137 *O. sativa* cultivars and 117 wild rice accessions. In wild rice, 45 accessions showed deletion and 73 accessions did not show it. The annual wild rices showed the deletion and the perennial ones did not. This kind of differentiation in *O. rufipogon* has been observed by several other authors using other techniques and procedures as well; e.g., Second (1982) using enzyme markers, Lu et al. (2002) using RFLP analysis, Cheng et al. (2003) based on interspersion patterns of SINEs, Zhou et al. (2003) based on microsatellite analyses, Londo et al. (2006) with DNA sequence variation from chloroplast and nuclear gene regions, Caicedo et al. (2007) using nucleotide polymorphism, Duan et al. (2007) using differences in DNA sequences and polymorphism, Rakshit et al. (2007) using DNA sequence polymorphism, and so on. There have been no contrary observations (also, McCouch et al. 2007).

Some authors have recently proposed dual or multiple origins for rice. This proposition now appears to be gaining more currency.

Yamanaka et al. (2003) analyzed the distribution of the retrotransposon, *p–SINE1-r*^{2}, a short interspersed nuclear element (SINE) at the waxy locus. They found that most annual species possessed this retrotransposon, while most perennial types lacked it. They inferred that the indica rices evolved from the annual wild rice forms and the japonica rices from the perennial wild rices.

Vitte et al. (2004) came to a similar conclusion, but by using a different approach. They compared the transpositional history of 41 LTR retrotransposon families in the genomes of the two rice cultivars, Nipponbare (japonica) and 93–11 (indica), both of whose complete sequences are available (cf. Goff et al. 2002, Yu et al. 2002). Using a genomic paleontology approach, they estimated that the two genomes diverged at least 200,000 years ago. They confirmed this with a survey of insertion polymorphism in a wide range of both japonica and indica landraces. The results indicated independent origins of japonica and indica rices.

There are other authors also who have proposed multiple origins for Asian rice. Kovach et al. (2007) observed that: "the diverse geography and ecology of the broad region that is home to *O. rufipogon* (east, southeast, and south Asia) suggests that this ancestral species was broadly adapted to diverse climates and would have provided ample genetic diversity from which *O. sativa* could have been domesticated on numerous occasions".

They stated this on the basis of genome-wide studies of variation which demonstrated that indica and japonica rices arose from genetically distinct gene pools within a common ancestor, *O. rufipogon*. Earlier, Morishima (2001) too had concluded from her extensive studies that "rice domestication has been a diffuse process in both space and time".

Sang and Ge (2007) have attempted to strike a balance between single and multiple origins of Asian rice by explaining the two processes differently and giving them different names, viz., the snow-balling model and the combination model. Their Model 1 referred to the single origin process. In this, a primitive form of *O. sativa* was initially selected by the early farmers, who then introduced it to other regions. There, they introgressed with the locally available wild rices that were adapted to the local physical and environmental situations. Out of these arose japonica and indica rices as hybrid derivatives with different genomic backgrounds of the wild populations. They called this process the snow-balling model. Model 2 is the multiple origin model. From the locally adapted and diverged wild rices, farmers selected the cultivable forms that were best suited to the environment of the region. Alleles for the various traits were fixed in locally adapted ecotypes, which came to be known as japonicas and indicas. This model was designated the Combination Model.

Vaughan et al. (2008a,b) proposed that Asian rice had evolved only once, and did not have dual or multiple origin. This is contrary to their own earlier viewpoints. They based their present proposal on the recent finding of Li et al. (2006), and Konishi et al. (2006), that the same shattering genes, *sh4* and *gSH1*, are present in both japonica and indica rices. Vaughan et al. argued that since the chances of the same mutation occurring independently and at the same locus in two rice groups are very small, this mutation must have occurred only once, and then diffused into other regions of Asia through human migration and movement. They proposed that it occurred initially in japonica rice in the lower Yangtze river region in eastern central China – the region from where the oldest rice archaeological remains have been obtained – as has also been proposed by some other workers (Fuller 2011, Liu and Chen 2012).

Vaughan et al.'s assumptions appear to be presumptuous. First, even though the chances of the same mutation occurring more than once may be small, but this is not impossible, especially when the initial material is under intensive selection pressure in the same direction. The non-shattering character would have been the most sought-after gain for the early farmers during the domestication of rice. There is good evidence for the movement of people across the eastern Himalayas from prehistoric times, between what is presently northeastern India and southern China (southern Yunnan, eastern Tibet) and of people from the Yangtze region moving south and eastwards to the Philippines and mainland southeast Asia.

Molina et al. (2011) conducted detailed studies on the evolutionary history of rice using two methods, and concluded that *O. sativa* has undergone only a single domestication. The two methods they employed were: (a) a diffusion-based approach to demographic modeling of SNP-data; and (b) a Bayesian evolutionary approach for phylogenetic analysis using previously published data sets. Both the analyses supported a single origin. The authors indentified SNPs through direct sequencing of >250 kb sequences from 630 fragments of three chromosomes (nos. 9, 10, and 12) of both wild and cultivated rices. They obtained 255.9 ± 2.09 kb sequence data for each accession consisting of 2,800 SNPs for indica, 2,070 SNPs for tropical japonica, and 7,274 SNPs for *O. rufipogon*. In addition, the authors used the data sets of Zhu and Ge (2006) and Londo et al. (2006), since the data that Molina et al. had generated were insufficient to support the BLAST software analysis employed (Table 6.7). (Incidentally, Londo et al. (2006) had used this very data to infer a multiple origin for Asian rice!)

Using the molecular clock approach, Molina et al. (2011) further proposed that *O. sativa* split from *O. rufipogon* 13,500 yr BP and the indica and japonica rices diversified 6,700 yr BP (Table 6.7).

This analysis once again shows the uncertain and contrary results that are revealed when diverse biometrical methodologies are employed for analysis. This may also be the consequence of excessive reliance on highly involved method for biological analyses. It is difficult to come to any other conclusion when the data used in Molina's study and their conclusions are compared with those obtained earlier.

Fuller et al. (2010) have recently proposed a novel pathway for the origin of Asian rice. In their hypothesis, japonica was developed in the Yangtze river basin 6,500–6,000 ya with selection for the principal domestication traits. At that time, south Asia had an inferred non-domestication feedback

Table 6.7 Results of BEAST re-analyses of Previously Published Phylogenetic Datasets*

Dataset	No. of Loci	Total no. of Sites	No. of Consensus Topologies+	Single Origin Support++	Indica and Japonica Not Monophyletic†
Zhu and Ge (2005)	4	2,917	11	7.7%	90.1%
Londo et al. (2006)	3	3,232	9	3.9%	92.9%
Tang et al. (2006)	11	9,775	3	96.3%	3.9%
Rakshit et al. (2007)	22	19,789	1	100%	0%
Zhu et al. (2007)	10	8,031	3	100%	0%
Yu et al. (2011)	5	19,780	3	100%	0%

*Reproduced from Molina et al. (2011) with permission
+Of 19 possible topologies (including polytomies)
++Indica and tropical japonica monophyletic
†Percentage values indicate the proportion of trees of a given topology in the posterior probability distribution of trees

system for protoindica. Later, this hybridized with domesticated japonica rice. Out of this was selected indica rice about two millennia later. Based on linguistic evidence, they proposed that cannabis represented the earliest and best archaeological evidence of introduction from China to India. Also introduced alongside cannabis were peach, apricot, and harvest knife. They were introduced to the Middle East region across central Asia by a northern route and also into the Indus region. This started around 2,400–2,200 BC and was well established by 2,000 BC. Japonica rice would also have been brought in along with these items. The authors stressed that the introductions were made through trade, and not through demic diffusion. Incidentally, the suggested period of introductions corresponds with the Indus valley civilization (3,200–1,400 BCE), which was based primarily on trade. They observed further that there was some evidence for rice cultivation in the Indus valley during 2,500–2,000 BC. This would have used proto-indica, and its hybridization with the introduced domesticated japonica rice was hypothesized to take place after 2,000 BC. This coincided also with the spread of upland rainfed rice from southwest China (Yunnan) to northeast India across the Himalayas. In further support of this hypothesis, Fuller et al. pointed out that the Chinese common millet *Panicum miliaceum* was introduced into Yemen and Sudan via the central Asian trade route during 2,000–1,700 BC (Boivin and Fuller 2009). Fuller et al. (2010) then proposed the following sequence of events for the formation of indica rice: *O. nivara* (prostrate type) → protoindica crossed with japonica (possessing the following domestication-related genes: *sh4, PROG1, rc, sw 5, QSH1, RC, SV5, wxy*) → indica/proto-aus forms *O. nivara* (weedy type).

While the proposition about the origin of japonica rice may be in order, it is felt that actual evidence is required to support the hypothesis about the origin of indica rice.

Huang et al. (2012) have recently proposed that japonica rice was first domesticated from a certain population of *O. rufipogon* around the middle area of the Pearl River (southern China), and that indica rice was subsequently developed from crosses between japonica rice and local wild rice as the initial varieties spread to southeast and south Asia. These authors generated genome sequences from 446 geographically diverse *O. rufipogon* and 1,083 indica and japonica varieties, and constructed a comprehensive map of rice genome variation. They then identified 55 selective sweeps that occurred during domestication. After this, they conducted in-depth analyses of the domestication sweeps and genome-wide patterns. They analyzed the domestication-associated traits through high resolution genetic mapping. The authors further studied the geographic variation of *O. rufipogon* using chloroplast and low-copy nuclear markers. They detected a possible differentiation within *O. rufipogon*, but this was not consistent with their geographic distribution. Hence, to characterize the differentiation critically, they surveyed sequence variation at 42 genome-wide sequence-tagged sites (STS) in 108 *O. rufipogon* accessions, which represented the full range of the species' distribution. They analyzed the data using Bayesian clustering, principal component analyses and AMOVA. They concluded that there are two genetically distinct forms of *O. rufipogon*, Ruf-I and Ruf-II. These two forms exhibited a clinal variation pattern from northeast to southwest of the species' distribution. Ruf-I is found mainly in China and Indo-China peninsula and showed genetic similarity with indica rices. Ruf-II is found mainly in south Asia and the Indo-China peninsula. It was found to be closely related to local cultivated rice varieties. The japonica was not similar to either of the two *O. rufipogon* groups. From their study, Huang et al. (2012) concluded that their results supported a single origin of domesticated *O. sativa* in China.

It is clear from the above discussions that the viewpoints expressed by various authors using different methods have produced very diverse and very contradictory results. This calls for a fresh look at this subject.

EPILOGUE

The previous six chapters of this volume had looked at: (1) the present position of rice in the world as a major food crop; (2) the history and spread of rice in the world; (3) phylogeny of rice and its genus, *Oryza;* (4) interrelationships among species and taxa in the genus *Oryza;* (5) origin of the African rice; and lastly (6) origin of the Asian rice.

Rice is one of the major food crops showing the highest rates of increase in productivity and production (the others being cassava and banana). This trend is expected to continue. In the Preface to this volume, Dr. M. S. Swaminathan stressed the increasing significance of rice as a climate-resilient crop in this age of global warming. Intensive R & D efforts are underway to develop new high-yielding cultivars possessing additional attributes such as biofortification and resistance to various biotic and abiotic stresses, especially to floods and heat. Studies have been taken up to even convert the present C-3 photosynthesis pathway of rice to the more productive C-4 pathway. We can expect that these ongoing efforts will be able to meet the increasing demand for rice in the future.

In Chapter 2, we reviewed the history of the introduction and spread of rice in different continents and regions of the world from its centers of origin in Asia. Rice was first introduced into southeast Asia from the lower Yangtze basin (China) about 50,000 years ago. The earliest archaeological remains of both the wild and cultivated rices have been found in the lower Yangtze river basin in east China. From here, rice spread northwards and southwards to south and southeast Asia, both by diffusion and by movement of people. These people carried with them both rice and the skills for cultivating it. However, in New Guinea, the west Pacific, and Australia, rice and its cultivation reached only about 600 years ago, carried there by a migrant population.

The assumption hitherto had been that rice originated independently in south Asia. From south Asia, it diffused westwards to southwest Asia about 2,500 years ago. However, it has been recently suggested – although based on rather weak indirect evidence – that rice might have been brought to this region about 2,200–2,000 BCE from the Yangtze river basin by a western route north of the Himalayas. This rice was a proto-indica form, it hybridized locally with the available wild rices and formed the present-day indica rices of south Asia. Until some supporting archaeological, botanical, or other direct evidence is obtained from central Asia, it is difficult to take note of this proposition.

Most of the rice in Africa is grown in Egypt, Madagascar, and west Africa. Rice is thought to have been introduced into Egypt in the 7th century CE by the Arabs, and into Madagascar by the people of southeast Asia migrating across the Indian Ocean, by the 5th century CE at the latest, with the earliest suggested date of 5th century BCE. However, hard evidence is not yet available for the earliest suggested date.

In west Africa, the situation is unique and different. This region of Africa produces the most rice, but also imports more than half of its rice requirements. The African rice, *O. glaberrima,* is indigenous to this region, but Asian rice, *O. sativa,* has now nearly replaced African rice in cultivation. This has happened because all the developmental activities in the region, which were taking place since the First World War (1914–1918), was promoting only the Asian rice. While Asian rice possesses a better ability to respond to modern agricultural practices and has a higher yielding capacity, African rice possesses several superior agronomic attributes of local relevance. It has superior competitive ability over weeds, and is better adapted to low soil fertility, high soil acidity, uncertain moisture regimes, and abrupt changes in environmental conditions. After the near-complete displacement of African rice, the Africa Rice Center (Africa Rice Center, Cotonou, Benin) has successfully released hybrid derivatives of Asian and African rices (*O. sativa* x *O. glaberrima*) designated NERICA (**N**ew **R**ice for **A**frica), and introduced them into cultivation. However, the first series of improved cultivars released by the Center has only about 10% *O. glaberrima* blood. From now on, new NERICAs will be known as ARICA (**A**dvanced **R**ices for **A**frica).

African rice, *O. glaberrima*, originated in west Africa itself. Like the Asian rice, it was generally assumed that it was introduced here by the European traders and fortune seekers, after they arrived here in the mid-17th century. Various authors have since proposed, based on numerous travel accounts, that Asian rice was first introduced into west Africa possibly by the Arabs during the 8–12th centuries CE. Three routes have been suggested for this introduction: from the north across the Sahara via the then active trans-Saharan trade route, from the east coast across central Africa, and from the northeast through the Ethiopa–Eritrea–Somalia region. The trans-Saharan trade routes were very active during about 1500–1000 BCE. Present-day Libya (encompassing Cyrenaica, Fezzan, etc.) used to support active agricultural societies during prehistoric times (from about 900 BCE). Similarly, the coastal region of eastern Africa from Eritrea, the Horn of Africa, and up to the present-day northern Mozambique, also used to see much human activity from Roman times, beginning around the 5th century BCE. This continued during the Arab Muslim period until the arrival of the first Europeans in the late 15th century CE. Further south, the earliest people from southeast and south Asia are thought to have landed in Madagascar at various times from the 5th century BCE to 5th century CE. There are even suggestions of some southeast Asian influences in west Africa, attested by Jones (1959), based on the use of the xylophone, and by Rush (2008). on the basis of cultural contacts. Hence, Asian rice could easily have been introduced into eastern Africa, and possibly also western Africa, long before the arrival of Europeans during the mid-15th century.

The introduction of rice into the Americas is an analogous situation. Here too, the widely held belief has been that rice was introduced into the Americas by Europeans after its discovery by Christopher Columbus in 1498. More recently, several scholars have uncovered convincing historical evidence that the first introductions of rice into the Americas were carried out by the African slaves, most of whom were brought into America from west Africa from the 16th century onwards (Carney 2001, Fields-Black 2008, etc.).

Our knowledge-history of the introduction and spread of rice into Europe and the Americas is patchy in the detail, even though it took place only within the last 600 years.

Rice has shown tremendous resilience in its adaptability to comparatively extreme regimes of temperature, water/moisture levels, and soil as indicated by its spread worldwide.

The phylogeny of the genus *Oryza* is covered in Chapter 3. The genus is part of the tribe Oryzeae. This tribe includes 8–10 genera and ca. 70 species. While GPWG (2001) included the tribe Oryzeae in the subfamily Ehrhartoideae, some authors (Clayton and Renvoize 1986, Mabberley 2006) have included it in Bambusoideae: the most primitive subfamily of the family Poaceae. These latter authors do not recognize the subfamily Ehrhartoideae.

Within the tribe Oryzeae, the genus *Leersia* has a sister relationship with the *Oryza* genus (Fig. 3.2). Incidentally, some of the species now included in *Leersia* were previously included in *Oryza* until Launert (1965) transferred those species that did not have sterile lemmas into the genus *Leersia*. This transfer has since been universally endorsed. The remaining four genera of *Oryzeae* form the second clade, *Ehrharteae,* in the subfamily Ehrhartaoideae (Fig. 3.2).

The cluster containing *Oryza* and *Leersia* has been proposed to have separated from the other genera 20.5 mya. *Oryza* and *Leersia* diverged at

14.2 mya (Fig. 3.3, Guo and Ge 2005). Within the genus *Oryza, O. granulata* was the first species to differentiate. This happened 10.2 mya (Kellogg 2009).

The genus *Oryza* includes 18–26 species and taxa. They have a pantropical distribution: with three species and two genomes (AA, EE) in Australia; seven species and six genomes (AA, BB, CC, GG, HH, JJ, and two unknown) in New Guinea and New Caledonia; seven species and six genomes (AA, BB, CC, GG, HH, JJ) in Island Southeast Asia and Malayan Peninsula; three species and genomes each in east Asia (AA, CC, GG); seven species and four genomes (AA, BB, CC, GG) in south Asia; six species and four genomes (AA, BB, CC, FF) in Africa; and four species and three genomes (AA, CC, DD) in Central and South America.

Some authors had initially attributed the global distribution of *Oryza* species to the break-up of Gondwanaland, but this does not fit with our present understanding of both the timings of Gondwana break-up and the Poaceae phylogeny (Table 3.1). For instance, South America became a separate continent 30 mya, the Indian plate collided with Asia 45 mya, and the Australia–New Guinea separation took place during the period 55–15 mya. However, the divergence of Poaceae took place in the period 77–55 mya, and according to various estimations, the genus *Leersia* separated from *Oryza* only around 14.2 mya, and species differentiation within *Oryza* began only 10.2 mya. At the same time, since the pantropical distribution of *Oryza* species is a reality, we can only assume that primitive forms of Poaceae/Ehrhartoideae might have been distributed along with the Gondwanan break-up and the drift of the various landmasses that followed this event/process. Thereafter, parallel evolution of genera and species might have taken place in the various land masses, which subsequently evolved into the present-day continents. The last continents to develop their respective identities were South America (30 mya) and New Guinea–Australia (15 mya) (Table 3.1). Vaughan speculated that long-distance dispersal by birds could explain the disjunct distribution of some *Oryza* species, but it may be stretching the imagination too far to explain the wide occurrence of species like *O. rufipogon, O. eichingeri, O. punctata, O. officinalis,* etc. That is, until hard evidence of specific bird flights are available.

Species relationships within the genus *Oryza* is covered in Chapter 4. *Oryza* is a small genus consisting of about 18 species and 6–10 subspecies taxa. This genus has been always considered to be a difficult one to classify.

Since *Oryza* is a small genus, it is unnecessary to include any subgeneric classification; it serves no purpose. Instead, the species and taxa in the genus have been grouped into 4–6 species complexes for convenience of discussion. These groups have been identified more or less on the basis of the genomic composition of the species and taxa.

In the 8–10 revisions of the genus carried out so far, there has been unanimity about the status of 13 species – *O. sativa, O. glaberrima, O. australiensis, O. barthii, O. latifolia, O. longiglumis, O. longistaminata, O. meyeriana, O. minuta, O. officinalis, O. ridleyi, O. rufipogon*, and *O. schlechteri*. The species status has been questioned in the case of nine taxa – *O. abromeitiana, O. alta, O. glumipatula, O. granulata, O. inandamanica, O. malampuzhensis, O. meridionalis, O. neocaledonica*, and *O. rhizomatis*. In this volume, some of them have been listed as subspecies of one or the other of the "valid" species. Only detailed morphological and molecular biology studies, using carefully chosen material and methods, and pooled analyses of published work, can resolve their status. The few studies performed on some, such as *O. malampuzhensis,* have not helped the matters. This is mainly because such studies have not been carried forward to a conclusive end.

A recurring problem that is seen in intertaxa and inter-ecotype (japonica, indica, for instance) hybrids is the widespread occurrence in their progenies of low levels of natural crossings, and the consequent prevalence of semisterility, mainly male sterility. This appears to be nature's way of preserving the genetic integrity of the female taxon, even while permitting a limited introduction of alien genes into it for providing the necessary wherewithal. This introduction gives the taxon an opportunity to better adapt itself to changing environmental conditions. Incidentally, Sano and his associates have studied the phenomenon of sterility in rices extensively, and they have attributed it to genic factors.

Stebbins (1959) observed more than 50 years ago that "occasional hybridization between recognizable species is the rule in flowering plants". This phenomenon is more widespread when cultivated plants are involved. Ellstrand (2003) has discussed this aspect in detail, and termed it "dangerous liaisons", although in a different context. Such natural crossings occur in rice, because of the highly disturbed and constantly changing habitats of the cultivated crop.

There are only about 12–14 *Oryza* species that possess characters distinct enough to be identified visually without using any metrical measurements. The detailed experimental taxonomy work done by Tateoka showed that in all the species complexes of *Oryza*, majority of the species/taxa could only be identified only by a combination of 3–5 characters, and even then, there remained 10–15% of samples (both herbarium sheets and live collections) that showed intergrades and could not therefore be identified with certainty. In the Latifolia complex, out of 34 samples of *O. minuta* and *O. officinalis*, 27 samples could be identified by a combination of five characters and the identity of the remaining seven samples could not be determined. *O. minuta* (2n = 48) has only a very restricted distribution in the Philippines and New Guinea and *O. officinalis* (2n = 24) is widely distributed in south, southeast, and east Asia. In *O. eichingeri* and *O. punctacta*, the author studied 73 samples (23 live + 50 herbarium sheets). Of these, three sheets and two live samples were intermediate in features. The two species, *O. eichingeri* and *O. punctata,* were distinguishable only by using a combination of five characters. Here too, *O. eichingeri* has a restricted distribution – and that also discrete – central Africa, Srilanka, while *O. punctata* is widely distributed in middle Africa (eastern, central, and west).

The situation is analagous in the case of the three tropical American species, *O. alta, O. latifolia,* and *O. grandiglumis. O. grandiglumis* is easily distinguished by the presence of sterile lemmas that are similar in size to the fertile lemmas. *O. alta* and *O. latifolia* are identifiable only by a combination of characters. Here also, there were a few specimens that possessed intermediate characters.

A similar situation is seen in the four southeast Asian taxa, *O. granulata, O. meyeriana, O. longiglumis,* and *O. abromeitiana.* As the name implies, *O. longiglumis* is characterized by long sterile lemmas that are longer than the spikelet length. Other than this, all the four taxa possess overlapping characters with regard to spikelet dimensions, awn length, etc.

The situation appears to be similar also in the case of *O. glaberrima* and its allied taxa, *O. barthii* and *O. stapfii. O. glaberrima* even shows some morphological similarity with Asian rice. Incidentally, Lolo and Second (1988) had observed *O. glaberrima*-like plants on the Konkan-Kerala coast (India), and from elsewhere in India, there have been reports of *O. glaberrima*-like plants (Assam, Jeypore Tract) also, Morinaga and Kuriyama (1955, 1958). It was suggested earlier in this volume that it was the apparent similarity between the two cultivated species that caused the delay in the first description of African rice, *O. glaberrima,* which occurred in 1848 – 95 years after *O. sativa* was described, and in recording its presence the second (29 yr) and third (44 yr) time.

We have a comparable situation with the Asian wild rices, but in a more acute form. They are variously designated as *O. nivara, O. fatua,* or just fatua or spontanea rices. Since this matter has been discussed at length in Chapter 6, we shall not go into it again here.

What does the foregoing account indicate? That widespread hybridization is taking place in nature within the species and taxa of all the species complexes in the genus *Oryza?* (That which Ellstrand (2003) indicated as "dangerous liaisons", though in a slightly different context.) The species identities of *O. meridionalis* and *O. rhizomatis* have been questioned, as has those of *O. inandamanica* and *O. malampuzhensis*. This leaves us with *O. australiensis, O. brachyantha, O. longistaminata, O. schelechteri*, and *O. neocaledonica*.

No studies are known to have been done on *O. neocaledonica* after it was first described in 1994 (Morat et al. 2004). *O. schlerchteri* was assumed to have become extinct after Pilger had described it in 1907, as no examples of this species had been collected since that date. Then, Vaughan (1994) collected *O. schlechteri* again, from the same location in New Guinea from where Pilger had collected the first samples. In the only study done on this species subsequently, Naredo et al. (1993) studied it to see if this taxon belonged to the genus *Oryza* or to *Leersia*. They confirmed its generic status as *Oryza*.

In species relationship studies, we posed nine queries that were unresolvable through taxonomy. They related to the taxonomic identities of *O. abromeitiana, O. alta, O. glumipatula, O. granulata, O. malampuzhensis, O. meridionalis, O. nivara, O. rhisomatis*, and *Portersia coarctata*.

A large number of phylogenetic studies have been carried out on *Oryza* by different groups using various procedures. In this book, we reviewed about 15 representative studies. Three of the studies used two or three diverse techniques using the same experimental material. Even in these, no two studies gave similar results. The results of other authors also show a wide variation.

It was expected that molecular biology studies would help resolve most of these phylogenetic relationships – these methodologies have been used successfully to resolve many phylogenetic issues regarding the grass family (GPWG 2001) and angiosperms (APWG 2009). The groups investigating these issues pooled the data generated by different authors and conducted pooled analyses. GPWG folded after publishing the 2001 report, but APWG continues to be active, and the present revision of their report is the third published to date, APG III (2009). We may attempt a similar procedure in *Oryza* also. The studies of Molina et al. (2011) are a pointer in this regard, even if they do not provide an ultimate resolution of the problem.

A review of various studies also suggests that greater care needs to be taken in the choice of materials and in the robustness of methodologies. The intraspecific variability of *Oryza* species appears to be higher than in other similar self-fertilized genera, as is seen in the few results published on within-species variability. Linkage disequilibrium (LD) appears to be widely prevalent in *Oryza* (Agrama and Eizenga 2008). This may be adding to the difficulties in delimiting several species and taxa in the genus. Natural crossings among the species in *Oryza* also appear to be higher than in other similarly-placed genera. Hence, it is advisable to use more samples per species/taxon, and make sure that they represent the full range of their natural distribution. The results appear to be more robust in the couple of studies in which such precautions had been taken.

It is usually assumed that all the rice cultivated in west Africa is of the African rice, *O. glaberrima*, but the evidence presented here indicates that this is not so. There are good reasons to assume that at least part of the rice that was cultivated in west Africa was *O. sativa*. The belated description of *O. glaberrima* (in 1855, vs. *O. sativa* in 1753), the long time-lapse between the collection of its first three samples (29 and 44 years), the fact that the first herbarium specimen of *O. glaberrima* was pasted along with one specimen of *O. sativa* and the sheet labeled as *O. sativa*, the observation that African rice shows characteristics of a recently differentiated species (the apparent newness of *O. glaberrima* [Buddenhagen 1978]), and the general observation that its genetic variability is "extremely low" (e.g., Li et al. 2011) are all pointers to *O. glaberrima* being a neospecies.

Cultivation of *O. glaberrima* has been confined to west Africa. Presently, it has all but been replaced by the Asian rice, *O. sativa,* because of the policy actions of the two European colonial powers of west Africa of the time, UK and France, who promoted only *O. sativa* in the development activities initiated in the region since the First World War (1914–1918). In the last two decades, the African Rice Center has initiated a program to combine the superior attributes of the two cultivated rices by interspecific hybridization. This has led to the release of NERICA (**N**ew **R**ice for **A**frica) rices. From now on, Nerica will be relabeled as ARICA (**A**dvanced **R**ice for **A**frica).

Presently, Africa is the second biggest consumer of rice after Asia, but still, this is only 19.6 kg/head/year when the whole continent is considered (comparable values for Asia are 77.9 kg and for the world, 52.9 kg). Currently, Africa has the highest compound growth rate in production (31.2%), in domestic supply (38.4%), and in per capita supply (15.8%) of rice for the period 1961 – 2007 (Table 1.3). However, the continent is now importing >60% of its annual requirements. Among the different regions of Africa, western Africa has the highest per capita rice supply (consumption), at 32.8 kg/cap/yr. In Africa, there are six countries whose annual per capita rice consumption is comparable to that of a typical south, southeast, or east Asian country. They are: Guinea, 128.2 kg; Madagascar, 102.1 kg; Sierra Leone, 92.4 kg; Guinea-Bissau, 84.7 kg; Liberia, 73.0 kg; and Cote d' Ivorie, 61.0 kg (see tables in Chapter 1). West Africa in general, and the Guinea coast and Inner Niger Delta (IND) in particular, have had long, established traditions of rice culture among the several language/ethnic groups. They do not seem to have been documented properly so far.

Roland Porteres has been the only author to have botanically studied the rice in Africa consistently and continuously. He published his first paper in 1945, and his last (posthumously and jointly with Prof J. Barrau) appeared in 1981. He was the sole author of all his papers (except the last, for the multi-volume, *The UNESCO History of Africa*). Porteres' early writings dealt with both the cultivated rice species and covered their variability, parallelism of the two species in their characteristics and variability, the records of African rice in South America, nomenclatural similarities between the rices of Madagascar and southeast and south Asia, and so on. Possibly because Porteres was the only author writing on African rice (except for very brief and specific mentions by Angladette, Roschevicz, and Chevalier), his views on its origin have persisted and continued to be repeated in the literature, even including the prestigious (US) National Academy of Sciences publication, *Lost Crops of Africa* (1996), and the present-day blogs of the Africa Rice Center. This is in spite of the fact that several of his views on the origin of African rice had been severely criticized by leading archaeologists, anthropologists, and philologists, especially at the Third Conference on African History and Archaeology (Gray 1962). Very briefly, Porteres' proposition was that African rice originated from the weedy wild rice, *O. barthii*, its center of origin was the Guinea coast, with two secondary centers of diversity located in the Guinean hills and Inland Niger Delta, and that it originated 3,500 years ago. He based his theories on the basis of his botanical studies, some linguistic evidences, the Neolithic state of the Mande people at that period, who (according to Porteres) domesticated African rice, and its association with the Senegambian megaliths. We have discussed this aspect in some detail in Chapter 5.

It had been proposed 40 years ago that *O. glaberrima* might have originated from *O. sativa* by mutation, and that *O. barthii,* instead of being the progenitor of African rice, was the product of natural hybridization between the two cultivated rices (Nayar 1973). As a prequel to this, it was pointed out that Asian rice was possibly introduced into west Africa during the 8th–12th centuries CE. Asian rice was introduced into Madagascar from Asia (mainly the southeast, also from the south) by the 5th century CE, at the latest. The only comment on this proposal has been from Second (1986), who stated that at least part of African rice could have originated from

O. sativa. Considerable supporting evidence, albeit indirect, for this hypothesis has been put forward recently. The archaeological findings of rice from the region have not been very definitive, mainly because of the poor state of preservation of the material, or that they were of wild rices.

What we need in the present situation is direct experimentation to test the veracity of these proposals. It is also necessary to initiate crop improvement work on *O. glaberrima* itself to increase its yield and further "top it up" or improve the already available positive attributes – its greater resilience to uncertain weather conditions, superior competitive ability over weeds, while retaining its red pericarp and other positive attributes.

Sakai et al. (2011) has recently reported the results of the whole genome sequencing of *O. glaberrima* using a shot gun sequencing method. It yielded 206.3 Mbp of 437,642 sequence-reads corresponding to about 0.6 x coverage of 357 Mbp of the species. Using subtractive and methalation filtration, they selected 22.5% of total nucleotides. This subset was compared with a similar one of *O. sativa* (japonica variety). Patterns of amino acid changes did not differ between the two species in terms of biochemical properties. Further, genomes of *O. glaberrima* generally stored a larger synonymous–nonsynonymous substitution ratio. This suggested to the authors that it had "undergone a genome-wide relaxation of purifying selection".

Li et al. (2011) used multi-locus DNA sequences of 14 unlinked genes located on different chromosomes to compare the levels and patterns of nuclear variation in African rice (20 accessions) and the related wild species, *O. barthii* (20 plants). The authors studied nucleotide diversity based on total sequences using the program DnaSP version 4.10.8. They then constructed genealogical trees using NJ, MP, and ML methods. The nucleotide diversity was very low in both the species. That in *O. glaberrima* was only one-fourth of that in *O. barthii* – 0.0006 and 0.00025, respectively, in both the entire region and silent sites. They attributed the much lower variation observed in *O. glaberrima* to the genetic bottleneck that it would have undergone during the domestication process. It is however possible that the higher nucleotide variation in *O. barthii* could be more easily explained if it is considered as a hybrid derivative of *O. sativa* and *O. glaberrima*.

The origin and evolution of Asian rice, *O. sativa*, was considered a more or less "settled" issue until recently. Asian rice is the rice universally cultivated across the world, the food crop that is the staff of life for the majority of humankind, and certainly the life-supporting food of more than half the human population living in east, southeast, and south Asia, parts of tropical Africa, and the West Indies (cf., tables in Chapter 1). However, the extensive studies on the topic carried out in the last two decades have produced wide differences and diverse points of view regarding the place and mode of domestication of Asian rice. This has prompted some workers to describe rice domestication as a "puzzle", as having a "complex history", and so on.

The situation may be the result of various authors employing disparate material and methodologies for their experimentation and analyses. Also, the early foundation work done on basic aspects of the domestication processes does not appear to have been adequately weighted. This information may help to interpret the results in a wider perspective.

There is broad agreement that Asian rice *O. sativa* evolved from the very widely occurring perennial wild rice, *O. rufipogon*. It is a moot question if this happened directly or from an annual taxon. The material intermediate between *O. sativa* and *O. rufipogon* is variously termed as spontaneous, *O. nivara, O. sativa* var. fatua, fatua rices, red rices, and so on. Even if this was the case, it is well nigh be impossible to distinguish them from the widely occurring weedy rices. Further, in tropical plants, especially graminaceous ones, it is not always possible to distinguish annual plants from perennial forms. Most varieties of Asian rice can be "perennated" if they are provided with the basic growing requirements of water and nutrients after the first harvest. This ability appears to be more marked in the medium and long duration landraces. A few decades ago, there was even a movement to cultivate a ratoon rice crop, since this form of crop production could reduce cultivation costs substantially, even though such crops yielded only one-half to two-thirds of the "mother" crop.

The available evidence shows that the earliest domestication of rice occurred 6–7 millennia ago in the lower Yangtze basin in eastern China. This was preceded for about two millennia by harvesting, and possibly also growing, of wild rices for grain harvesting, as revealed by archaeological studies. Whether the wild rices were cultivated, or their natural stands were just protected for grain collection is also a moot question. The evidence indicates that in that period, the population had a choice of alternate foods including millets, so they might not have felt an immediacy to domesticate rice and increase the productivity of wild rices.

From the core area of lower Yangtze River basin, rice was possibly carried by migrating people of various language and ethnic groups, southwards and eastwards to southeast Asia and the Philippines. They carried with them also the know-how for rice production. This spread of rice and its cultivation might have been also facilitated by diffusion.

Huang et al.'s (2012b) recent study has shown that japonica rice was first domesticated from a local strain of *O. rufipogon* around the middle area of the Pearl river in south China. This japonica rice spread from here to southeast and south Asia, where they crossed with the local *O. rufipogon* strains from which arose indica strains. This has added a new element of uncertainty on the Asian rice domestication.

It was suggested four decades ago that Asian rice had a "diffuse origin in both space and time" (Nayar 1973). This is because both early and indigent people (until quite recently) used to find it worthwhile to harvest grains from wild rices, and because of its very wide occurrence.

The above human habit, together with other supporting evidence – such as its very high varietal diversity, the wide range of ecological situations under which rice is cultivated, and the diverse preferences of various culture or ethnic groups (such as for red-pericarped rice, glutinous rice, long-grained rice, thick-grained rices, small grained rice, medicinal rice, scented rice, etc.) – implies that Asian rice was domesticated several times, at different periods, and in different regions by different language/ethnic/cultural groups. Some of these possible regions are the eastern Himalayan region comprising Yunnan, northeast India, Bhutan, and Sikkim; the Ganga plains, including the contiguous lower Himalayas; the Konkan-Kerala coast; southeast India; and the Laos-Myanmar hills. Added to this is the observation that resistance to seed shattering is imparted in Asian rice by a single gene *sh4*, with 69% "strength". This attribute has been pinpointed to variation in just a *single* base pair. Hence, the first defining step towards domestication of rice was apparently less arduous than in other major crops. Consequently, the probability of this happening more than once in both space and time would have been higher.

Recently, several characters have been designated as being domestication-related. The non-shedding nature of spikelets can be considered as the most critical and necessary among them. The acquisition of this character alone would have given the early, first farmers an adequate advantage and incentive to harvest the grains and use them for further crop production, and possibly also, further crop improvement. Some of the other characters which could be linked to rice domestication are synchronous flowering and tillering, increased self-fertilization, determinacy, bigger panicles, bolder grains, and awnlessness. All the other characters mentioned in connection with rice domestication may have only regional, local, or cultural relevance, based on the preference of various culture or ethnic groups and ecological situations. These include glutinous rice, white-pericarped rice, long slender grained rice, very variable duration rices (70–330 days), flood resistant rice, very small grained rice, scented rice, medicinal rice, and so on. Hence, these latter characters need not be considered as components of the so-called "domestication syndrome", it is indicated.

Literally, a "raging controversy" is presently going on about the number of times that the Asian rice has been domesticated – whether once, twice, or several times. Historical (including archaeological), anthropological, biological, and circumstantial factors indicate that rice has been domesticated several times in time and space. The currently available molecular biology evidence has been unable to resolve this issue. We shall consider this aspect in some detail here. These studies and several others have already been discussed in earlier chapters. For this dicsussion, we shall revisit three pieces of work.

Sang and Ge (2007) have termed the domestication of Asian rice as "puzzling". The difference of opinion surrounding the domestication event of Asian rice – whether once or more than once – and the mode of evolution of japonica and indica rices appears to be the reason behind this statement. To reconcile this situation, they proposed two models for domestication; the snowballing model and the combination model. These have been discussed already. Based on the lower SNP density (only one-tenth) of the region containing the *sh4* gene, they suggested also that a strong selection sweep took place at the major shattering locus. This led them to propose that the *sh4* gene has only a single origin.

Sweeney and McCouch (2007) had termed the domestication process of Asian rice as a "complex" one. This description too is apparently based on the ongoing controversy about the origin of Asian rice – whether single or multiple origin. These authors have made a case for the latter alternative using the results of the phylogenetic studies done by Wang et al. (1992) using RFLPs, isozyme (Second 1982), sequence haplotype analysis done at three genetic loci (Londo et al. 2006), patterns of retrotranspon insertion (Ma and Bennetsen 2004), and genomic paleontology studies (Vitte et al. 2004) to support their contention. They observed that QTLs for domestication traits tended to cluster within certain regions of the rice genome. Sweeney and McCouch noted also the observations made by various authors (Chang 1976, Second 1982, Wang et al. 1992, Cheng et al. 2003, Garris et al. 2005) that the progenitor species of rice must have contained two groups of "different suites of *O. rufipogon* genes and corresponding mutations influencing domestication traits would have been selected within different ecotypes or subpopulations", and japonica and indica rices would have evolved from them.

Sweeney and McCouch (2007) further observed that progenies of interspecific hybrids between *O. rufipogon* and *O. sativa* often showed a reduction in the intensity of expression of wild characters. This indicated to them either that the japonica and indica varieties might be sharing some domestication alleles, or that independent mutations might have occurred in the two ecotypes within the same domestication locus, and they would not then naturally complement each other, when crossed. The authors found further support for multiple origin of rice in the studies that showed that different domestication genes were under selection in different populations. This was indicated by the QTL studies of some authors (Xiao et al. 1998, Moncada et al. 2001, Septiningsinh et al. 2003, Thomson et al. 2003, McCouch et al. 2006, Viao et al. 2006). In these studies, populations derived from crosses between a single wild accession and diverse cultivars had often identified different QTLs for various domestication traits.

In order to identify the geographical locations of various domestication events, Londo et al. (2006) examined DNA sequence variation at three genetic loci in 161 collections of *O. rufipogon* and 203 accessions of *O. sativa*, both samples representing their respective ranges of distribution. Then, they studied the sequence of three genes (*alp B–rbc L, p–VATPase*, and *SAM*) and compared indica and japonica haplotypes with those from the geographically diverse *O. rufipogon* accessions. Their data indicated that India and IndoChina represented the ancestral center of diversity of *O. rufipogon*. Their results suggested to them further that Asian rice was domesticated at least twice from different *O. rufipogon* populations, and these gave rise to indica and japonica rices. Some japonica rices however did not share a haplotype with any of the *O. rufipogon* accessions. This they stated might be because either their putative ancestral *O. rufipogon* populations had not been sampled, or because the relevant wild rices were now extinct. They then suggested that ecotype differentiation was a consequence of significant geographical separation, which in turn had resulted in developing sterility barriers.

In contrast to the above observations, Molina et al. (2011) came to the opposite conclusion – that Asian rice had only a single origin, in the lower Yangtze river basin in China. Domesticated rice then spread worldwide, both by diffusion and carried by migrating people. They concluded this from their own experimentation and by re-analyzing the earlier published data of five authors using different analytical methods. We shall now go into this study in some detail.

Molina et al. first resequenced 630 gene fragments of chromosomes 8, 10, and 12 collected from a diverse set of wild and domesticated rice accessions. Then, utilizing SNP patterns, they identified 20 putative selective sweeps in cultivated rice. Demographic modeling based on these SNP data and a diffusion-based approach gave strong support for a single domestication event. Further, a Bayesian phylogenetic analysis, implementing the multispecies coalescent and using previously published phylogenetic sequence data, also pointed to a single origin for rice. The authors placed the origin of rice domestication at 8,200–13,500 BP and the indica-japonica divergence at 3,900–6,700 BP, based on the biological clock method. They further inferred that the differentiation of japonica and indica rices took place after the origin of *O. sativa*.

Molina et al. observed that "concatenation of data could lead to misleading phylogenetic relationships, and conflicting gene trees could be formed because of coalescent stochasticity", and these would lead to problems in species delimitation. They stated further that phylogenetic analyses using multispecies coalescent (MSC) could detect signals of species differentiation even before their trees became reciprocally monophyletic (Knowles and Carstens 2007, Liu et al. 2009, Degnan and Rosenberg 2009).

Molina et al. stated that when they re-analyzed data of studies (in which the authors had observed multiple origins of rice) using *BEAST analysis (Bayesian Evolutionary Analysis by Sampling Trees, *BEAST V1.6 Species Tree Ancestral Reconstruction), 90% of the datasets showed "disputed monophyly". All of them had used fewer than five loci in their studies. In four other studies, in which the authors had used more than five loci, the data showed strong support for monophyly when their data were re-analyzed using the *BEAST program.

What can we make out of these seemingly irreconcilable results? What is revealing is the recovery of completely opposite results when different programs were used for the analysis, as was the case in the five instances studied by Molina et al. (Zhu and Ge 2005, Londo et al. 2006, Tang et al. 2006, Zhu et al. 2007, Rakshit et al. 2007). The most obvious explanation would be that the lack of robustness of methodology – as indicated more than once in this volume – can cast doubts on the conclusions. But, one cannot be sure about this either, when selection takes place during domestication.

Unlike in nature, where evolution takes place in response to the environmental conditions experienced by the organisms, selection under domestication is a different "ball game" altogether. As was so eloquently stated 60 years ago by one of the best known botanists of the mid-20th century, I. H. Burkill (1953), the "habits of man" override many rules of nature and plant breeding in this case. Varietal selection can be directed for most characters based on the personal preferences of the breeder. There are several instances of the operation of such preferences in rice. While japonica and indica rices appear to have been formed as a result of environmental adaptation (and are therefore best designated as "ecotypes"), this cannot be said of the selection of all the forms of specialty rices by different cultural/ethnic/religious societies in southeast, south, and east Asia.

This section will close with a brief mention of the strong influence of human preferences on varietal diversity in the Konkan-Malabar coast of southwest India (8–21°N; south Asia). This coastal region is a long, narrow strip of land, ca. 1,000 km long and 30–80 km wide (Nayar 2011) hemmed in between the western Ghats in the east and Arabian Sea in the west. This area is very undulating, and overall the land slopes westwards to the sea. The region receives 1,500–4,000 mm precipitation annually. It is now recognized by the IUCN (International Union for Conservation of Nature and Natural Resources) as one of the 32 biodiversity hotspots of the world (cf. Nayar 2011).

Rice is the staple food crop of the region. Because of the highly variable physiography, rice is grown here under a very wide variety of environmental conditions, ranging in elevation from 1500 m above msl to 3 m below msl as in Kuttanad. Kuttanad has been recognized recently by the FAO as a Globally Important Agricultural Heritage Site.

Generally, three crops can grow in the Kerala-Konkan region, this area being in the tropics. This region has since long been suggested as one of the centers of diversity and origin of rice (Watt 1892, Ramiah 1937, Nair et al. 1963, Vaughan and Muralidharan 1988, Lolo and Second 1988, etc.).

The people here have a strong preference for parboiled, red-pericarped rice. As mentioned earlier, of the 63 improved cultivars released in Kerala, which forms the southern part of the region, as many as 61 cultivars possess a red pericarp. Again as mentioned already, the release of Red Triveni possessing red pericarp rice as a substitute for the already improved cultivar Triveni (white pericarp) is a good example of the authorities responding to the preferences of the local population (Rosamma et al. 1991).

In conclusion, notwithstanding the strong evidence put forward by Molina et al. (2011) for a single origin of Asian rice in the Yangtze basin of China and its subsequent spread throughout the world, the considerable evidence now available for a multiple origin of Asian rice cannot be overlooked. The archaeological evidence for rice domestication obtained from the Ganga valley in north India, as pointed out earlier, sets up a reasonably strong case for the gradual ennoblement of wild rice. This is hard to ignore. There is also good evidence of rice domestication from several other archaeological sites in south Asia (Table 6.2). We should take into consideration also the strong, genetic and molecular biology evidences presented by Sweeney and McCouch (2007). In the case of cultivated crops, the habits and preferences of the local humans may be the prime determining factor in the nature, direction, and speed of crop domestication.

DIVERGENCE TIMINGS WITHIN THE GENUS *ORYZA*

The last aspect that will be considered here is the wide differences in divergence timings reported for subgeneric differentiation in *Oryza,* and the further divergence of *O. sativa* into indica and japonica rices. There is fair agreement about the time of differentiation at the supergeneric level (Tables 3.2 and 3.3).

The clade containing the genera *Oryza* and *Leersia* separated from a second clade containing all the other genera of the tribe Oryzeae ~20.5 mya. *Oryza* and *Leersia* separated ~14.2 mya (Guo and Ge 2005). This observation was based on DNA sequences from chloroplast, mitochondrial, and nuclear genomes. The most basal split in *Oryza* took place ~9 mya, producing *O. granulata* (Fig. 3.3). This was followed by the differentiation of *O. brachyantha, O. australiensis, O. eichingeri, O. rhizomatis/O. officinalis, O. punctata, O. sativa/O. rufipogon, O. meridionalis, O. longistaminata, O. glaberrima* and *O. barthii,* in that order (Guo and Ge 2005).

Zhu and Ge (2005) studied the phylogenetic relationships of eight A-genome species based on sequences of four nuclear genes and MITE-insertions. The divergence time of the lineage was estimated by a molecular clock using the formula $T = K/2r$ (r = absolute rate of substitution/site/year; K = number of substitutions/site between homologous sequences). The four primary datasets were pooled and analyzed. *O. meridionalis* was the earliest diverged lineage. Indica and japonica formed two separate monophyletic groups suggesting their polyphyletic origin. The authors estimated that the A-genome group diverged ca. 2.0 mya, and indica and japonica separated ca. 0.4 mya.

Ma and Bennetzen (2004) used the nuclear DNA of African rice, *O. glaberrima,* as reference material to study the timing, nature, mechanisms, and specificities of recent sequence evolution in indica and japonica rices. Based on the presence/absence of LTR transposons, they estimated the time of indica–japonica divergence as ~0.44 mya and that between *O. sativa* and *O. glaberrima* at ~0.64 mya.

Vitte et al. (2004) have estimated that *O. sativa* differentiated from *O. rufipogon* ca. 200,000 ya (0.2 mya) based on a genome paleontology approach.

In contrast to the timings reported by the above authors, Molina et al. (2011) estimated the time of the onset of domestication at 8,200 BP, the split of *O. sativa* from *O. rufipogon* commencing at 13,500 BP and the japonica-indica split at 6,700–3,900 yBP. The authors used both a molecular clock of 6.5 x 10^{-9} substitutions/site/yr and also estimated the divergence from chromosome scan data. The higher values given above in the range values were obtained with the latter method.

What can we make out of these "estimates"? They vary so very widely that it is not possible to get a consensual idea about the divergence timings. Even the two estimates of Guo and Ge (2005) and Zhu and Ge (2005) do not appear to show much correspondence. Incidentally, the estimates provided by Molina et al. (2011) correspond with the archaeological findings. Clearly, detailed studies and discussions are called in order for us to have a more standardized methodology.

REFERENCES

APG, 2009. An update of the Angiosperm Phylogeny Group classification for the orders and families of flowering plants. APG III. Botanical Journal of the Linnean Society 161, 105–121.

Africa Research Centre (WARDA) (Various Years): Annual Reports. www.warda.org.

Aggarwal, R.K., Brar, D.S., Nandi, S., Huang, N., Khush, G.S., 1999. Phylogenetic relationships among *Oryza* species revealed by AFLP markers. Theoretical and Applied Genetics 98, 1320–1328.

Agrawal, P.K., 2003. Molecular approaches in evolutionary studies and speciation in genus *Oryza* – present status. In: Nanda, J.S., Sharma, S.D. (Eds.), Monograph on Genus *Oryza*. Oxford & IBH Publishing Co., New Delhi, pp. 139–170.

Ahmadi, N., Glassmann, J.C., Rabary, E., 1991. Traditional highland rices originating from the intersubspecific recombination on Madagascar. Rice Genetics II, 67–79 (IRRI, Los Banos, Philippines).

Ahn, S.H., Bollich, C.N., Tanksley, S.D., 1992. RFLP tagging of a gene for aroma in rice. Theoretical and Applied Rice 84, 825–828.

Ahn, S.-M., 2010. The emergence of rice agriculture in Korea archaeobotanical perspectives. Archaeological and Anthropological Sciences 2, 89–98.

Ahuja, S.C., Ahuja, V., 2010. Rice in social and cultural life of people. In: Sharma, S.D. (Ed.), Rice: Origin, Antiquity, and History. CRC Press, Boca Raton, FL., USA, pp. 39–54.

Akagi, H.Y., Yokozeki, A., Inagaki, A., Fujimura, T., 1996. Microsatellite DNA markers for rice chromosomes. Theoretical and Applied Genetics 93, 1071–1077.

Akasaka, M., Ushiki, J., Iwata, H., Ishikawa, R., Ishii, T., 2009. Genetic relationships and diversity of weedy rice (*Oryza sativa* L.) and cultivated rice varieties in Okayama Prefecture, Japan. Breeding Science 59. 401–109.

Allchin, R., Allchin, B., 1993. The Birth of Indian Civilization. 338 pp. Penguin India, New Delhi.

Allchin, R., Allchin, B., 1997. Origins of a Civilization. Viking, New Delhi.

Allston, R.E.W., 1846. The rice plant. De Bow's Review 1, 320–356 (cf., FAO2007).

Alpern, S.B., 2008. Exotic plants of west Africa: where they come from and when? History in Africa 35, 63–102.

An, Z., 1989. Prehistoric agriculture in China. In: Harries, D.R., Hillman, G.C. (Eds.), Foraging and Farming: The Evolution of Plant Domestication. Unwin Hyman, London, pp. 643–650.

Anderson, E., 1949. Introgressive Hybridization. Hafner, New York.

Anderson, E., 1951. Discussion on rice. Indian Journal of Genetics and Plant Breeding 11, 114–115.

Anderson, E., 1953. Introgressive hybridization. Biological Reviews 28, 280–350.

Atchinson, J.C., Ali, J.R., Davis, A.M., 2007. When and where did India and Asia collide? Journal of Geophysical Research 112, B05423. http://dx.doi.org/10.1029/2006JB004706.

Backer, C.A., 1946. Wild species of *Oryza* in Malayan Archipelago. Blumea (Suppl. 3), 45–55.

Badawi, A.T., Maximos, M.A., Aidy, I.R., Olaoye, R.A., Sharma, S.D., 2010. History of rice in Africa. In: Sharma, S.D. (Ed.), Rice: Origin, Antiquity, and History. Science Publishers, Enfield, NH, USA, pp. 372–431.

Baker, H.G., Stebbins, G.L. (Eds.), 1965. The Genetics of Colonizing Species. Academic Press, New York, NY, USA.

Bao, Y., Lu, B.R., Ge, S., 2005. Identification of genomic constitutions of *Oryza* species with the B and C genomes by the PCR–RFLP method. Genetic Resources and Crop Evolution 52, 69–76.

Barbier, J.M., Mouret, J.C., 1993. La Camargue: une region de production avec dis atouts mais ussi des contraintes. In: Riz-du debuche a la culture. ICTF & CFR, pp. 21–23(cf., Ferrero and Vidotto (2010)).

Bautista, N.S., Solis, R., Kamijima, O., Ishii, T., 2001. RAPD, RFLP and SSLP analysis of phylogenetic relationships between cultivated and wild species of rice. Genes and Genetic Systems 76, 71–79.

Bedaux, R., Macdonald, K., Person, A., Polet, J., Sanogo, K., Schmidt, A., Sidibe, S., 2001. The Dia Archaeological Project: rescuing cultural heritage in the Inland Niger Delta (Mali). Antiquity 75, 837–848.

Bellwood, P., Glover, I., 2004. Foundations for an archaeological society. In: Glover, I., Bellwood, P. (Eds.), Southeast Asia: From Prehistory to History. Routledge Curzon, Abingdon, UK, pp. 4–18.

Bellwood, P., 1996. The origin and spread of agriculture in the Indo-Pacific region. In: Harris, D.R. (Ed.), The Origin and Spread of Agriculture and Pastoralism in Eurasia. UCL Press, London, pp. 469–498.

Bellwood, P., 2004a. First Farmers: The Origin of Agricultural Societies. Blackwell, Oxford.

Bellwood, P., 2004b. The origins and dispersal of agricultural communities in southeast Asia. In: Glover, I., Bellwood, P. (Eds.), Southeast Asia: From Prehistory to History. Routledge Curzon, Abingdon, UK, pp. 21–40.

Bellwood, P., 2007. Prehistory of the Indo-Malaysian Archipelago, Third ed. ANU E-Press, Canberra.

Bellwood, P., 2010. Language Families and History of Human Migration. 18 pp. Australian National University, Canberra.

Bellwood, P., 2011. The chequred prehistory of rice movement southwards as a domestic cereal from the Yangtze to the equator. Rice 4, 93–103.

Bhalerao, S.G., 1928. The wild rice (*Oryza sativa*) of the Bombay Presidency. Poona Agricultural College Magazine 20, 23–25.

Bhalerao, S.G., 1930. The Grain Shedding Character and its Importance. Bulletin No. 205. Agricultural Research Institute, Pusa, India.

Biswal, J., Sharma, S.D., 1987. Taxonomy and phylogeny of *Oryza collina*. Oryza 24, 24–29.

Blench, P., 2007. New palaeozoological evidence for the settlement of Madagascar. Azania 42, 69–82.

Boivin, N., Fuller, D.Q., 2009. Shell middens, ships, and seeds: exploring coastal subsistence, maritime trade and the dispersal of domesticates in and around ancient Arabian peninsula. Journal of World Prehistory 22, 113–180.

Bor, N.L., 1960. The Grasses of Burma, Ceylon, India, and Pakistan (Excluding Bambuseae). Pergamon Press, Oxford.

Bouchenak-Khelladi, Y., Salamin, N., Savolainen, V., Forest, F., Van Der Bank, M., Chase, M.W., 2008. Large multigene phylogenetic trees of the grasses (Poceae): progress towards complete tribal and generic taxon sampling. Molecular Phylogenetics and Evolution 47, 488–505.

Bouchenek-Khelladi, Y., Verboom, G.A., Savolainen, V., Hodkinson, T.R., 2010. Biogeography of the grasses (Poaceae) a phylogenetic approach to reveal evolutionary history in geographical space and time. Botanical Journal of the Linnen Society 162, 543–557.

Bourke, R.M., Allen, M.G., Salisbury, J.G. (Eds.), 2000. Food Security for Papua New Guinea. ACIAR, Canberra, Australia.

Bradshaw, A.D., 1965. Evolutionary significance of phenotypic plasticity in plants. Advances in Genetics 13, 115–155.

Breunig, P., Neumann, K., 2002. From hunters and gathers to food producers. In: Hassan, F.A. (Ed.), Droughts, Food, and Culture. Kluwer, Dordrecht, The Netherlands, pp. 123–155.

Breunig, P., Neumann, K., Van neer, W., 1996. New research on the Holocene settlement and environment of the Chad basin in Nigeria. African Agricultural Review 13, 111–145.

Bromham, L., Penny, D., 2003. The modern molecular clock. Nature Genetics 4, 216–224.

Buddehhagen, I.W., Persley, G.J. (Eds.), 1978. Rice in Africa. Academic Press, London, pp. 356.

Buddenhagen, I.W., 1978. Rice ecosystems in Africa. In: Buddenhagen, I.W., Persley, G.J. (Eds.), Rice in Africa. Academic Press, New York, NY, USA, pp. 11–28.

Cai, H.W., Morishima, H., 2002. QTL clusters reflect character associations in wild and cultivated rice. Theoretical and Applied Genetics 104, 1217–1228.

Cai, X.L., Wang, Z.Y., Xing, Y.Y., Zhao, J.L., Hong, M.M., 1998. Aberrant splicing of intron 1 leads to the heterogeneous 5′ UTR and decreased expression in waxy gene in rice cultivars of intermediate amylose content. The Plant Journal 14, 459–465.

Cai, H.W., Akimoto, M., Morishima, H., 2008. Genetic diversity in wild relatives of rice and domestication events. In: Hirano, H.Y., Hirai, A., Sano, Y., Sasaki, T. (Eds.), Rice Biology in the Genomic Era. Springer, Berlin, pp. 261–276.

Caicedo, A.L., Williamson, S.H., Hernandez, R.D., Boyko, A., Fledel-Alton, A., York, T.L., Polato, N.R., Olsen, K.M., Nielson, R., McCouch, S.R., Bustamante, C.D., Purugannan, M.D., 2007. Genome–wide patterns of nucleotide polymorphism in domesticated rice. PLoS Genetics 3, 1745–1756.

Capinpin, J.M., Pancho, J.V., 1961. Wild species of rice in the Philippines. Philippine Agriculturist 44, 523–524.

Cappers, R. (Ed.), 2006. Roman Footprints at Berenike. University of California, Los Angeles, CA, USA, pp. 229.

Cappers, R. (Ed.), 2007. Fields of Change. Proceedings of the 4th International Workshop of African Archaeobotany. Groningen Institute of Archaeology, Groningen, The Netherlands, p. 214.

Carney, J.A., Rosomoff, R.N., 2009. In the Shadow of Slavery: Africa's Botanical Legacy in the Atlantic World. 280 pp. University of California Press, Berkeley, CA, USA.

Carney, J.A., 2001. Black Rice: The African Origins of Rice Cultivation in the Americas. 240 pp. Harvard University Press, Cambridge, MA, USA.

Carpenter, A.J., 1978. The history of rice in Africa. In: Buddenhagen, I.W., Persley, G.J. (Eds.), Rice in Africa. Academic Press, New York, pp. 3–26.

Casson, L., 1989. The Periplus Maris Erythracei. Princeton University Press, Princeton, NJ, USA.

Castillo, C., 2011. Rice in Thailand: the archaeobotanical investigations. Rice 4, 114–120.

Chang, T.T., 1964. Present Knowledge of Rice Genetics and Cytogenetics. IRRI, Los Banos.

Chang, T.T., 1976a. Rice, *O. sativa and O. glaberrima*. In: Simmonds, N.W. (Ed.), Evolution of Crop Plants. Longman, London, pp. 98–104.

Chang, T.T., 1976b. The origin, evolution, cultivation, dissemination and diversification of Asian and African rices. Euphytica 25, 425–441.

Chang, T.T., 1976c. The rice cultures. Philosophical Transactions of the Royal Society of London, B 275 (936), 143–157.

Chang, T.T., 1985. Crop history and genetic conservation: rice – a case study. Iowa State Journal of Research 59, 425–455. 457–496.

Charbolin, R., 1977. Rice in west Africa. In: Leakey, C.L.A., Wills, J.B. (Eds.), Food Crops of the Lowland Tropics. Oxford University Press, Oxford, pp. 7–25.

Chatterjee, D., 1947. Botany of wild and cultivated rices. Nature (London) 160, 234–237.

Chatterjee, D., 1948. A modified key and enumeration of the species of *Oryza* Linn. Indian Journal of Agricultural Sciences 18, 185–192.

Chatterjee, D., 1951. Note on the origin and distribution of wild and cultivatd rices. Indian Journal of Genetics and Plant Breeding 11, 18–22.

Chen, W.B., Santo, Y.-I., Nakamura, I., Nakai, H., 1993. Indica–Japonica differentiation in Chinese rice land races. Euphytica 74, 195–201.

Chen, X., Temnykh, S., Xu, Y., Cho, Y.G., McCouch, S.R., 1997. Development of micro satellite framework map providing genome wide coverage in rice (*Oryza sativa* L.). Theoretical and Applied Genetics 95, 553–567.

Chen, M.-H., Bergman, C., Pinson, S., Fjellstrom, R., 2008. *Waxy* gene haplotypes. Associations with apparent amylase content and the effect by the environment in an international rice germplasm collection. Journal of Cereal Science 47, 536–545.

Cheng, C.Y., Tsuchimoto, S., Ohtsubo, H., Ohtsubo, E., 2002. Evolutionary relationships among rice species with AA genome based on SINE insertion analysis. Genes and Genetic Systems 77, 323–334.

Cheng, C.Y., Motohashi, R., Tsuchimoto, S., Ohstubo, H., Ohtsubo, E., 2003. Polyphyletic origin of cultivated rice: bases on the interspatial pattern or SINES. Molecular Biology and Evolution 20, 67–75.

Chevalier, A., 1932. Nouvelle contribution a l'etude systemative des *Oryza*. New contribution to the systematic study of the genus *Oryza* (in French). Revue Internationale de Botanique Applique & Agriculture Tropicale 12, 1014–1032.

Chowdhury, K.A., Ghosh, S.S., 1953. Rice in ancient India. Science and Culture 19, 207–209.

Chowdhury, K.A., 1965. Plant remains from pre- and proto-historic sites and their scientific significance. Science and Culture 31, 177–178.

Chu, Y.E., Oka, H.I., 1970. The genetic base of crossing barriers between *Oryza perennis* subsp. *barthii* and its related taxa. Evolution 24, 135–144.

Chu, Y.E., Morishima, H., Oka, H.I., 1969. Partial self-incompatibility found in *Oryza perennis* ssp. *barthii*. Japanese Journal of Genetics 44, 225–229.

Clark, P.U., Dyke, A.S., Shakun, J.D., Carlson, A.E., Clark, J., Wohlfarth, B., Mitrovica, J.Y., Hosteter, S.W., McCabe, A.M., 2009. The last glacial maximum. Science 325, 710–714.

Clark, P.U., Shakun, J.D., Baker, P.A., Bartlein, P.J., Brewer, S., Brook, E., Carlson, A.E., Cheng, H., Kaufman, D.S., Liu, Z., Marchitto, T.M., Mix, A.C., Morril, C., Otto-Bliesner, B.L., Pahnke, K., Russel, J.M., Whitlock, C., Adkins, J.E., Blois, J.L., Clark, J., Colman, S.M., Curry, W.B., Flower, B.P., He, F., Johnson, T.C., Lynch-stieglitz, J., Markgraf, V., McManus, J., Mitrovica, J.X., Moreno, P.I., Williams, J.W., 2012. Global climate evolution during the last deglaciation. Proceedings of the National Academy of Sciences of the United States of America. http://dx.doi.org/10.1073/pnas11166 19109.

Clayton, W.D., Renvoize, S.A., 1986. Genera Graminum, Grasses of the World. Royal Botanic Garden, Kew, UK.

Clayton, W.D., Vorontsova, M., Harman, K.T., Williamson, H., 2010. GrassBase – the Online World Grass Flora. Oryza. RBG, Kew, U. K.

Clayton, W.D., 1968. Studies in the Gramineae–17. Kew Bulletin 21, 485–488.

Clayton, W.D., 1970. Gramineae (Part 1). In: Milne-Redhead, E., Polhill, R.M. (Eds.), Flora of Topical East Africa. Crown Agents for Overseas Governments, London, p. 176.

Clayton, W.D., 1975. Chorology of the genera of Gramineae. Kew Bulletin 30, 111–132.

Coclanis, P.A., 2010. The rice industry of the United States. In: Sharma, S.D. (Ed.), Rice: Origin, Antiquity, and History. Science Publishers, Enfield, NH, USA, pp. 411–431.

Collins, R.O., Burns, J.M., 2007. A History of Sub-Saharan Africa. Cambridge University Press, Cambridge.

Connah, G., 1976. The Daima sequence and the prehistoric chronology of the Lake Chad region of Nigeria. Journal of African History 17, 321–352.

Connah, G., 2001. African Civilizations. Cambridge University Press, Cambridge.

Coyne, J.A., Orr, H.A., 2004. Speciation. Sinauer Associates, Sunderland, Massachusetts.

Cranston, K.A., Hurwitz, B., Sanderson, M.J., Ware, D., Wing, R.A., Stein, L., 2010. Phylogenetic analysis of BAC-end sequence libraries in *Oryza* (Poaceae). Systematic Botany 35, 512–523.

Crawford, G.W., Lee, G., 2003. Agricultural origins in the Korean peninsula. Antiquity 77 (295), 87–95.

Crawford, G.W., 2005. The origins of rice agriculture: recent progress in East Asia. Antiquity 46, 309–317.

Crawford, G.W., 2006. The Jomon in early agriculture discourse. World Archaeology 40, 445–465.

Crawford, G.W., 2011. Early rice exploitation in the lower Yangtzi valley: what are we missing? The Holocene. http://dx.doi.org/10.1177/0959683611424177.

Crepet, W.L., Feldman, G.D., 1991. The earliest remains of grasses in the fossil record. American Journal of Botany 78, 1010–1014.

Crosby Jr., A.W., 2003. In: The Columbian Exchange, 30th Anniversary ed. Praeger, Westport, CT., USA.

Dally, A.M., Second, G., 1990. Chloroplast DNA diversity in wild and cultivated species of rice (Genus *Oryza*, Section *Oryza*): cladistic mutation and genetic distance analysis. Theoretical and Applied Genetics 80, 209–222.

De Candolle, A., 1886. Origin of Cultivated Plants, Reprint, second ed. Hafner, New York. 1967.

Denham, T.P., Haberle, S., Lentfer, C., 2004. New evidence and revised interpretations of early agriculture in highland New Guinea. Antiquity 78, 839–857.

Denham, T.P., Haberle, S.G., Lentfer, C., Fullagar, R., Field, J., Therin, M., Porch, N., Winsborough, B., 2007. Origins of agriculture at Kuk swamp in the Highlands of New Guinea. Science 301, 189–193.

Denham, T., Iriarte, J., Vrydaghs, L., 2007. Rethinking Agriculture: Archaeological and Ethnoarchaeological Perspectives. Left Coast Press, Walnut Creek, CA, USA.

Dingkuhn, M., Jones, M.P., Johnson, D.E., Fofana, B., Sow, A., 1996. Towards new high yielding, weed-competitive rice plant types drawing from *Oryza sativa* and *O. glaberrima* genepools. In: Annual Report 1995. WARDA, Bouake, Cote d' Ivorie, pp. 27–35.

Dobzhansky, T., 1970. Genetics of the Evolutionary Process. Columbia University Press, New York.

Doyle, J.J., Gaut, B.S., 2000. Evolution of genes and taxa: a primer. Plant Molecular Biology 42, 1–23.

Dreni, L., Pilatone, A., Yun, D., Erreni, S., Pajoro, A., Caporali, E., Zhang, D., Kater, M.K., 2011. Functional analysis of all *AGAMOUS* subfamily members in rice reveals their role in reproductive organ identity determination and meristem determinacy. Plant Cell 23, 2850–2863.

Duan, S., Lu, B.R., Li, Z., Tong, J., Kong, J., Yao, W., Li, S., Zhu, Y., 2007. Phylogenetic analysis of AA-genome *Oryza* species based on chloroplast, mitochondrial, and nuclear DNA sequences. Biochemical Genetics 45, 113–129.

Duistermatt, H., 1987. A revision of *Oryza* (Gramineae) in Malaysia and Australia. Blumea 32, 157–193.

Ebana, K.T., Shibaya, T., Wu, J.-H., Matsuhara, K., Kanamori, H., Yamane, H., Yamanouchi, V., Mizubayashi, T., Kono, I., Shomura, A., Ito, S., Ando, T., Hori, K., Matsumoto, T., Yano, M., 2011. Uncovering of major genetic factors generating naturally occurring variation in heading date among Asian rice cultivars. Theoretical and Applied Genetics 122, 1199–1210.

Eiguchi, M., Sano, R., Hirano, H.-Y., Sano, Y., 1993. Genetic and development bases for phenotype plasticity in deep water rice. Journal of Heredity 84, 201–205.

Ellstrand, N.C., 2003. Dangerous Liaisons? When Cultivated Plants Mate with Their Wild Relatives. The Johns Hopkins University Press, Baltimore, MD, USA.

Erygin, P.S., Natalin, N.B., 1968. Rice (in Russian) Moscow. (cf., Lu and Chang (1980)).

Fagan, B., 1969. Early trade and raw materials in south central Africa. Journal of African History 10, 1–13.

Falvey, L., 2010. History of rice in southeast Asia and Australia. In: Sharma, S.D. (Ed.), Rice: Origin, Antiquity, and History. Science Publishers, Enfield, NH, USA, pp. 183–224.

FAO, 1993. Rice in Human Nutrition: B O Juliano. Food & Nutrition Series # 26. FAO, Rome.

FAO, 1995. Dimensions of Need: Staple Foods, What do People Eat? FAO, Rome.

FAO, 1999. Global Workshop on Red Rice Control, Varadero, Cuba. Proceedings. FAO, Rome.

FAO, 2001. Glossary of Biotechnology for Food and Agriculture. FAO, Rome.

FAO, 2004. International Year of Rice: Rice is Life. www.fao.org.

FAO, 2007. Weedy Rices – Origin, Biology, Ecology, and Control. 156 pp. Plant Production & Protection Paper # 188. FAO, Rome.

FAOSTAT, 2009. Production of Crops. Various Years. Available at: http://faostat.fao.org.

Felsenstein, J., 1985. Phylogenies and the comparative method. American Naturalist 125, 1–15.

Ferrero, A., Vidotto, F., 2010. History of rice in Europe. In: Sharma, S.D. (Ed.), Rice: Origin, Antiquity, and History. Science Publishers, Enfield, NH, USA, pp. 341–372.

Fields-Black, E.L., 2008. Deep Roots: Rice Farmers in West Africa and the African Diaspora. Indiana University Press, Bloomington, IN., USA.

Fogliatto, S., Vidotto, F., Ferrero, A., 2011. Morphological characterization of Italian weed rice populations. Weed Research. http://dx.doi.org/10/1111/j. 1365 – 3180. 2011. 00890x.

Fujita, D., Tagle, A.G., Ebron, L.A., Fukuta, Y., Kobayashi, N., 2012. Characterization of near isogenic lines carrying for QTL for high spikelet number with the genetic background of an indica rice variety, IR 64. Breeding Science 62, 18–26.

Fuller, D.Q., Qin, L., Qin, L., 2007. Presumed domestication? Evidence for wild rice cultivation and domestication in the fifth millennium BC of the lower Yangtze region. Antiquity 81 (312), 316–331.

Fuller, D.Q., Qin, L., Harvey, E., 2008. Evidence for a late onset of agriculture in the lower Yangtze region and challenges for an archaeobotany of rice. In: Sanchez-Mazaz, A., Blench, R., Ross, M.D., Peiros, I., Lin, M. (Eds.), Past Human Migrations in East Asia: Matching Archaeology, Lingusitics, and Genetics. Routledge, London, pp. 40–83.

Fuller, D.Q., Qin, L., Zheng, Y.-F., Zhao, Z., Chen, X., Hosoya, L.O., Sun, G.-P., 2009. The domestication process and domestication rate in rice: spikelet bases from the lower Yangtze. Science 323, 1607–1610.

Fuller, D.Q., Sato, Y.-I., Castillo, C., Qin, L., Weiskopf, A.R., Kingwell-Banham, E.J., Song, J., Ahn, S.-M., Van Etten, J., 2010. Consilience of genetics and archaeobotany in the entangled history of rice. Archaeological and Anthropological Science 2, 115–131.

Fuller, D.Q., 2002a. Fifty years of archaeobotanical studies in India: Laying a solid foundation. Indian Archaeology in Retrospect III. pp. 247–364. Indian Council of Historical Research, New Delhi.

Fuller, D.Q., 2011. Pathways to Asian civilizations. Tracing the origin and spread of rice and rice cultures. Rice 4, 78–92.

Furukawa, T., Maekawa, M., Oki, T., Suda, I., Shimada, H., Takamure, I., Kadowaki, K.-I., 2006. The *Rc* and *Rd* genes are involved in proanthocyanidin synthesis in rice pericarp. Plant Journal 49, 91–102.

Futuyma, D.J., Mayer, G.C., 1980. Nonallopatric speciation in animals. Systematic Zoology 29, 254–271.

G.P.G. (GRASS PHYLOGENY WORKING GROUP), 2001. Phylogeny and subfamilial classification of the grasses (Poaceae). Annals of the Missouri Botanic Garden 88, 373–457.

Gao, H., Zheng, X.-M., Fei, G., Chen, J., Jin, M., Ren, Y., Wu, W., Zhou, K., Sheng, P., Zhou, F., Jiang, L., Wang, J., Zhang, X., Guo, X., Wang, J.L., Cheng, Z., Wu, C., Wang, H., Wan, J.-M., 2013. *Edh 4* encodes a novel and *Oryza* genus-specific regulator of photoperiod flowering in rice. PLOS Genetics 9 (2), e1003281.

Garris, A.J., Tai, T.H., Coburn, J., Kresovich, S., McCouch, S., 2005. Genetic structure and diversity in *Oryza sativa* L. Genetics 169, 1631–1638.

Gaut, B.S., 2002. Evolutionary dynamics of grass genomes. New Phytologist 154, 15–28.

Ge, S., Sang, T., Lu, B.R., Hong, D.Y., 1999. Phylogeny of rice genomes with emphasis on origins of allotetraploid species. Proceedings of the National Academy of Sciences of the United States of America 96, 14400–14405.

Ge, S., Li, A., Lu, B.R., Zhang, Z.S., Hong, D.Y., 2002. A phylogeny of the rice tribe Oryzeae (Poaceae) based on *matk* sequence data. American Journal of Botany 89, 1967–1972.

Ghesquiere, B., Sequier, A.J., Second, G., Lorieux, M., 1997. First steps towards a rational use of African rice, *Oryza* glaberrima, in rice breeding through a 'contig line' concept. Euphytica 96, 31–39.

Ghosh, S.S., 1961. Further records of rice (*Oryza* ssp.) from ancient India. Indian Forester 87, 295–301.

Gibb, H.A.R., 1929. Reprint 2001. Ibn Batutta: Travels in Asia and Africa. Manohar Publishers, New Delhi, India.

Gifford-Gonzalez, D., 2005. Pastorlaism and its consequences. In: Stahl, A.B. (Ed.), African Archaeology. Blackwell Publishing, Oxford, UK, pp. 187–224.

Glaszmann, J.C., 1986. A varietal classification of Asian cultivated rice (*Oryza* sativa L.) based on isozyme polymorphism. Rice Genetics, 83–90 (IRRI, Los Banos, Philippines).

Glaszmann, J.C., 1987. Isozymes and classification of rice varieties. Theoretical and Applied Genetics 74, 21–30.

Glover, I., Bellwood, P., 2004. Southeast Asia: From Prehistory to History. Routledge Curzon, Abington, UK.

Goff, S.A., and 54 others, 2002. A draft sequence of the rice genome (*O. sativa* L. ssp. japonica). Science 296, 92–100.

Golomb, L., 1976. The origin, spread, and persistence of glutinous rice as a staple crop in Mainland southeast Asia. In: Golomb, L. (Ed.), From Subsistence to Symbol: How Glutinous Rice Has Stuck. Siam Society Library, Bangkok, Thailand, pp. 1–12.

Golson, J., 1977. No room at the top: agricultural intensification in the New Guinea Highlands. In: Allen, J., Golson, J., Jones, J. (Eds.), Sunda and Sahul. Academic Press, London, pp. 601–638.

Gopalakrishnan, R., Nayar, N.M., Sampath, S., 1964. Cytogenetical studies of two amphidiploids in the genus *Oryza*. Euphytica 13, 57–64.

Gorman, C.F., 1969. Hoabinhian: a pebble tool complex with early plant association in southeast Asia. Science 163, 671–673.

Gorman, C.F., 1972. Excavations at Spirit Cave, North Thailand. Asian Perspectives 13, 79–107.

Govindaswamy, S., Krishnamurthy, A., 1958a. On the occurrence of japonica type of rice grains in Jeypore tract. Rice Genetics Newsletter 6 (4), 22–24.

Govindaswamy, S., Krishnamurthy, A., 1958b. A note on the occurrence of a new species of *Oryza* in Jeypore tract. Science and Culture 24, 234–236.

Grant, V., 1981. Plant Speciation, second ed. Columbia University Press, New York.

Gray, R., 1962. A report on the Conference. Journal of African History 3, 175–191.

Grillo, M.A., Li, C., Fowlkes, A.M., Briggeman, T.M., Zhou, A., Schemske, D.W., Sang, T., 2009. Genetic architecture for the adaptive origin of annual wild rice, *Oryza nivara*. Evolution 63, 870–883.

Grist, D.H., 1975. Rice, fifth ed. Longman, Green & Co., London.

Griswold, D., 2006. Grain Drain: The Hidden Cost of US Rice Subsidies. Cato Institute, Washington D. C., USA (cf., Coclains, P. H. 2010).

Gu, X.-Y., Kianian, S.F., Foley, M.E., 2004. Phenotypic selection for dormancy induced a set of adaptive haplotypes from weedy into cultivated rice. Genetics 171, 695–704.

Guo, Y.L., Ge, S., 2005. Molecular phylogeny of Oryzeae (Poaceae) based on DNA sequences from chloroplast, mitochondrial, and nuclear genomes. American Journal of Botany 92, 1548–1558.

Gustchin, G.G., 1938. Le riz: origin et histoire de sa culture. Rizet Rizicultura 12, 61–96.

Gustin, G.G., 1938. Le riz: origin et historic de couture. Riz et Riziculture 12, 61–96.

Hammer, K., 1984. Das Domestikationsyndrome. Kulturpflanze 32, 11–34.

Hanebuth, T., Stattegger, K., Groots, P.M., 2000. Rapid flooding of the Sunda Shelf. A late glacial sea-level record. Science 288, 1033–1035.

Harlan, J.R., De Wet, J.M.J., 1971. Toward a rational classification of cultivated plants. Taxon 20, 509–517.

Harlan, J.R., Dewet, J.M.J., 1965. Some thoughts about weeds. Economic Botany 19, 16–24.

Harlan, J.R., De Wet, J.M.J., Stemler, B.L. (Eds.), 1976. Origins of African Plant Domestication. Mouton Publishers, The Hague.

Harlan, J.R., 1965. The possible role at weedy races in the evolution of cultivated plants. Euphytica 14, 173–176.

Harlan, J.R., 1970. The evolution of cultivated plants. In: Frankel, O.H., Bennett, E. (Eds.), Genetic Resources of Plants. Their Exploration and Conservation. Blackwell, Oxford, pp. 19–32.

Harlan, J.R., 1971. Agricultural origins: centres and noncentres. Science 174, 468–474.

Harlan, J.R., 1992. Crops and Man, second ed. American Society of Agronomy, Madison, WI., USA.

Harland, S.C., 1936. The genetical conception of the species. Biological Reviews 11, 83–112.

Harper, J.L., 1977. Population Biology of Plants. Academic Press, New York.

Harries, D.R., 1976. Traditional systems of plant food production and the origins of agriculture in west Africa. In: Harlan, J.R., et al. (Ed.), Origins of African Plant Domestication. Mouton, The Hague, pp. 311–356.

Hassan, F.A., 1997. Holocene palaeoclimates of Africa. African Archaeological Review 14, 213–230.

Hassan, F.A., 2002a. Conclusion. In: Hassan, F.A. (Ed.), Droughts, Food, and Culture. Kluwer Dordrecht, The Netherlands, pp. 321–333.

Hassan, F.A., 2002b. Palaeoclimate, food and climate change in Africa. In: Hassan, F.A. (Ed.), Droughts, Food, and Culture. Kluwer, Dordrecht, The Netherlands, pp. 11–26.

Hassan, F.A., 2002c. Introduction. In: Hassan, F.A. (Ed.), Droughts, Food, and Culture. Kluwer, Academic, Dordrecht, The Netherlands, pp. 1–9.

Hassan, F.A. (Ed.), 2002d. Droughts, Food and Cultures. Kluwer Academic, Dordrecht, The Netherlands, p. 347.

Hawley, J.C. (Ed.), 2008. India in Africa, Africa in India. Indiana University Press, Bloomington, IN., USA.

Heu, M.H., Moon, H.P., 2010. History of rice culture in Korea. In: Sharma, S.D. (Ed.), Rice: Origin, Antiquity, and History. Science Publishers, Enfield, NH, USA, pp. 115–153.

Higham, C.F.W., Lu, T., 1998. The origins and dispersal of rice cultivation. Antiquity 72, 867–877.

Higham, C., Maloney, B., 1989. Coastal adaptation, sedentism, and domestication. a model for socioeconomic intensification in prehistoric southeast Asia. In: Harries, D.R., Hillman, G.C. (Eds.), Foraging and Farming. The Evolution of Plant Exploitation. Unwin Hyman, London, pp. 650–666.

Higham, C., 2004. Mainland southeast Asia from the Neolithic to the stone age. In: Glover, I., Bellwood, P. (Eds.), Southeast Asia: From Prehistory to History. Routledge Curzon, Abington, UK, pp. 41–67.

Hill, M.H., 1978. Dating Senegambian megaliths. Current Anthropology 19, 604–605.

Hinata, K., Oka, H.I., 1962a. A survey of hybrid sterility relationships in the Asian forms of *Oryza perennis* and *O. sativa*. Japanese Journal of Genetics 37, 314–328.

Hinata, K., Oka, H.I., 1962b. Some conditions on the evolutionary dynamics of cultivated rice. Japanese Journal of Genetics 37, 329–342.

Hirano, Y.-K., Eiguchi, M., Sano, Y., 1998. A single base change altered the regulation of the *Waxy* gene at the post-transcriptional level during the domestication of rice. Molecular Biology and Evolution 15, 978–987.

Hitchcock, A.S., 1935. Oryza. In: North American Flora. Part 5 (Poales), vol. 17, New York Botanic Garden, Bronx, NY, USA, pp. 355–356.

Hooker, J.D., 1897. Flora of British India. Botanical Survey of India, Calcutta, vol. 7.

Hu, C.H., 1960. Karyological studies in haploid plants–IV. Chromosome morphology and intragenome pairing in haploid plants of *Oryza glaberrima* Steud. as compared with those of *O. sativa* L. Cytologia 25, 437–449.

Huang, P., Molina, J., Flowers, J.M., Rubinstein, S., Jackson, S.A., Purugannan, M.D., Schaal, B.A., 2012. Phylogeography of Asian wild rice, *Oryza rufipogon*: a genome-wide view. Molecular Ecology 21, 4593–4604.

Hurles, M.E., Sykes, B.C., Jobling, M.A., Forster, P., 2005. The dual origin of the Malagasy in Island Southeast Asia and east Africa: evidence from maternal and paternal lineages. Human Genetics 76, 894–901.

Hutchinson, S.G., 1965. Crop plants evolution, a discussion. In: Hutchinson, J.B. (Ed.), Essays in Crop Plants Evolution. Cambridge University Press, Cambridge, pp. 165–181.

IM (1978): cf., Ahn, S. (2010).

Insoll, T., Shaw, T., 1977. Gao and Igbo–Ukwa: beads, interregional trade, and beyond. African Archaeological Review 14, 9–23.

IRAT–ORSTOM, 1977. Meeting on African Rice Species, Paris. 25–26 January 1977. IRAT–ORSTOM, Paris.

IRRI, 1964a. Report of committee appointed to attempt a standard classification and nomenclature of the genus *Oryza*. In: Rice Genetics and Cytogenetics. IRRI, Los Banos, The Philippines, pp. 251–252.

IRRI, 1964b. Rice Genetics and Cytogenetics. In: Proceedings of a Symposium, 4–8 February 1963. IRRI, Los Banos Philippines, pp. 253–254.

IRRI, 1986. Rice Genetics I. IRRI, Los Banos, Philippines.

IRRI, 2000. Wild and weedy rice ecosystems in Asia – a review. In: Proceedings of an International Symposium, Vietnam. 1998. IRRI, Los Banos, Philippines.

Ishii, T., Terachi, T., Mori, N., Tsunewaki, K., 1993a. Comparative study on the chloroplast, mitochondrial, and nuclear genome differentiation in two cultivated rice species, *Oryza sativa* and *O. glaberrima* by RFLP analysis. Theoretical and Applied Genetics 85, 88–96.

Ishii, T., Nakano, T., Maeda, H., Kamijima, O., 1996. Phylogenetic relationships in A genome species of rice as revealed by RAPD analysis. Genes & Genetic Systems 71, 195–201.

Ishii, T., Xu, Y., McCouch, S.R., 2001. Nuclear- and chloroplast-microsatellite variation in A-genome species of rice. Genome 44, 658–666.

Iwamoto, M., Nagashima, H., Nagamine, T., Goh, H.I., Hogo, K., 1999. *P-SINE 1*- like intron of Cat A catalase homology and phylogenetic relationships among AA_genome *Oryza* and related species. Theoretical and Applied Genetics 98, 853–861.

Iyama, S., Morishima, H., (Compiled by), 1980. The Ancestors of Cultivated Rices and Their Evolution. National Institute of Genetics, Misima, Japan.
Jacobs, B.F., Kingston, J.D., Jacobs, L.L., 1999. The origin of grass dominated ecosystems. Annals of the Missouri Botanical Garden 86, 590–644.
Ji, H.S., Chu, S.H., Jiang, W., Cho, Y.-I., Hahn, J.-H., Eun, M.-Y., McCouch, S.R., Koh, H.J., 2006. Characterization and mapping of a shattering mutant in rice that corresponds to a block of domestication genes. Genetics 173, 995–1005.
Jin, J., Huang, W., Gao, J.-P., Yang, J., Shi, M., Zhu, M.-Z., Luo, D., Lin, H.-X., 2008. Genetic control of rice plant architecture under domestication. Nature Genetics 40, 1365–1369.
Johnston, B.F., 1958. The Staple Food Economics of Western Tropical Africa. Stanford University, Press, Stanford, CA, USA.
Joseph, L., Kuriachan, P., Kalyanaraman, K., 1999. Collection and evaluation of the tetraploid *Oryza officinalis* Wall. ex Watt (*O. malampuzhensis* Krish.et Chand.) endemic to Western Ghats, India. Genetic Resources and Crop Evolution 46, 531–541.
Joshi, S.P., Gupta, V.S., Aggarwal, R.K., Ranjekar, P.K., Brar, D.S., 2000. Genetic diversity and phylogenetic relationship as revealed by inter simple sequence repeat (ISSR) polymorphism in the genus *Oryza*. Theoretical and Applied Genetics 100, 1311–1320.
Kajale, M.D., 1989. Mesolithic exploitation of wild plants in Sri Lanka archaeobotanical study at the cave site of Beli–Lena. In: Harries, D.R., Hillman, G.C. (Eds.), Foraging and Farming: The Evolution of Plant Exploitation. Unwin Hyman, London, pp. 270–281.
Kajale, M.D., 1991. Current status of Indian paleo-ethnobotany: introduced and indigenous food plants with a discussion of the early historical and evolutionary development of Indian agriculture and agricultural systems in general. In: Renfew, J. (Ed.), New Light on Early Paleo-ethnobotany. Edinburgh University Press, Edinburgh, pp. 155–189.
Kane, N.C., Baack, E.J., 2007. The origins of weedy rice. Molecular Ecology. http://dx.doi.org/10.1111/j. 1365–294X. 2007. 03520.x.
Kaneda, C., 2010. History of rice in Japan. In: Sharma, S.D. (Ed.), Rice, Origin, Antiquity, and History. Science Publishers, Enfield, NH, USA, pp. 154–182.
Kanno, A., Hirai, A., 1992. Comparative studies of the structure of chloroplast DNA from four species of *Oryza*: cloning and physical maps. Theoretical and Applied Genetics 83, 791–798.
Kasem, S., 2012. Evaluating grain food potential of wild relatives of rice (PhD thesis). Southern Cross University, Lismore, NSW, Australia.
Katayama, T.C., 1963. A survey of botanical studies on the genus *Oryza*, especially of photoperiodic studies. Seiken Ziho 15, 98–109.
Kato, S., Kosaka, H., Hara, S., 1928. On the affinity of rice varieties as shown by the fertility of rice plants. Science Bulletin, Faculty of Agriculture, Kyushu Imperial University 3, 132–142.
Kellog, E.A., 2009. The evolutionary history of Ehrhartoideae, Oryzeae, and *Oryza*. Rice 2, 1–14.
Khush, G.S., Multani, D.S., Vergera, G.V., Brar, B.S., 1990. Taxonomic status of *Oryza inandamanica*. Rice Genetics Newsletter 7, 88–89.
Khush, G.S., Brar, D.S., Virk, P.S., Tang, S.X., Malik, S.S., Busto, G.A., Lee, Y.T., Mc Nally, R., Trinh, L.N., Jiang, Y., Shata, M.A.M., 2003. Classifying Rice Germplasm by Isozyme Polymorphism and Origin of Cultivated Rice. Discussion Paper # 46. IRRI, Los Banos.
Khush, G.S., 1997. Origin, dispersal, cultivation and variations of rice. Plant Molecular Biology 35, 25–34.
Khush, G.S., 2000. Taxonomy and origin of rices. In: Singh, R.K., et al. (Ed.), Aromatic Rices. IRRI, Los Banos, pp. 5–14.
Kihara, H., Katayama, T.C., 1964. Occurrence of Indica and Japonica Types among Sikkimese Rice Varieties. pp. 3–11. Annual Report No. 14. National Institute of Genetics, Misima, Japan.

Kihara, H., Nakao, S., 1960. The rice plant in Sikkim: a report on the scientific mission. Seiken Ziho 11, 46–54.

Kihara, H., 1958. Consideration on the origin of cultivated rice. Seiken Ziho 10, 68–83.

King, R.C., Stansfield, W.D., Mulligan, P.K., 2006. A Dictionary of Genetics, Seventh ed. Oxford University Press, Oxford.

Kinoshita, T., 1995. Report of the Committee on Gene Symbolization, nomenclature, and Linkage Groups. Rice Genetics Newsletter 12, 9–153.

Klee, M., Zach, B., 1999. The exploitation of wild and domesticated food plants at settlement mounds in northeast Nigeria. In: van der Veen, M. (Ed.), Exploitation of Plant Resources in Ancient Africa. Plenum Publishers, New York, pp. 81–88.

Klee, M., Zach, B., Neumann, K., 2000. Four thousand years of plant exploitation in the Chad basin of northeast Nigeria. III. Vegetation History and Archaeobotany 69, 223–237.

Koide, Y., Onishi, K., Kanazawa, A., Sano, Y., 2008. In: Hirano, H.Y., et al. (Ed.), Genetics of speciation in rice. Rice Biology in the Genomics Era, Springer, Berlin, pp. 247–260.

Konishi, S., Izawa., T., Lin, S.Y., Ebana, K., Fukuta, Y., Sasaki, T., Yano, M., 2006. An SNP caused loss of seed shattering during rice domestication. Science 312, 1392–1396.

Kovach, M.J., Sweeney, M.T., McCouch, S.R., 2007. New insights into the history of rice domestication. Trends in Genetics 23, 578–587.

Kovach, M.J., Kalingaclon, M.N., Fitzgerald, M.A., McCouch, S.R., 2009. The origin and evolution of fragrance in rice. Proceedings of the National Academy of Sciences of the United States of America 106, 14444–14449.

Krishnaswamy, N., Chandrasekharan, P., 1967. Note on a naturally occurring tetraploid species of *Oryza*. Science and Culture 23, 307–310.

Krishnaswamy, N., Chandrasekharan, P., 1968. A new species of *Oryza* L. Madras Agricultural Journal 45, 471–472.

Kumar, T., 1988. History of Rice in India. Mythology, Culture, Agriculture. Gyan Publishing House, Delhi.

Kurata, N., 2008. Chromosomes and genome evolution in rice. In: Hirano, H.Y., et al. (Ed.), Rice Biology in the Genomics Era. Springer, Berlin, pp. 235–246.

Kusimba, C.M., Kusimba, S.B., 2005. Mosaics and interactions. East Africa. In: Stahl, A.B. (Ed.), African Archaeology. Blackwell, Oxford, pp. 392–419.

Lambeck, K., Chappel, J., 2001. Sea level change through the Last Glacial Cycle. Science 292, 679–686.

Langenin, S.A., Clay, K., Grace, J.B., 1990. The incidence and effects of hybridization between cultivated rice and its related weedy red rice. Evolution 44, 1000–1009.

Launert, E., 1965. A survey of the genus *Leersia* in Africa (Gramineae, Oryzoideae, Oryzeae). Senckenbergiana Biologie 46, 129–148.

Levin, D.A., 2000. The Origin, Expansion, and Demise of Plant Species. Oxford University Press, Oxford, UK.

Lewicki, T., 1974. West African Food in the Middle Ages. Cambridge University Press, Cambridge.

Li, C.D., Ni, P.X., Francki, M., Hunter, A., Zhang, Y., Schibeci, D., Li, H., Tarr, A., Wang, J., Cakir, M., Yu, J., Bellgard, M., Lance, R., Appels, R., 2004. Genes controlling seed dormancy and pre-harvest sprouting in a rice-wheat-barley comparison. Journal of Functional and Integrative Genomics 4, 84–93.

Li, C.G., Zhou, A.L., Zang, T., 2006. Genetic analysis or rice domestication syndrome with the wild annual species, *Oryza nivara*. New Phytologist 170, 185–194.

Li, Z.-M., Zheng, X.-M., Ge, S., 2011. Genetic diversity and domestication history of African rice as inferred from multiple gene sequences. Theoretical and Applied Genetics 123, 21–31.

Lin, S.Y., Sasaki, T., Yano, M., 1998. Mapping quantitative trait loci controlling seed dormancy and heading date in rice using backcross inbred lines. Theoretical and Applied Genetics 96, 997–1003.

Lin, Z.W., Griffith, M.E., Li, X.R., Zhu, Z.F., Tan, L., Fu, Y.C., Zhang, W.X., Wang, X.K., Xie, D.X., Sun, C.Q., 2007. Origin of seed shattering in rice (*Oryza sativa* L.). Planta 226, 11–20.

Linares, O.F., 2002. African rice (*Oryza glaberrima*): history and future potential. Proceedings of the National Academy of Sciences of the United States of America 99, 16360–16365.

Linder, H.P., 1986a. The evolutionary history of the Poales/Restionales – a hypothesis. Kew Bulletin 42, 297–318.

Litke, R., 1968. Uber den Nachweis tertiarer Gramineen. Monats Deutschen Akademische Wissenoschaft Berlin 10, 462–471.

Littlefield, D.C., 1981. Rice and Slaves. Lousiana State University Press, Baton Rouge, LA.

Liu, L., Chen, X., 2012. The Archeology of China. Cambridge University Press, Cambridge, UK.

Londo, J.P., Schaal, B.A., 2007. Origins and population genetics of the weedy red rice in the USA. Molecular Ecology. http://dx.doi.org/10.1111/j.1365–294x.2007.03489.x.

Londo, J.P., Chiang, Y.C., Hung, K.H., Chiang, T.Y., Schaal, B., 2006. Phylogeography of Asian wild rice, *Oryza rufipogon*, reveals multiple independent domestications of cultivated rice, *Oryza sativa*. Proceedings of the National Academy of Sciences of the United States of America 103, 9578–9583.

Lorieux, M., Petrov, M., Huang, N., Guideroni, E., Ghesquire, A., 1996. Aroma in rice: genetic analysis of a quantitative trait. Theoretical and Applied Genetics 93, 1145–1151.

Lorieux, M., Ndjiondjop, M.N., Ghesqure, A., 1999. A first interspecific *Oryza sativa x O. glaberrima* microsatellite-based genetic linkage map. Theoretical and Applied Genetics 100, 593–601.

Lu, J.J., Chang, T.T., 1980. Rice in its temporal and spatial perspectives. In: Luh, B.S. (Ed.), Rice Production and Utilization. AVI, Westford, CT., USA, pp. 1–74.

Lu, B.R., Ge, S., 2004. *Oryza coarctata*: the name that best reflects the relationships of *Porteresia coarctata* (Poaceae: Oryzeae). Nordic Journal of Botany 23, 555–558.

Lu, B.R., Zheng, K.L., Qian, H.R., Zhuang, J.Y., 2002. Genetic differentiation of wild relatives of rice as assessed by RFLP analysis. Theoretical and Applied Genetics 106, 101–106.

Lu, B.R., 1999. Taxonomy of the genus *Oryza* (Poaceae): historical perspective and current status. International Rice Research Notes 24 (3), 4–8.

Lu, T.L.-D., 2006. The occurrence of cereal cultivation in China. Asian Perspectives 41, 1–14.

Ma, J., Bennetzen, J.L., 2004. Rapid recent growth and divergence of rice nuclear genomes. Proceedings of the National Academy of Sciences of the United States of America 101, 12404–12410.

Mabberley, D.J., 2006. Mabberley's Plant-Book, third ed. Cambridge University Press, Cambridge.

Mangelsdorf, P.C., 1965. The evolution of maize. In: Hutchinsion, J.B. (Ed.), Essays in Crop Plants Evolution. Cambridge University Press, pp. 23–49.

Marshall, F., Hildebrand, E., 2002. Cattle before crops. The beginning of food production in Africa. Journal of World Prehistory 16, 99–143.

Matsuo, T., 1952. Genecological Studies on Cultivated Rice. Bulletin, National Institute of Agricultural Sciences Japan D3. 111 pp. NIAS, Ibaraki, Japan.

Maurizio, A., 1926. Vegetable Food Stuffs and Agriculture, a Historical Survey. Warsaw. (cf. Lewitski 1974).

Mayr, E., 1942. Systematics and the Origin of Species. Columbia University Press, New York.

Mayr, E., 1970. Animal Species and Evolution. Harvard University Press, Cambridge, MA, USA.

McEachern, S., 2005. Two thousand years of west African history. In: Stahl, A.B. (Ed.), African Archaeology. Blackwell Publishing, Oxford, UK, pp. 442–466.

McIntosh, R.J., McIntosh, S.K., 1981. The Inland Niger Delta before the empire of Mali: evidence from Jenne-Jeno. Journal of African History 22, 1–22.

McIntosh, S.K., McIntosh, R.J., 1988. From stone to metal: new perspectives on the later prehistory of west Africa. Journal of World Prehistory 2, 89–133.

McIntosh, S.K., McIntosh, R.J., 1993. Cities without citadels: understanding urban origins along the middle Niger. In: Shaw, T., Sinclair, P.J.J., Andah, B., Okopoko, A.I. (Eds.), Archaeology of Africa. Food, Metals and Towns. Routledge, London, pp. 622–641.

McIntosh, S.K., 1995. Excavations at Jenne-Jeno, Hambarketolo and Kaniana (Inland Niger Delta Mali), the 1981 Season. University of California Press, Berkeley, CA.

McIntosh, S.K., 2006. The Holocene prehistory of west Africa. In: Akeyampong, E.K. (Ed.), Themes in West African History. Ohio University Press, Athens, OH, USA, pp. 11–32.

McPherson, K., 1993. The Indian Ocean: A History of People and the Sea. Oxford University Press, New Delhi.

Miamoto, K., 2009. Searching for the Origins of Agriculture (in Japanese). . Yoshikawa, Tokyo (cf., J. Whitman 2011: (Rice 4: 149–158. 2011).

Mikami, I., Uwatoko, N., Ikeda, Y., Yamaguchi, J., Hirano, H.Y., Suzuki, Y., Sano, Y., 2008. Allelic diversification at the *wx* locus in landraces of Asian rice. Theoretical and Applied Genetics 116, 979–989.

Mochizuki, K., Ohtsubo, H., Hirano, H., Sano, Y., Ohtsubo, E., 1993. Classification and relationships or rice accessions with AA genome by identification of transposable elements at nine loci. Japanese Journal of Genetics 68, 205–217.

Mokhtar, G. (Ed.), 1981. General History of Africa. II. Ancient Civilizations of Africa. UNESCO, Paris, p. 785.

Molina, M., Sikora, M., Garnol, N., Flowers, J.M., Rubenstein, S., Reynolds, A., Huang, P., Jackson, S., Schaal, B.A., Bustamante, C.D., Boyko, A.R., Purugannan, M.D., 2011. Molecular evidence for a single evolutionary origin of domesticated rice. Proceedings of the National Academy of Sciences of the United States of America 108, 8351–8356.

Morat, P., Deroin, T., Coudere, H., 1994. Presence in New Caledonia of an endemic species in the genus *Oryza* L (Gramineae). Bulletin du Museum National d'Histoire Naturalle Series 16, Section B, Adansonia 1, 3–10.

Morinaga, T., Kuriyama, H., 1955. Japonica type of rice in the subcontinent of India and Java. Japanese Journal of Breeding 5, 149–153.

Morinaga, T., Kuriyama, H., 1957. Cytogenetical studies on *Oryza sativa* L. IX. The F1 hybrid of *O. sativa* L. and *O. glaberrima* Steud. Japanese Journal of Breeding 7, 57–65.

Morinaga, T., Kuriyama, H., 1958. Intermediate type of rice in the subcontinent of India and Java. Japanese Journal of Breeding 7, 253–259.

Morinaga, T., 1954. Classification of rice varieties on the basis of affinity. Japanese Journal of Breeding 4 (Suppl.), 1–14.

Morishima, H., Oka, H.I., 1981. Phylogenetic differentiation of cultivated rice. 22 Numerical evaluation of the indica-japonica differentiation. Japanese Journal of Breeding 31, 402–413.

Morishima, H., Oka, H.I., Chang, W.T., 1961. Directions of differentiation in populations of wild rice *Oryza perennis* and *O. sativa f. spontanea*. Evolution 15, 326–339.

Morishima, H., Hinata, K., Oka, H.I., 1962. Comparison between two cultivated rice species, *Oryza sativa* L. and *O. glaberrima* Steud. Japanese Journal of Breeding 12, 153–165.

Morishima, H., Hinata, K.H., Oka, H.I., 1963a. Comparison of modes evolution of cultivated forms from two wild rice species, *O. breviligulata* and *O. perennis*. Evolution 17, 170–181.

Morishima, H., Hinata, K.H., Oka, H.I., 1963b. Comparison of modes of evolution of cultivated forms from two wild rice species, *O. breviligulata* and *O. perennis*. Evolution 17, 170–181.

Morishima, H., Shimamoto, Y., Sano, Y., Sato, Y.I., 1984. Observations on Wild and Cultivated Rices in Thailand for Ecological-genetic Study. National Institute of Genetics, Misima Japan.

Morishima, H., Sano., Y., Oka, H.I., 1992. Evolutionary studies in cultivated rices and its wild relatives. Oxford Surveys on Evolutionary Biology 8, 135–184.

Morishima, H., 1969. Phenetic similarity and phologenetic relationships among strains of *Oryza perennis* estimated by methods of numerical taxonomy. Evolution 23, 429–443.

Morishima, H., 2001. Evolution and domestication or rice. Rice Genetics IV, 63–78 (IRRI, Los Banos, Philippines).

Murdoch, G.P., 1959. Africa, its People and Their Culture History. McGraw Hill, New York, USA.

Murray, S.S., 2004. Searching for the origins of African rice domestication. Antiquity 78 (300), 5.

Murray, S.S., 2005. The Rise of African Rice Farming and the Economic use of Plants in the Upper Middle Niger Della (Mali). PhD Dissertation. Department of Anthropology, University of Wisconsin, Madison, WI, USA. 539 pp.

Murray, S.B., 2007. Identifying African rice domestication in the Middle Niger Delta (Mali). In: Cappers, R. (Ed.), Fields of Change. Groningen Institute of Archaeology, Groningen, The Netherlands, pp. 53–61.

Naidu, P.D., 1999. A review on Holocene climate changes in the Indian subcontinent. Memoir Geological Society of India 42, 303–314.

Nair, N.R., Nayar, N.M., Sahadevan, P.C., 1964. On the occurrence of *Oryza* species in south–western India and its significance. Current Science 33, 515–517.

Nair, V.G., John, S., Manju, P., 2011. Compendium of Crop Varieties of Kerala. Kerala Agricultural University, Vellayani, Trivandrum, India.

Nakagahra, M., Akihama, T., Hayashi, K., 1975. Genetic variation and geographic cline of esterase isozymes in native rice varieties. Japanese Journal of Genetics 51, 373–382.

Nakagahra, M., 1978. The differentiation, classification and centre of genetic diversity of cultivated rice by isozyme analysis. Tropical Agriculture Research Series 11, 77–82.

Nakamura, I., Chen, W.B., Sato, V.I., 1991. Analysis of chloroplast DNA from ancient rice seeds. Annual Report, National Institute of Genetics, Japan 42, 108–109.

Nakamura, S.I., 2010. The origin of rice cultivation in the lower Yangtze region, China. Archaeological and Anthropological Science 2, 107–113.

Naredo, M.E.B., Vaughan, D.A., Cruz, F.S., 1993. Comparative spikelet morphology of *Oryza schlechteri* Pilger and related species of *Leersia* and *Oryza* (Poaceae). Journal of Plant Research 106, 109–112.

NAS, 1990. Sea-level Change. National Academy of Sciences, Washington, D. C., USA.

Nayar, N.M., Gopalakrishnan, R., Sampath, S., 1966. Genetic studies in four interspecific hybrids of the genus *Oryza*. Euphytica 15, 184–194.

Nayar, N.M., 1958. Studies on the Origin of Cultivated Rice *Oryza sativa* L (Associate IARI Thesis). Indian Agricultural Research Institute, New Delhi.

Nayar, N.M., 1967. Prevalence of self incompatibility in *Oryza barthii* Cheval.: its bearing on the evolution of rice and related taxa. Genetica 38, 521–527.

Nayar, N.M., 1973. Origin and cytogenetics of rice. Advances in Genetics 17, 153–292.

Nayar, N.M., 2010. The history and genetic transformation of the African rice, *Oryza glaberrima* Steud. (Gramineae). Current Science 99, 1681–1689.

Nayar, N.M., 2012. Evolution of the African rice: a historical and biological perspective. Crop Science 52, 505–516.

Nesbitt, M., Simpson, J., Svanberg, I., 2010. History of rice in Western and Central Asia (1947–2007). In: Sharma, S.D. (Ed.), Rice, Origin, Antiquity, and History. Science Publishers, Enfield, NH, USA, pp. 308–340.

Neumann, K., 2003. The late emergence of agriculture in sub-Saharan Africa. In: Neumann, K., Butler, A., Kahlheber, S. (Eds.), Food, Fuels, and Fields. Heinrich Barth Institute, Cologne, Germany, pp. 71–92.

Neumann, K., 2005. The romance of farming: plant cultivation and domestication in Africa. In: Stahl, A.B. (Ed.), African Archaeology. Blackwell Publishing, Oxford, UK, pp. 249–275.

Ng, N.Q., Hawkes, J.G., Williams, J.T., Chang, T.T., 1981. The recognition of a new species of rice (*Oryza*) from Australia. Botanical Journal, Linnean Society 82, 327–330.

Nicolson, S., 1976. A climate chronology for Africa (PhD Dessertation). University of Wisconsin, Madison, WI, USA.

Niklas, K.J., 1997. The Evolutionary Biology of Plants. University of Chicago Press, Chicago, IL, USA.

Nishikawa, T., Vaughan, D.A., Kadowaki, K., 2005. Phylogenetic analysis of *Oryza* species, based on simple sequence repeats and their flanking nucleotide sequences from the mitochondrial and chloroplast genomes. Theoretical and Applied Genetics 110, 696–705.

NRC, 1996. Lost Crops of Africa. Grains African Rice, pp. 17–37, vol. I, National Research Council, National Academy Press, Washington D. C., USA.

Nuitjen, E., Van Treuren, R., Struik, P.C., Mokuwa, A., Okry, F., Teeken, B., 2009. Evidence for the emergence of new rice types of interspecific hybrid origin in west African farmers fields. PLOS One 4 (10), e 7335. http://dx.doi.org/10.1371/journal.pone 0007335.

Ohmido, N., Fukui, K., 1995. Cytological studies of African cultivated rice, *Oryza glaberrima.* Theoretical and Applied Genetics 91, 212–217.

Oka, H.I., Chang, W.T., 1961. Hybrids swarms between wild and cultivated rice species *Oryza perennis* and *O. sativa*. Evolution 15, 418–430.

Oka, H.I., Chang, W.T., 1962. Rice varieties intermediate between wild and cultivated forms and the origin of the japonica rice. Botanical Bulletin Academia Sinica 3, 109–131.

Oka, H.I., Chang, W.T., 1964. Evolution of responses to growing conditions in wild and cultivated rice forms. Botanical Bulletin Academia Sinica 5, 120–138.

Oka, H.I., Morishima, H., 1967. Variations in the breeding systems of a wild rice *Oryza perennis*. Evolution 21, 249–258.

Oka, H.I., Sano, Y., 1981. Differentiation of *Oryza perennis* populations in adaptive strategy. In: Problems in General Genetics. Proceedings of the XIV International Genetics Congress, Moscow. Mir Publications, Moscow, pp. 65–85.

Oka, H.I., Chang, W.T., 1959. The impact of cultivation on populations of wild rice, *Oryza sativa* f. spontanea. Phyton 13, 105–117.

Oka, H.I., Chang, W.T., 1962. Rice varieties intermediate between wild and cultivated forms and the origin of the japonica type. Botanical Bulletin of Academia Sinica 3, 109–131.

Oka, H.I., Chang, W.T., 1963. Observations of Wild and Cultivated Rice Species in Africa. Report of Trip from Sierra Leone and Tchad, 73 pp. National Institute of Genetics, Misima, Japan.

Oka, H.I., Morinaga, H., Sano, Y., Koizumi, T., 1978. Observations of Rice Species and Accompanying Savannah Plants on the Southern Fringe of Sahara Desert: Report of Study Tour in West Africa. 86 pp. National Institute of Genetics, Misima, Japan.

Oka, H.I., 1956. Observations on Natural Populations of Formosan Wild Rice. pp. 46–47. Annual Report No. 6. National Institute of Genetics, Misima.

Oka, H.I., 1961. Report of Trip for Investigation of Rice in Latin American Countries. National Institute of Genetics, Misima, Japan.

Oka, H.I., 1964a. Pattern of interspecific relationships and evolutionary dynamics in *Oryza*. In: Rice Genetics and Cytogenetics. Elsevier, Amsterdam, pp. 71–90.

Oka, H.I., 1988. Origin of Cultivated Rice. Elsevier, Amsterdam, The Netherlands.

Olsen, K.M., Purugannan, M.D., 2002. Molecular evidence on the origin and evolution of glutinous rice. Genetics 162, 941–950.

Olsen, K.M., Caicedo, A.L., Polato, N., McClung, A., McCouch, S., Purugganan, M.D., 2006. Selection under domestication: evidence for a sweep in the rice *Waxy* genomic region. Genetics 173, 975–983.

Olsen, K.M., Caicedo, A.L., Jia, Y.L., 2007. Evolutionary genomics of weedy rice in the USA. Journal of Integrative Plant Biology 49, 811–816.

Park, K.C., Kim, N.H., Cho, Y.S., Kang, K.H., Lee, J.K., Kim, N.S., 2003. Genetic variations of AA genome *Oryza* species measured by MITE–AFLP. Theoretical and Applied Genetics 107, 203–209.

Park, K.C., Kim, N.H., Lee, J.-H., Kim, N.-S., 2003. Genetic variation in *Oryza* species detected by MITE–AFLP. Genes and Genetic Systems 78, 235–243.

Parker, G., 2008. The Makings of Roman India. Cambridge University Press, Cambridge, UK.

Parsons, B.J., Newbury, H.J., Jackson, M.I., Ford–Lloyd, B.Y., 1997. The genetic structure and conservation of aus, aman and boro rices from Bangladesh. Genetic Resources and Crop Evolution 46, 587–598.

Paterson, A.H., Bowers, J.E., Chapman, B.A., 2004. Ancient polyploidization predating divergence of the cereals, and its consequences for comparative genomics. Proceedings of the National Academy of Sciences of the United States of America 101, 9903–9908.

Pearsell, D.M., Piperno, D.R., Dinan, E.H., Umlauf, M., Zhao, Z.J., Benfer Jr., R.A., 1995. Distinguishing *Oryza* species through phytolith analysis: results of preliminary research. Economic Botany 49, 183–193.

Pellig, R., 2008. Garamantian agriculture: the plant remains from Jarma, Fazzan. Libyan Studies 39, 41–71.

Pelling, R., 2005. Garamantan agriculture and its significance in a wider North African context: the evidence of plant remains from Fazzan project. Journal of North African Studies 10, 397–411.

Pental, D., Barnes, S.R., 1985. Interrelationship of cultivated rices *Oryza sativa* and *O. glaberrima* with *O. perennis* complex. Analysis of Fraction 1 protein and some repeated DNA sequences. Theoretical and Applied Genetics 70, 185–191.

Pereira, J.A., Guimaraes, E.P., 2010. History of rice in Latin America. In: Sharma, S.D. (Ed.), Rice: Origin, Antiquity, and History. Science Publishers, Enfield, NH, USA, pp. 432–451.

Piqueras, J., Boira, J.V., 2005. Spain rice in history. In: Ferroro, A., Scansetti, M. (Eds.), Rice Landscapes of Life, pp. 177–191. Ediziou Mercurio Vercelli, Italy (cf., Ferrero, A. and F. Vidotto 2010).

Pokharia, A.K., Pal, J.N., Srivastava, A., 2009. Plant macroremains from Neolithic Jhusi in Ganga plain: evidence for grain-based agriculture. Current Science 97, 564–572.

Pokharia, A.K., 2008. Record of macrobotanical remains from the Aravalli hill, Ojiyana, Rajasthan evidence for agriculture-based subsistence economy. Current Science 94, 612–622.

Pokharia, A.K., 2011. Palaeoethnobotany at Lahurdewa: a contribution to the 2nd millennium B C agriculture of the Ganga Plain. India Current Science 101, 1569–1578.

Porteres, R., 1945. Sur la segregation geogrpique des genes de l'*O.glaberrima* dans l'Ouest African et sur les centres de cette espece. Comptes Rendes du Akademic Science 221, 152–153.

Porteres, R., 1949a. Le riz vivace de l'Afrique. Agronomic Tropicale (Paris) 4, 5–24.

Porteres, R., 1949b. Les Guyanes centre secondaire nouveau de variation des riz cultives (*Oryza sativa* L.). Agronomic Tropicale 4, 379–404.

Porteres, R., 1950. Articulation intraspecifique homoloque et origine monophyletique de chacune des especes *Oryza sativa* L. et *O. glaberrima* Steud. Revieu Bitanique Applique Agronomie Tropicale 30, 147–157.

Porteres, R., 1955a. Historique sur les premiers echantillons d'*Oryza glaberrima* Steud. recueillis en Afrique. Journal of Agronomie Tropicale Botanique Applique 2, 535–537.

Porteres, R., 1955b. Presence ancienne d'une variete cultivates *O. glaberrima* Steud. en Guyane Francaise. Journal of Agronomie Tropicale Botanique Applique 2, 680.

Porteres, R., 1956. Taxonomie agrobotanique des riz cultives O. sativa Linn. et O. glaberrima Stued. Journal of Agronomic Tropicale Botanique Applique 3, 314–384. 541–580, 627–700, 821–856.

Porteres, R., 1959. Les appellations des cereals en Afrique. IX. Les riz. Journal of Agronomic Tropicale Botanique Applique 6, 176–220. 252–265.

Porteres, R., 1962. Berceaux agricoles primaries sur le Continent African. Journal of African History 3, 195–210.

Porteres, R., 1970. Primary cradles of agriculture in the African continent. pp. 43–58. In: Fage, J.D., Oliver, R.A. (Eds.), Papers in African Prehistory. First appeared in French in Journal of African History, 3 pp. 195–210. Cambridge University Press, Cambridge, UK, 1962.

Porteres, R., 1976. African cereals: *Eleusine, fonio*, black fonio, teff, *Brachiaria, Paspalum*, Pennisetum, and African rice. In: Harlan, J.R., De Wet, J.M.J., Stemler, A.B.L. (Eds.), Origins of African Plant Domestication. Mouton Publishers, The Hague, pp. 409–452.

Posnansky, M., McIntosh, R.J., 1976. New carbon dates from north and west Africa. Journal of African History 17, 161–195.

Possehl, G.L., 1997. The transformation of the Indus civilization. Journal of World Prehistory 11, 425–472.

Possehl, G.L., 2002a. Harappan Civilization. A Contemporary Perspective. Alta Mira Press/Vistar Publications, New Delhi.

Potts, D.T., 1991. A Note on rice cultivation in Mesopotamia and Susano. NABU(Nouvelles Assyriologiques Breves et Utilitaries. 1991: 1 – 2 (cf. Nestbitt, M. et al. (2010)).

Prance, G., Nesbitt, M., 2005. The Cultural History of Plants. Routledge, London.

Prasad, V., Stroemberg, C.A.E., Alimohammadian, H., Sahni, A., 2005. Dinosaur coprolites and the early evolution of grasses and grazers. Science 310, 1177–1180.

Premathileke, R., 2006. The emergence of early agriculture in the Horton Plains, central Sri Lanka linked to late Pleistocene and early Holocene climate changes. Pragdhara 18, 90–104.

Prodoehl, A., 1922. Oryzeae monographie describuntur. Botanische Archieve 1, 211–224. 231–255.

Purseglove, J.W., 1976. The origins and migrations of crops in tropical Africa. In: Harlan, J.R., de Wet, J.M.J., Stemler, A.B.L. (Eds.), Origins of African Plants Domestication. Mouten Publishers, The Hague, pp. 291–301.

Pyraman, K., 1989. New evidence of plant exploitation and environment during the Hohabinhian (Late Stone Age) from Ban Cao caves, Thailand. In: Harries, D.R., Hillman, G.C. (Eds.), Foraging and Farming: The Evolutionary of Plant Exploitation. Unwin Hyman, London, pp. 282–291.

Rakshit, S., Rakshit, A., Matsumura, H., Hasegawa, Y., ITO, A., Ishii, T., Miyashita, N.T., Terauchi, R., 2007. Largescale DNA polymorphism study of *Oryza sativa* and *O. rufipogon* reveals the origin and divergence of rice. Theoretical and Applied Genetics 114, 731–743.

Ramiah, K., Ghose, R.L.M., 1951. Origin and distribution of rice. Indian Journal of Genetics and Plant Breeding 11, 7–13.

Ramiah, K., Rao, M.B.V.N., 1953. Rice Breeding and Genetics. Indian Council of Agricultural Research, New Delhi.

Ramiah, K., 1937. Rice in Madras. Government Press, Madras, India.

Raven, P.H., Axelrcd, D.I., 1974. Past continental movements. Annals of the Missouri Botanical Garden 61, 539–674.

Ray, H.P., 1993. The Archaeology of Seafaring in Ancient South Asia. 335 pp. Cambridge University Press, Cambridge, UK.

Ray, H.P., 2002. The archaeology of early historical maritime India. In: In: Sattar, S., Korisettar, R. (Eds.), Indian Archaeology in Retrospect, vol. IV. Manohar Publishers, New Delhi, pp. 341–353.

Ren, F.G., Lu, B.R., Li, S., Huang, J., Zhu, Y., 2004. A comparative study of genetic relationships among the AA-genome *Oryza* species using RAPD and SSR markers. Theoretical and Applied Genetics 108, 113–120.

Richards, P., 1996. Culture and community values in the selection and maintenance of African rice. In: Brush, D.S., Stabinksy, D. (Eds.), Valuing Local Knowledge: Indigenous People and Intellectual Property Rights. Island Press, Washington, DC, USA, pp. 209–229.

Ridley, H.N., 1925. 1922–1925. The Flora of Malay Peninsula, vols 5, Reeve & Co, London.

Rieseberg, L.H., 1995. The role of hybridization in evolution: old wine in new skins. American Journal of Botany 82, 944–953.

Rieseberg, L.H., 1995. The role of hybridization in evolution: Old wine in new skins. American Journal of Botany 82, 944–953.

Roschevicz, R.J., 1931. A contribution to the knowledge of rice. Applied Botany, Genetics and Plant Breeding Bulletin 27, 3–133.

Roy, S.C., 1921. A preliminary classification of wild rices of C. P. and Berar. Agricultural Journal of India 16, 365–380.

Sahadevan, P.C., 1967. Rice in Kerala. Agriculture Department Kerala, Trivandrum, India.

Sakamoto, S., 1977. Hsien rice, keng rice, and glutinous rice. In: Matsuo, T., et al. (Ed.), Science of the Rice Plant. Genetics, vol. 3. Food and Agriculture Research Centre, Tokyo, pp. 143–153.

Sampath, S., Govindaswamy, S., 1958. Wild rices of Orissa, their relationship to cultivated varieties. Rice Newsletter 6 (3), 17–20.

Sampath, S., Rao, M.B.V.N., 1951. Interrelationships between species in genus *Oryza*. Indian Journal of Genetics and Plant Breeding 11, 14–17.

Sampath, S., 1962. The genus *Oryza*: its taxonomy and interrelationships. Oryza 1, 1–29.

Sanderson, M.J., 1997. A nonparametric approach for estimating divergence timings in the absence of tree constancy. Molecular Biological Evolution 14, 1218–1231.

Sang, T., Ge, S., 2007a. Genetics and phylogenetics of rice. Current Opinion in Genetics & Development 17, 1–16.

Sang, T., Ge, S., 2007b. The puzzle of rice domestication. Journal of Integrative Plant Biology 49, 760–768.

Sano, Y., Morishima, H., 1982. Variation in resource allocation and adaptive strategy of a wild rice, *Oryza perennis* Moench. Botanical Gazette 143, 518–523.

Sano, Y., Chu, Y.E., Oka, H.I., 1980. Genetic studies of speciation in cultivated rice. 2. Character variations in backcross derivatives between *O. sativa* and *O. glaberrimaa*: M–V linkage and key characters. Japanese Journal of Genetics 55, 19–39.

Sano, Y., 1978. Analysis of genes controlling the sterility between *Oryza sativa* and *O. glaberrima*. pp. 116. Annual Report No. 28. National Institute of Genetics, Misima, Japan.

Sano, Y., 1986. Sterility barriers between *Oryza sativa* and *O. glaberrima*. In: IRRI: Rice Genetics. IRRI, Los Banos, Philippines, pp. 109–118.

Sano, Y., 1989. The direction of gene flow between two co–occurring rice species, *Oryza sativa* and *O. glaberrima*. Heredity 63, 353–357.

Sano, Y., 1990. The genic nature of gamete eliminature in rice. Genetics 125, 183–191.

Sano, Y., 1997. Inheritance of sterility. In: Matsuo, T., Futsuhara, Y., Kikuchi, F., Yamaguchi, H. (Eds.), Science of the Rice Plant. Genetics, vol. III. Food and Agriculture Policy Research Centre, Tokyo, pp. 367–376.

Saraswat, K.S., Pokharia, A.K., 2004. Plant economy at Lahurdewa: A preliminary contemptation (Abstract) National Seminar on the Archaeology of the Ganga Plain, Lucknow, U. P., India.

Saraswat, K.S., 2005. Agricultural background of the early farming communities in the Middle Ganga Plain. Pragdhara 15, 145–177.

Sarkar, R., Raina, S.N., 1992. Assessment of genomic relationships in the genus *Oryza* L. based on seed protein profile analysis. Theoretical and Applied Genetics 85, 127–132.

Sarla, N., Swamy, B.P.M., 2005. *Oryza glaberrima*, a source for the improvement of *Oryza sativa*. Current Science 89, 955–963.

Sato, Y.I., 1987. Character association patterns among japonica rice varieties of Asian origin. In: Hsieh, S.C. (Ed.), Crop Exploration and Utilization of Genetic Resources(Taichung District Agricultural Improvement Station, Changhua, Taiwan).

Sato, T., 2005. State and Rural Society in Medieval Islam. Brill, The Netherlands (cf., Badawi et al. (2010)).

Sattar, M., Sharma, S.D., Pokharia, A.K., 2010. History of rice in south Asia (up to 1947). In: Sharma, S.D. (Ed.), Rice: Origin, Antiquity, and History. Science Publishers, Enfield, N. H., USA, pp. 225–271.

Sauer, C.O., 1952. Agricultural Origins and Dispersal. 110 pp. American Geographical Society, New York.

Saxena, A., Prasad, V., Singh, I.B., Chauhan, M.S., Hasan, R., 2006. On the Holocene records of phytoloths of wild and cultivated rice from Ganga plain evidence for rice-based agciculture. Current Science 90, 1547–1552.

Saxena, A., Prasad, V., Singh, I.B., 2013. Holocene–palaeocene reconstruction from the phytoliths of the lake-full sequence of Ganga plain. Current Science 104, 1054–1062.

Schlichting, C.D., 1986. The evolution of phenotypic plasticity in plants. Annual Review of Ecology and Systematics 17, 667–693.

Schoff, W.A., 1912. (Translated and Annotated by): The Periplus of the Erythrean Sea Longmans. Green, and Co., London.

Second, G., 1982. Origin of the genic diversity of cultivated rice (*Oryza* spp.): study of the polymorphism scored at 40 isozyme loci. Japanese Journal of Genetics 57, 25–57.

Second, G., 1985a. Eurasian versus Gondwanian origin of the genus *Oryza*. East Asian Tertiary/Quarternary Newsletter 2, 53.

Second, G., 1985b. Geographic origins, genetic diversity, and the molecular clock hypothesis in the Oryzeae. In: Jacquard., P. (Ed.), Genetic Differentiation and Dispersal in Plants. Springer Verlag, Berlin, pp. 42–56.

Second, G., 1985c. New insight into the genome differentiation in *Oryza* L. through isozyme studies. In: Sharma, Ak, Sharma, A. (Eds.), Advances in Chromosome and Cell Research. Oxford & IBH Publishing Company Ltd., New Delhi, pp. 45–78.

Second, G., 1986. Isozymes and phylogenetic relationships in *Oryza*. Rice Genetics, 27–39 (IRRT, Los Banos, The Philippines).

Seetharaman, R., Ghoral, D.P., 1976. Occurrence of types with characters of O. glaberrima in Assam Rice Collection. Current Science 45, 67.

Semon, M., Nielsen, R., Jones, M.P., McCouch, S.R., 2005. The population structure of African cultivated rice, *O. glaberrima* Steud. Evidence for elevated levels of linkage disequilibrium caused by admixture with *O. sativa* and ecological adaptation. Genetics 169, 1639–1647.

Senaratna, J.E., 1956. The Grasses of Ceylon. Government Press, Colombo.

Seshu, D.V., Sorrels, M.E., 1986. Genetic studies on seed dormancy in rice. Rice Genetics I, 369–382 (IRRI, Los Banos, Philippines).

Sharma, G.R., Manda, D., 1980. Excavation in Mahagara, 1977–78 (A Neolithic Settlement in the Belan Valley). Archeology of the Vindhyas and the Ganga Valley 6. Department of Ancient History, Culture and Archeology. University of Allahabad, UP, India.

Sharma, S.D., Shastry, S.V.S., 1965b. Taxonomic studies in genus *Oryza* L. III. *O. rufipogon* Griff. *Sensu stricto* and *O. nivara* Sharma et Shastry nom. nov. Indian Journal of Genetic and Plant Breeding 25, 157–167.

Sharma, S.D., Shastry, S.V.S., 1965d. Taxonomic studies in genus *Oryza* L. VI. A modified classification. Indian Journal of Genetic and Plant Breeding 25, 173–178.

Sharma, G.R., Misra, V.D., Mandal, D., Misra, B.B., Pal, J.N., 1980. Beginnings of Agriculture. Abinash Prakashan, Allahabad, UP, India.

Sharma, G.R., 1985. From hunting and food gathering to domestication of plants and animals in the Belan and Ganga valleys. In: Misra, V.S., Bellwood, P. (Eds.), Recent Advances in Indo-Pacific Prehistory. Oxford & IBH, New Delhi, pp. 369–371.

Sharma, S.D., 2003a. Species of genus *Oryza* and their interrelationships. pp. 73–112. In: Nanda, J.S., Sharma, S.D. (Eds.), Monograph on Genus *Oryza*. Oxford & IBH Publishing, New Delhi.

Sharma, S.D., 2003b. Origin of cultivated rices. pp 311–330. In: Nanda, J.S., Sharma, S.D. (Eds.), Monograph on Genus *Oryza*. Oxford & IBH Publishing Co. Post. Ltd., New Delhi.

Sharma, S.D., 2010a. Domestication and diaspora of rice. In: Sharma, S.D. (Ed.), Rice, Origin, Antiquity, and History. Science Publishers, Enfield, NH, USA, pp. 1–24.

Sharma, S.D. (Ed.), 2010b. Rice, Origin, Antiquity, and History. Science Publishers, Enfield, NH, USA, p. 569.

Sheriff, A.M.H., 1981. The east African coast and its role in maritime trade. In: Mokhtar, G. (Ed.), General History of Africa. II. Ancient Civilizations. UNESCO, Paris, pp. 551–567.

Shimizu, H., Maruoka, M., Ichikawa, N., Baruah, A.R., Uwattoko, N., Sano, Y., Onishi, K., 2010. Genetic control of phenotypic plasticity in Asian cultivated and wild rice in response to nutrient and density change. Genome 53, 211–223.

Sidebotham, S.E., 2011. Berenike and the Ancient Maritime Spice Route. University of California Press, Berkeley, CA, USA.

Singh, R.K., Singh, U.S., Khush, G.S. (Eds.), 1995. Aromatic Rices. Oxford & IBH Publishing, New Delhi.

Singh, I.B., 2005. Climate change and human history in Ganga plain during late Pleistocene-Holocene. Palaeobotanist 54, 1–12.

Singh, U., 2008. A History of Ancient and Early Medieval India: From the Stone Age to the 12th Century. Pearson Education, New Delhi, India.

Smith, D.E., Harrison, S., Firth, C.R., Jordan, J.T., 2011a. The early Holocene sea level rise. Quaternary Science Reviews 30, 1846–1860.

Smith, B.D., 1998. The Emergence of Agriculture. Scientific American Library, New York.

Sokal, R.R., Rohlf, F.J., 1981. Biometry, second ed. Freeman, San Francisco, CA, USA.

Sokal, R.R., Sneath, P.H.A., 1963. Principles of Numerical Taxonomy. Freeman, San Francisco, CA, USA.

Sokal, R.R., 1958. Quantification of systematic relationships and phylogenetic trends. Proceedings of the 10th International Congress of Entomology 1, 400–416 (cf., Morishima and Oka (1960)).

Solheim II, W.G., 1969. Early Bronze Age in northeastern Thailand. Current Anthropology 9, 59–62.

Soltis, D.E., Soltis, P.S., 2000. Contributions of plant molecular systematics to studies of molecular evolution. Plant Molecular Biology 42, 45–75.

Soltis, D.E., Soltis, P.S., 2003. The role of phylogenetics in comparative genetics. Plant Physiology 132, 1790–1800.

Soltis, P.S., Soltis, D.E., Doyle, J.J. (Eds.), 1992. Molecular Systematics of Plants. Chapman and Hall, London.

Soltis, D.E., Soltis, P.S., Doyle, J.J. (Eds.), 1998. In: Molecular Systematics of Plants II. DNA Sequencing. Kluwer, Dordrecht, Netherlands.

Sreejayan, U., Sureshkumar, V., Varghese, G., Jacob, T.M., Thomas, G., 2010. Stratification and population structure of the genetic resources of ancient medicinal landrace njavara. Genetic Resources and Crop Evolution. http://dx.doi.org/10/1007/S10722-010-96-13-1.

Stebbins, G.L., 1950. Variation and Evolution in Plants. Columbia University Press, New York, USA.

Stebbins, G.L., 1981. Co-evolution of grasses and herbivores. Annales of the Missouri Botanic Gardens 68, 75–86.

Stephens, S.G., 1950. The internal mechanism of speciation in *Gossypium*. Botanical Review 15, 115–149.

Stephens, S.G., 1951. Possible significance of duplication in evolution. Advances in Genetics 4, 247–265.

Sugimoto, K., Takeuchi, Y., Ebana, K., Miyao, A., Hirochika, H., Nihara, Ishiyama, K., Kobayashi, M., Ban, Y., Hattori, T., Yano, M., 2010. Molecular cloning of *Sdr 4*, a regulator involved in seed dormancy and domestication of rice. Proceedings of the National Academy of Sciences of the United States of America 107, 5792–5797.

Sutton, J.E.G., 2001. Igbo-Ukwa and the Nile. African Archaeological Review 18, 49–62.

Swallen, J.R., 1936. *Oryza.* In: Publication No. 461. Botany of the Maya Area. Carnegie Institute of Washington, Washington DC, USA, pp. 155–156.

Swallen, J.R., 1943. Gramineae. *Oryza.* pp. 159–161 Annales of the Missouri Botanical Garden 30, 104–280.

Swallen, J.R., 1955. Grasses of Guatemala. Chicago National History Museum, Chicago, Ill, USA.

Sweeney, M., McCouch, S., 2007. The complex history of the domestication of rice. Annals of Botany 100, 951–957.

Sweeney, M.T., Thomson, M.J., Pfeil, B.E., McCouch, S., 2006. Caught red–handed: *Rc* encodes a basic helix–loop–helix protein conditioning red pericarp in rice. Plant Cell 8, 283–294.

Sweeney, M.T., Thomson, M.J., Cho, Y.G., Park, Y.J., Williamson, S.H., Butamante, C.D., McCouch, S.R., 2007. Global dissemination of a single mutation conferring white pericarp in rice. PLoS Genetics 3 (8), e133.

Takahashi, R., 1955. The origin and evolution of cultivated barley. Advances in Genetics 8, 227–262.

Tan, L., Li, X., Liu, F., Sun, X., Li, C., Zhu, Z., 2008. Control of a key transition from prostrate to erect growth in rice domestication. Nature Genetics 40, 1360–1364.

Tanaka, K., Honda, T., Ishikawa, R., 2010. Rice archaeological remains and the possibility of DNA archaeology. Archaeological & Anthropological Sciences 2, 69–78.

Tang, S.X., Xuan, S.W., 2010. Domestication of rice in China and its cultural heritage. In: Sharma, S.D. (Ed.), Rice, Origin, Antiquity, and History. Science Publishers, Enfield, NH, USA, pp. 85–114.

Tateoka, T., 1962a. Species of the genus *Oryza.* Natural Science Museum, Tokyo 29 (5/6), 73–78.

Tateoka, T., 1962b. Taxonomic studies of *Oryza.* I. *O. latifolia* complex. Botanical Magazine, Tokyo 75, 418–427.

Tateoka, T., 1962c. Taxonomic studies of *Oryza.* II. Several species complexes. Botanical Magazine, Tokyo 75, 455–461.

Tateoka, T., 1963. Taxonomic studies of *Oryza.* III. Key to the species and their enumeration. Botanical Magazine, Tokyo 76, 165–173.

Tateoka, T., 1965a. A taxonomic study of *Oryza eichingeri* and *O. punctata.* Botanical Bulletin Tokyo 78, 156–163.

Tateoka, T., 1965b. *Porteresia*, a new genus of Gramineae. Bulletin National Science Museum Tokyo 8, 405–406.

Tateoka, T., 1965c. Taxonomy and chromosome numbers of African representatives of the *Oryza officinalis* complex. Botanical Bulletin Tokyo 78, 198–201.

Terao, H., Mizushima, U., 1939. Some considerations on the classification of *Oryza sativa* L. into two subspecies, so-called *japonica* and *indica.* Japanese Journal of Genetics 10, 213–258.

Tewari, R., Srivastava, R.K., Saraswat, K.S., Singh, I.B., Singh, K.K., 2006. Early farming in Lahurdewa. Puratatwa 18, 347–373.

Thapar, R., 2003. The Penguin History of India from the Origins to AD 1300. Penguin Books, New Delhi.

The Plant List. 2001. www.theplantlist.org. (accessed 01.07. 2013).

Thoday, J.M., Gibson, J.B., 1962. Isolation by disruptive selection. Nature 193, 1164–1166.

Thomas, G., Joseph, L., Varghese, G., Sreejayan, Kalyanaraman, K., Kuriachan, P., 2001a. Analysis of phenotypic and genetic variations among populations of *Oryza malampuzhaensis* show evidence of attitude–dependent genetic changes. Canadian Journal of Botany 79, 1090–1098.

Thomas, G., Sreejayan, Joseph, L., Kuriachan, P., 2001b. Genetic variation and population structure in *Oryza malampuzhaensis* Krish. et Chand. endemic to Western Ghats, south India. Journal of Genetics 80, 141–148.

Thomasson, J.R., 1987. Fossil grasses, 1820–1986 and beyond. In: Soderstrom, T.R., Hilu, K.W., Campbell, C.S., Barkworth, M.E. (Eds.), Grass Systematics and Evolution. Smithsonian Institution, Washington, USA.

Tian, Z., Wang, X., Lee, R., Li, Y., Specht, J.E., Nelson, R.L., McClean, T.E., Qiu, L., Ma, J., 2010. Artificial selection for determinate growth habit in soybean. Proceedings of the National Academy of Sciences of the United States of America 107, 8563–8568.

Ting, Y., 1949. Origination (sic) of the rice cultivation in China. College of Agriculture, Sun Yat Sen University Agronomy Bulletin III (7), 18 (in Chinese).

Tofanelli, S., Bertoncini, S., Castri, L., Luiselli, D., Calafell, F., Donati, G., Paoli, G., 2009. On the origins and admixture of Malagasy: new evidence from high-resolution analyses of paternal and maternal lineages. Molecular Biology and Evolution 26, 2109–2124.

Uga, Y., Fukuta, Y., Cai, H.W., Iwata, H., Ohsawa, R., Morishima, H., Fujimura, T., 2003. Mapping QTL influencing rice floral morphology using recombinant inbred lines derived from a cross between *Oryza sativa* L. and *O. rufipogon* Griff. Theoretical and Applied Genetics 107, 218–226.

UNESCO, 2013. http://who.unesco.org: Stone Circles of Senegambia. (Ref. # 1228).

Van Der Veen, M., 1999. The Exploitation of Plant Resources in Ancient Africa. 283 pp. Plenum Publishers, New York.

Van Der Veen, M., 2003. Trade and diet at Roman and medieval Qnseir at-Qadim, Egypt. A preliminary report. In: Neumann, K., Butler, A., Kahlheber, S. (Eds.), Food, Fuel, and Fields: Progress in African Archaeobotany. Heinrich-Barth-Institut, Kologne, Germany, pp. 207–212.

Vasconcellos, J. de C.E., 1953. The Rice. Ministry of Finance, Lisbon (cf., Ferrero, A. and F. Vidotto 2010).

Vaughan, D.A., Kadowaki, K.I., Kaga, A., Tomooka, N., 2005. On the phylogeny and biogeography of the genus *Oryza*. Breeding Science 55, 113–122.

Vaughan, D.A., Ge, S., Kaga, A., Tomooka, N., 2008. Phylogeny and biogeography of the genus *Oryza*. In: Hirano, H.Y., Hirai, A., Sano, Y., Sasaki, T. (Eds.), Rice Biology in the Genomics Era. Springer, Berlin, pp. 219–234.

Vaughan, D.A., 1990. A new rhizomatous *Oryza* species (Poaceae) from Sri Lanka. Botanical Journal, Linnean Society 103, 159–163.

Vaughan, D.A., 1994. The Wild Relatives of Rice: A Genetic Resources Handbook. 137 pp. IRRI, Los Banos, Philippines.

Vaughan, D.A., 2003. Genepools of the genus *Oryza*. In: Nanda, J.S., Sharma, S.D. (Eds.), Monograph on Genus *Oryza*. Science Publishers, Enfield, NH, USA, pp. 113–138.

Verboom, G.A., Linder, H.P., Stock, W.D., 2003. Phylogenetics of the grass genus *Ehrharta*: evidence for radiation in the summer arid zone of the South African Cape. Evolution 57, 1008–1021.

Verin, P., 1981. Madagascar. In: Mokhtar, G. (Ed.), General History of Africa. II Ancient Civilizations. UNESCO, Paris, pp. 693–713.

Vernet, R., 2002. Climate during the late Holocene in the Sahara and the Sahel: evolution and consequences on human settlement. In: Hassan, F.A. (Ed.), Droughts, Food, and Culture: Ecological Change and Food Security in Africa Later Prehistory. Plenum Press, New York, pp. 47–63.

Vicentini, A., Barber, J.C., Aliscioni, S.A., Giussani, L.M., Kellog, E.A., 2008. The age of the grasses and clusters of origins of C_4 photosynthesis. Global Change Biology 14, 2963–2977.

Vitte, C., Ishii, T., Lamy, F., Brar, D., Panaud, O., 2004. Genomic paleontology provides evidence for two distinct origins of Asian rice (*Oryza sativa* L.) Molecular and General Genomics 272, 504–511.

Wang, S., Gong, D., 1999. Climate in China during the Four Special Periods in Holocene. 12 pp. Department of Geophysics, Peking University, Beijing 100 871, China.

Wang, Z.Y., Second, G., Tanksley, S.D., 1992. Polymorphism and phylogenetic relationships among species in the genus *Oryza* as determined by analysis of nuclear RFLPs. Theoretical and Applied Genetics 83, 565–581.

Wang, Z.Y., Zheng, F.Q., Shen, G.Z., Gao, J.P., Snutsad, D.P., Li, M.G., Zhang, J.L., Hong, M.M., 1995. The amylase content in rice endosperm is related to the post-transcriptional regulation of the *waxy* gene. Plant Journal 7, 613–622.

Watson, L., Dallwitz, M.J., 1992. Grass Genera of the World. CAB International, Wallingford, UK.

Watt, G.W., 1892. Rice. Dictionary of Economic Products of India, vol. 5, Government Printer, Calcutta. pp. 498–653.

WCSP, 2012. Oryza. World Checklist of Selected Plants. apps.kew.org/wesp Royal Botanic Garden, Kew, UK.

Weber, D.J., Rohilla, R., Singh, U.S., 2000. Biochemistry of aroma in scented rice. In: Singh, R.K., et al. (Ed.), Aromatic Rices. IRRI, Los Banos, pp. 29–46.

Weber, S., Lehman, H., Barela, T., Hawks, S., Harriman, D., 2010. Rice or millets: early farming strategies in prehistoric central Thailand. Archeological and Anthropological Sciences 2, 79–88.

Wei, M., 1978. Estimation of average hetero and genetic distance from a small number of individuals. Genetics 89, 583–590.

Welch, J.J., Bromham, L., 2005. Molecular dating when rates vary. Trends in Ecology and Evolution 20, 320–327.

Wenke, R.J., 1975 – 1976. Imperial investments and agricultural developments on Parthian and Sasaman Khuxestan: 150 BC to 640 AD. Mesopotamia 10/11, 31–221 (cf., Nesbitt, M. et al. 2010).

Wetlands International, 2013. Inland Niger Delta. www.wetlands.org (accessed 23.12.13.).

White, J.C., 1989. Ethno-ecological observations on wild and cultivated rice and yams in northeastern Thailand. In: Harries, D.R., Hillman, G.C. (Eds.), Foraging and Farming: The Evolution of Plant Exploitation. Unwin Hyman, London, pp. 153–158.

Whyte, R.O., 1972. The Gramineae, wild and cultivated, of monsoonal and equatorial Asia. Asian Perspectives 15, 127–151.

Wood, P.S., 1974. Black Majority. Negroes in Colonial South Carolina from 1670 through the Stono Rebellion. Knopf, New York (cf., Coclanis, P. H. 2010).

Wu, W.C., Lu, Y.G., Wang, G.C., 1990. A revision of the scientific and Chinese names of the verricose wild rice indigenous to China. Chinese Journal of Rice Science 4 (1), 33–37 (in Chinese with English summary).

Xie, K., Jiang, L., Lu, B., Yang, C., Li, L., Liu, X., Zhang, L., Zhao, Z., Wan, J., 2010. Identification of QTLs for seed dormancy in rice. Plant Breeding. http://dx.doi.org/10.1111/j.1439–0523. 2010. 01829.x.

Xiong, L.Z., Liu, K.D., Dai, X.K., Xu, C.G., Zhan, Q., 1999. Identification of genetic factors controlling domestication-revealed traits of rice using an F_2 population of a cross between *Oryza sativa* L. and *O. rufipogon*. Theoretical and Applied Genetics 98, 243–251.

Yamanaka, S., Nakamura, I., Nakai, H., Sato, Y.I., 2003. Dual origin of the cultivated rice based on molecular markers of newly collected annual and perennial strains of wild rice species, *Oryza nivara* and *O rufupogon*. Genetic Resources and Crop Evolution 50, 529–538.

Yamanaka, S., Nakamura, I., Watanabe, K.N., Sato, Y.I., 2004. Identification of SNPs in the *waxy* gene among glutinous rice cultivars and their evolutionary significance during the domestication process of rice. Theoretical and Applied Genetics 108, 1200–1204.

Yu, J., and 101 others, 2002. A draft sequence of the rice genome (*Oryza sativa* L. ssp. *indica*). Science 296, 79–92.

Zach, B., Klee, M., 2003. Four thousand years of plant exploitation in the Chad basin II. Discussion on the morphology of caryopses of domesticated *Pennisetum* and complete catalogue of the fruits and seeds of Kursekata. Vegetation History and Archaeobotany 12, 187–200.

Zhang, L.B., Zhu, Q.H., Wu, Z.Q., Ross-Ibarra, J., Gaut, B.S., Ge, S., Sang, T., 2009. Selection on grain shattering genes and rates of rice domestication. New Phytologist 184, 708–720.

Zhang, C.-C., Yuan, W.-Y., Zhang, Q.-E., 2012. RPL1, a gene involved in epigenetic processes regulates phenotypic plasticity in rice. Molecular Plant 5, 482–493.

Zhao, Z., Pearsall, D.M., Benfer, R.A., Piperno, D.R., 1998. Distinguishing rice (*Oryza sativa*, Poaceae) from wild *Oryza* species through phytolith analysis. II. Finalized method. Economic Botany 52, 134–145.

Zhao, K., Wright, M., Kimball, J., Eizenga, G., McClung, A., Kovac, M., Tyagi, W., Ali, M.L., Tung, C.W., Reynolds, A., Bustamante, C.D., McCouch, S.R., 2010. Genomic diversity and introgression in *O. sativa* reveal the impact of domestication and breeding on the rice genome. PLOS One 5 (5), e10780. 10.1371/journal. pone. 0010780.

Zhao, Z.J., 2010. New data and new issues for the study of origin of rice agriculture in China. Archaeological & Anthropological Sciences 2, 99–105.

Zhou, W., Xie, S., Mayers, P.A., Zheng, Y., 2005. Reconstruction of late glacial and Holocene climate evolution in southern China from geolipids and pollen in the Dingnan peat sequence. Organic Geochemistry 36, 1272–1284.

Zhou, H.F., Xie, Z.W., GE, S., 2003. Microsatellite analysis of genetic diversity and population genetic structure of a wild rice (*Oryza rufipogon* Griff.) in China. Theoretical and Applied Genetics 107, 332–339.

Zhu, Q.H., Ge, S., 2006. Phylogenetic relationships among A genome species of the genus *Oryza* revealed by intron sequences of four nuclear genes. New Phytologist 167, 249–265.

Zohary, D., Hopf, M., Weiss, E., 2012. Domestication of Plants in the Old World, 243 pp, fourth ed. Oxford University Press, Oxford.

Zohary, D., 1965. Colonizer species in the wheat group. In: Baker, H.G., Stebbins, G.L. (Eds.), The Genetics of Colonizing Species. Academic Press, New York, pp. 403–423.

Zuccolo, A., Sebastian, A., Yu, Y., Jackson, S., Rousley, S., Billheimer, D., Wing, R.A., 2010. Assessing the extent of substitution rate variation of retrotransposon long terminal repeat sequences in *Oryza sativa* and *O. glaberrima*. Rice 3, 242–250.

INDEX

Note: Page numbers followed by "f" denote figures; "t" tables.

T

U

V

W

Edwards Brothers Malloy
Ann Arbor MI. USA
April 11, 2014